DISCOVERING MOTION

An introduction to
the branch of physics
known as **MECHANICS**

A full curriculum designed for ages 12 and up
by Ellen Johnston McHenry

Discovering Motion; An Introduction to Mechanics

© 2024 Ellen Johnston McHenry

Published by Ellen McHenry's Basement Workshop
Pennsylvania, PA
ejm.basementworkshop@gmail.com

All rights reserved. No part of this book may be reproduced, copied, or transfered in digital format without the premission of the author. Howev,er purchasers of this book may make copies of the pages for use in their homeschool or their classroom.

ISBN 979-8-9868637-5-7

Printed in the U.S.A.

Other titles by this author:

The Elements; Ingredients of the Universe
The Chemical Elements Coloring and Activity Book
Carbon Chemistry
Dissect Your Dinner (food chemistry)
The Brain
Botany in 8 Lessons
Cells; An Introduction to the Anatomy and Physiology of Animal Cells
Protozoa; A Poseidon Adventure
Rocks and Dirt
Mapping the Body with Art
Mapping the World with Art

A NOTE ABOUT THIS BOOK

This book is a bit different from my other curricula. In most of my books, you read an entire chapter, then you do the activities that go with that chapter. In this book, most of the activities are written right into the chapter. As you are reading, the text will ask you to stop and do an activity. The reason for this is to help you discover the science for yourself as we go along, which will be much better than simply reading about it. Thus, you need to allow enough time to get through a chapter, since it is more than just reading.

This book can be used in a home setting or in a classroom. In my experience, it is easier to use it in a family or tutoring setting, but adaptations can be made for use in a classroom. For a weekly class, the students must be able to do at least some of the activities at home. The at-home activities can be the ones that don't require any special equipment.

If you are working with a group, you may want to have the students take turns reading the text, or you might decide to have one very proficient reader (or the teacher) do the aloud reading and ask the students to follow along in their own books. When you come to the cartoons, you might want to let the students take turns reading the lines for the fingerprint guys. Assign one reader to the fingerprint that is always on the left, another reader for the one on the right, and a third for any historical characters. Silly voices are fine, as long as they don't take the focus off what is being said.

Note that the teacher's section at the back of the book has not only the answer keys but also additional activity ideas if you need more activities.

THE MATH IN THIS BOOK

The math required to solve the problems in this book is mostly pre-algebra level. Students must be comfortable with fractions, decimals, squares and square roots. (Calculators can be used to do all the calculations.) Exponents are introduced in one of the chapters, just as a way to shorten the length of answers that have a lot of zeros on the end.

Students should be comfortable working with simple formulas like D=RT, and understand that formulas can be rearranged to solve for any variable, such as T=D/R. The only Algebra 1 topics included are the Cartesian plane and a very brief introduction to what cosine means.

Students who struggle in math can still use this book. In many cases, the math problems can be skipped without causing any critical gaps in knowledge. Students can go right on to the next section and pick up with the new concepts.

TOOLS AND MATERIALS YOU WILL NEED

Things you need to purchase from a science supply store or from Amazon:
1) Two good quality meter sticks (thickness of at least 5 mm or 3/16 inch, if possible)
2) Several spring scales   (You don't need a full set, but make sure you have at least these: 100 gram [1 N], 500 gram [5 N], and 1000 gram [10 N].)
3) Steel bearings (steel marbles) 15 mm size
4) Stopwatch (You can use the stopwatch feature on your phone, but if you are working with a class, having real stopwatches is very helpful.)

Supplies you might have around the house or can easily purchase locally:
1) Toothpicks
2) Thread
3) Washers (flat metal rings sold in hardware departments alongside screws)
4) Pins and/or pushpins
5) Paperclips
6) Straws
7) Rubber bands of various sizes
8) Bottle of water
9) Hacksaw blade (optional—you can also use a coat hanger) (activity 4.9)
10) Glass marbles
11) Cotton balls (or better yet, pom-poms from a craft store that will fit into your cannon in activity 8.10)
12) Very large sheets of paper, or roll of paper (though in a pinch you can substitute newspaper) (activity 8.5)

Consumable supplies you will need to buy:
1) Paper and cardstock (110 lb weight cardstock is good, as it will go through most home printers)
2) Glue stick, and/or white glue (not school glue—get the high quality white craft glue)
3) Clear tape and masking tape (or duct tape)
4) Poster paint (activity 8.5)

Items you probably already have around the house:
1) Scissors
2) Pennies
3) Cans of food
4) Hammer
5) Coat hanger
6) Colored pencils
7) Cardboard
8) Long board of some kind (even the leaf out of your dining room table)
9) Foil, waxed paper, plastic wrap, sand paper
10) Miscellaneous small balls (wood, plastic, rubber)
11) Cereal boxes
12) Calculator (or you can use the calculator on your computer or smart phone)
13) Cylindrical objects of various sizes

Optional items that are mentioned in some activities:
1) "Clackers"
2) Track car (a toy car that goes at a steady pace)
3) Skateboard
4) Spirograph® drawing toy
5) Gyroscope

CONTENTS

INTRODUCTION 1
    We meet our hosts and find out how they came to be in this book.

ADVENTURE 1: Balancing with Archimedes.. 5
    Center of mass, balancing, and the life of Archimedes

ADVENTURE 2: Balancing, lifting, and turning 23
    More about balancing, mechanical advantage, lever, simple machines, the "Lever Rap"

ADVENTURE 3: Swinging with Galileo 39
    Pendulums, Inverse Square Law, Foucault pendulums, resonance, oscillation, early life of Galileo

ADVENTURE 4: Inertia with Newton 57
    Inertia, life of Newton, Principia, Newton's First Law, mass, weight, inertial balance

ADVENTURE 5: Slowing and stopping with Da Vinci 69
    Static and kinetic friction, power of surface area, coefficient of friction, free body diagrams

ADVENTURE 6: Gravity with Galileo (and Newton and Einstein) 77
    Gravity, acceleration, speed versus velocity, air resistance and terminal velocity, Universal Law of Gravitation, big G, little g, Cavendish experiment, relativity and gravity, other models of gravity

ADVENTURE 7: Calculating and Colliding with Newton 97
    Newton's Second and Third Laws, F=ma, normal force, momentum, impulse, collisions

ADVENTURE 8: Flies and cannonballs 111
    Descartes, Cartesian grid, displacement, graphing linear motion, $d=1/2gt^2$, parabolic motion

ADVENTURE 9: Round and round we go 129
    Cycloids, uniform circular motion, centripetal force, $F_c=mv^2/r$, Marey's photography, angular momentum, torque, right-hand rule, L=rmv, angular velocity, moment of inertia, $L=I\omega$

ADVENTURE 10: Playing and working 161
    Kinetic and potential energy, PE=mgh, $KE=1/2mv^2$, W=Fd and $W=Fd(\cos\theta)$, joules and watts
    "The Roller Coaster Song"

EPILOGUE.. 191

SUPPLEMENTAL ACTIVITIES and ANSWER KEYS 195

OPTIONAL QUIZZES 235

BIBLIOGRAPHY 249

INTRODUCTION

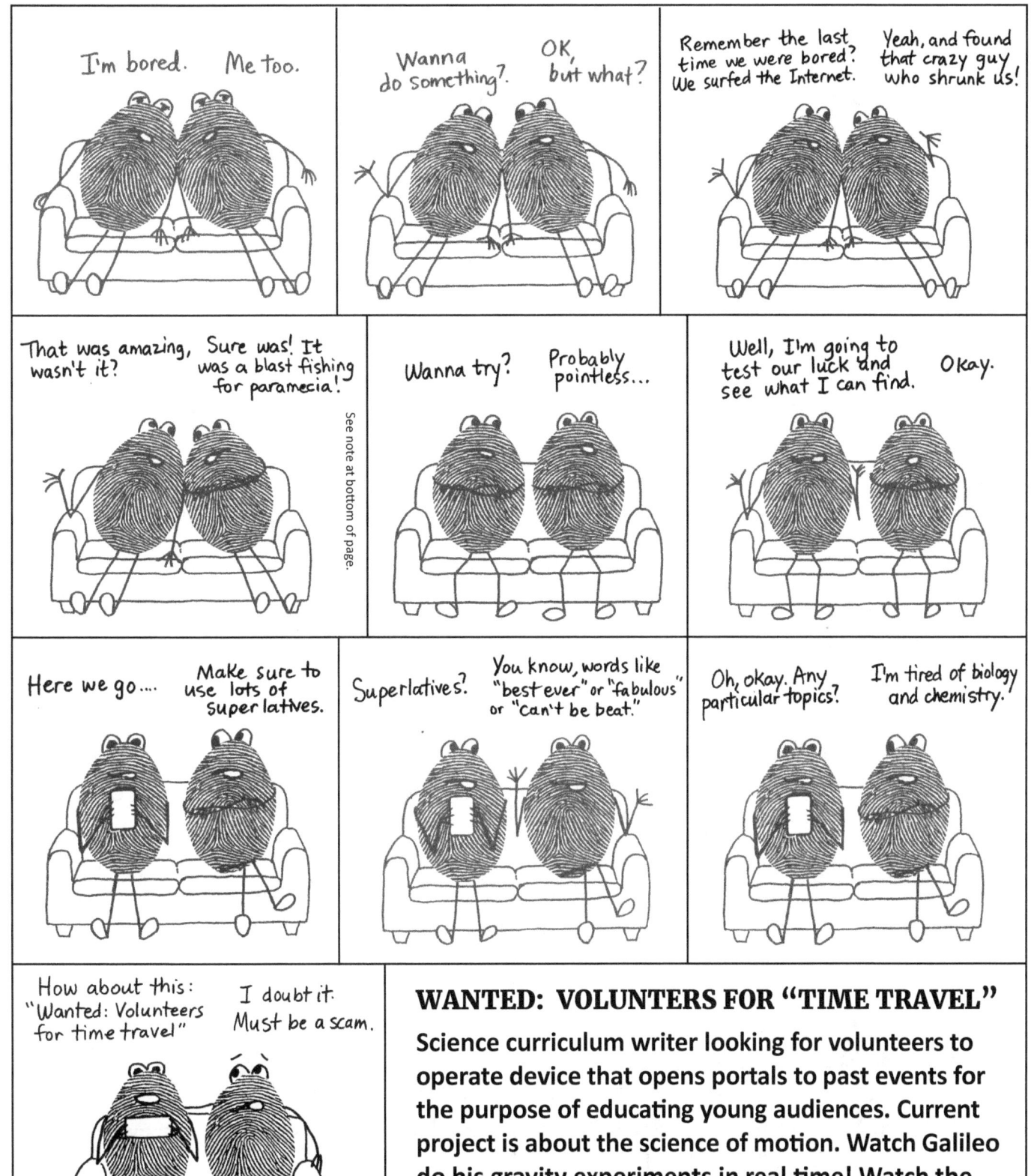

Their previous adventure fishing for protozoans is in "Protozoa: A Poseidon Adventure" by the same author.

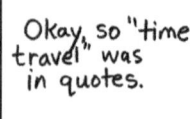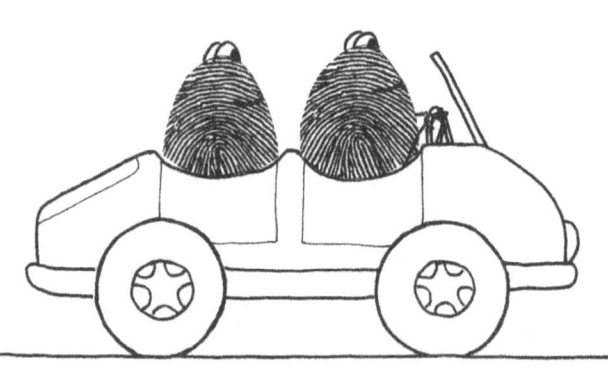

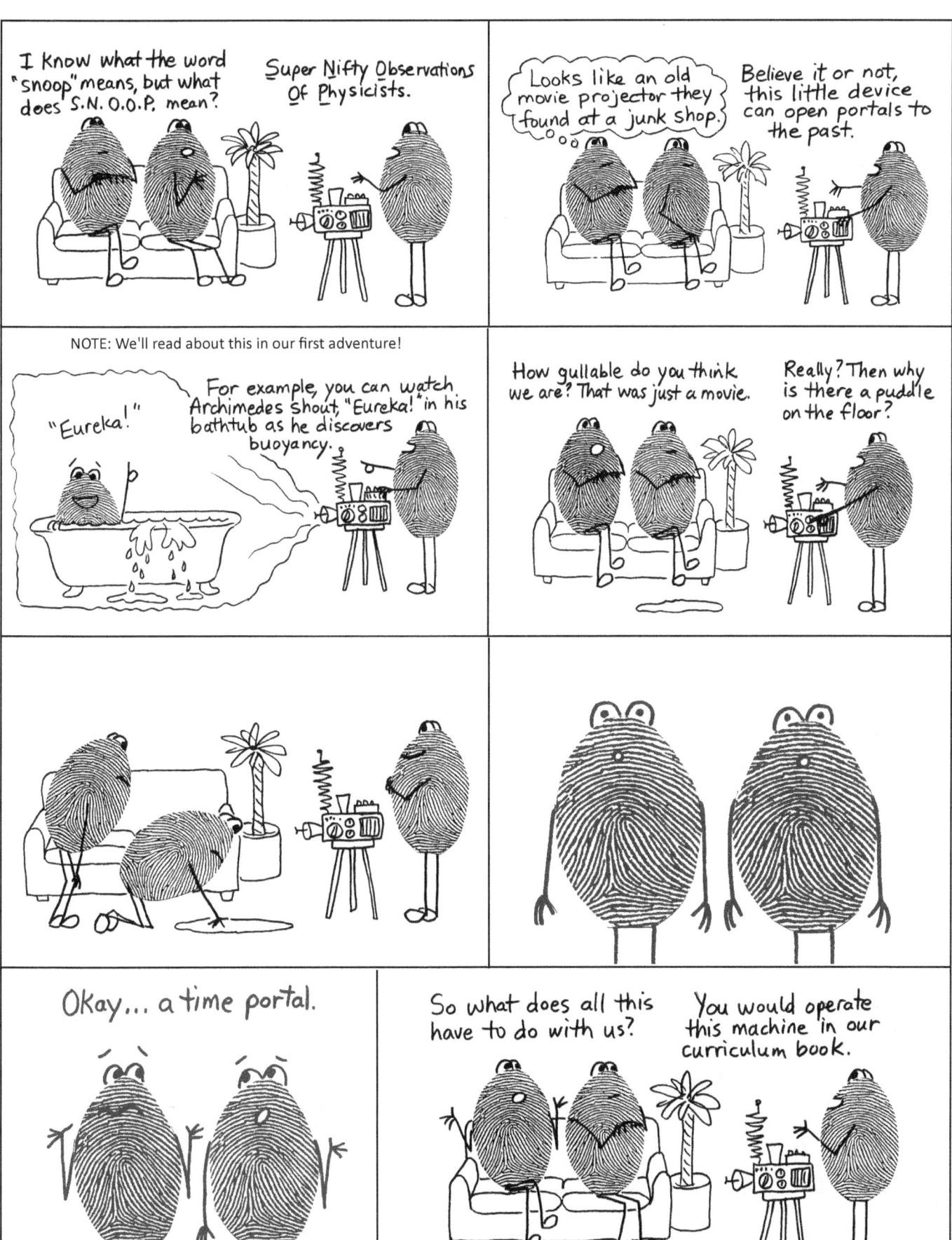

Adventure 1: Balancing with Archimedes

Our adventure begins about 2,200 years ago in the ancient Greek city of Syracuse, on the island of Sicily. (On a map, Sicily looks like the "ball" being kicked by the "boot" of Italy.) A brilliant mathematician named Archimedes *(ar-kih-ME-deez)* is experimenting with the way things balance. Archimedes wasn't the first person to think about how things balance, but he was the first to propose the idea that we now call "center of gravity." In Archimedes' day the word "gravity" had not yet been invented, so he didn't use that word.

Archimedes was interested in finding a mathematical way of describing how and why things balanced. It was easy to find the balance point for a wooden rod, but what about a flat shape such as a triangle or a trapezoid or a parallelogram? His final conclusions were written in a book called *On The Equilibrium of Planes*. ("Equilibrium" means "balancing," and in geometry, a plane is a flat surface.)

Finding the balance point ("center of gravity") of something can be quite important in our everyday lives. When you hang a picture on the wall, if you get the hanger off-center the picture will look crooked. If you are making a spinning top and don't get the handle perfectly centered, the top won't spin correctly. You can probably think of more examples just by looking at things around you right now.

You've probably used a ruler to find the center of a line or a square for a math assignment, or perhaps the center of a board for an art or craft project. Rulers are the most common tool for finding centers. But how accurate are your measurements? In this first activity, you will test your acumen (skill) with a ruler. Can you measure accurately enough to find the balance point for a rectangle? Sounds easy, but is it? Take a few minutes to do the following activity. (Archimedes must have done many experiments like this.)

ACTIVITY 1.1: Use a ruler to find the center of a rectangle, then test your results

You will need: *a ruler, a pencil, scissors, a piece of heavy card stock paper (or thin cardboard such as the side of a cereal box), a toothpick, half an apple (or potato), and a paper towel or paper napkin. Stick the toothpick into the apple (or potato or other firm object) to make a "balance stand."*

1) Use the ruler to draw a square on the piece of card stock or cardboard. The square can be about the size of your palm, or a little larger if you wish. (You can get help making a perfectly square corner by tracing the corner of a piece of paper.) Cut out the rectangle.
2) Use a ruler to find the center of the rectangle and mark it with a dot. The easiest way to do this is simply to align the ruler so that it connects two opposite corners, then draw a line. Do this to the other pair of opposite corners. The point where these lines cross is the center.
3) Place your balance stand on the corner of a table. This will allow you to look at the underside of the rectangle. Or, place the balance stand on top of something tall (but stable) that is sitting on the table.
4) Place the center dot of your rectangle right on the point of the toothpick, then let go. Does it balance? If not, move it around until you get it to balance. How far off were you? (NOTE: **DON'T** prick a hole in your shape using the sharp point of the toothpick! In fact, you can use scissors to trim off the tip of the toothpick so it is not quite so sharp.)

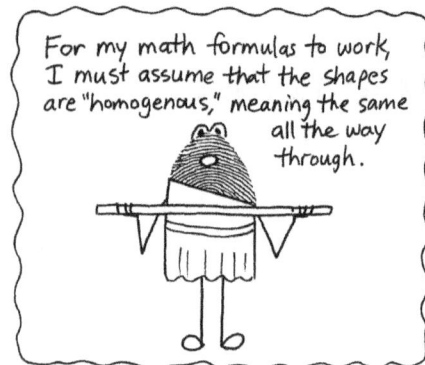

Here is another activity where you can do what Archimedes did. (We are skipping circles because it is pretty obvious that the balance point of a circle is at its center. If you draw a circle using a compass, you automatically get the balance point—the place where you stuck in the pointy part of the compass.)

ACTIVITY 1.2: Find the balance point of trapezoids and triangles
(A trapezoid is a 4-sided figure with at least two parallel sides. A parallelogram is a type of trapezoid.)

You will need: a ruler, a pencil, scissors, your balance stand, and another piece of card stock or thin cardboard

NOTE: If you are working in a group and are short on time, you can have each person in the group make one type of shape, then share them around for balancing.

TRAPEZOIDS:
1) Draw a rectangle, then turn it into parallelogram (which is a special type of trapezoid) by slanting two ends. Make sure the slanted ends are exactly parallel. Then draw another rectangle and turn it into a trapezoid by making those slanted ends into lines that are NOT parallel.
2) Cut out your shapes.
3) Try the technique you used in Activity #1, where you made lines from the corners that crossed in the middle. Is this the balance point for a parallelogram? For the trapezoid?
4) Now try using the ruler to determine the middle of each side. Connect the mid-points of opposite sides. Put a dot at the point where these two lines cross. Is this a better balance point?

TRIANGLES:
1) Draw some triangles on your card stock or cardboard. Make one "right" triangle (that has a 90-degree "square" corner), an isoceles triangle (two sides the same length, so it is symmetric), and one "scalene" triangle where each side is a different length. (Archimedes is holding a scalene triangle.)
2) Cut out your triangles.
3) Use the ruler to find the center of each edge. Mark it with a small dot.
4) Now use the ruler as a straight edge to make a connecting line between each center dot and the angle that is directly opposite to it. Make a dot at the place where all three lines cross.
5) Place each triangle on the balance stand and see if it will balance on that center dot.

The words that go with that diagram are even more impressive. Here is a sample:

Let there be a triangle, ABG, and let AD in it be against the middle of base BG. One must prove that the "center of weight" of ABG is on AD. For otherwise, still, if it is possible, let it be Q, and let QI be drawn through it parallel to BG. And, in fact, what remains of the repeatedly bisected DG will eventually be less than QI. Let each of BD, DG be divided into equals, and though the cuts parallel to AD let lines be drawn and let EZ, HK, LM be joined. They will, in fact, be parallel to BG. The center of weight of parallelogram MN is on US, the center of weight of KC on TU, and that of ZO on TD. Therefore, the center of weight of the magnitude composed from all of them is on straight-line SD. Let it be, in fact, R, and let RQ be joined and extended, and let GF be drawn parallel to AD. Triangle ADG, in fact, has this ratio to all the triangles inscribed up from AM, MK, KZ, ZG similar to ADG, that which GA has to AM, since AM, MK, ZG, KZ are equal. And since triangle ADB also has the same ratio to all the similar triangles inscribed up from AL, LH, HE, EB, which BA has to AL, therefore triangle ABG has this ratio to all the mentioned triangles, that which GA has to AM.

Thankfully, you won't have to do anything like that in this book. But let's take it up a notch and give you more challenging shapes to work with. What about very irregular shapes? Is there an easy way to find the balance point for any shape?

ACTIVITY 1.3: Find the balance point of some U.S. states

You will need: *copies of the state outlines provided at the end of this chapter (either printed onto heavy card stock or glued to thin cardboard), a pencil, scissors, a pin, a piece of thread, and a small nut or washer (or any small, heavy object with a hole)*
NOTE: If you are working from a paperback copy of this book and want digital patterns to print, go to www.ellenjmchenry.com/discovering-motion-printables
NOTE: Once again, make sure you don't prick a hole in these shapes using your toothpick! The shape must balance on its own.

1) Let's start with Utah. Just for minute, pretend that the extra bump at the top is not there and put a pencil dot at the center of the main rectangle. Place that dot on the point of the balance stand and see what happens. Now move Utah around until it balances. Put a dot on the correct balance point. Compare the two dots. How does that extra bump affect the balance point? Is the correct point closer to the bump or farther away from it?
2) Now try Nevada. Using what you learned from Utah, take a guess where the balance point might be and mark it with a dot.
   Then balance Utah and mark the correct balance point. How close were you?
3) Next, we'll try Texas, but for this one we'll show you a sneaky trick for finding the balance point. Cut a piece of thread (about 20 cm) and tie a nut or washer onto one end. At the other end, tie a small loop. Put the pin through the small loop, then prick the pin into one corner of Texas (any corner is fine), as close to the edge as you can possible get it. Wiggle the pin a bit so that Texas hangs freely, maybe even swinging just a bit. (If you have trouble with Texas wanting to fall off the pin, hold it against a wall, but very gently so that it still can swing.) When both Texas and the weight have come to rest, you will need to draw a line where the thread crosses Texas. You can do this easily by making a pencil mark right where the thread crosses the bottom edge, then using a ruler to connect that dot with the pin prick. Now move the pin to a different corner. Any corner will do. Repeat this process and you will get another line across Texas. Do this for at least two more corners so that have four or more lines. The intersection of these lines will be the center of mass for the shape of Texas. Place this center dot on your balancing stand and see if it balances. If it doesn't, you should only have to make a very small correction.

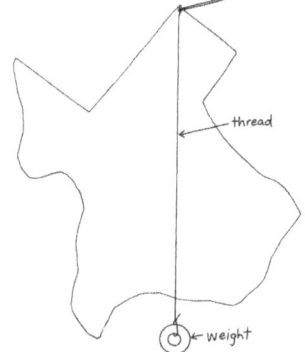

5) Try this method for the other states.
   CHALLENGE: Where is the balance point for Florida? Will Florida balance on your stand?

Let's stop and ponder something really strange. If a cardboard shape weighs 3 grams, all three of those grams are pressing down on the tip of your balance stand. If your balance stand was also a scale, it would register a weight of 3 grams. Yet only one small point of that piece of cardboard is touching the toothpick. Most of the cardboard is NOT touching the toothpick. It's like all of the mass of the entire shape is concentrated right at the place that is resting on the toothpick. How can this be? What is happening underneath all the rest of that shape— the parts that are not touching the toothpick? Don't they weigh something? This is the first of many strange concepts we'll come across in this book. For mathematical purposes, we can indeed assume that ALL of the mass of that shape is present right at that tiny point, and simply ingore the rest of the shape. Very odd indeed. And speaking of odd, the next activity will give us another strange idea to ponder.

In the next activity we'll begin using the term **center of mass**. We've been using the term "balance point" very loosely. We need to tighten up our definition so that we can evaluate an object where the balance point is not resting on the tip of our toothpick. Sometimes the term "center of gravity" is used in place of "center of mass." The two terms are interchangeable as long as we are on or near the earth. We'll leave the discussion of mass versus gravity for a future chapter.

ACTIVITY 1.4: Find the balance point of some letters

You will need: copies of the letter page provided at the end of this chapter (either printed onto heavy card stock or glued to thin cardboard), a pencil, scissors, a pin, a piece of thread, and a small nut or washer (or any small, heavy object with a hole)

1) Start with the letters F and J. See if you can balance them just by experimenting.
2) Now try the letter L. Does it balance? Mark a dot where the balance point is. Then snip off the portion marked "1" and try again. Now does it balance? Mark the balance point. Snip off the portion marked "2" and try again, marking the balance point. Finally, snip off "3." How does the balance point change? What direction does it go? How are you changing the center of mass by snipping off parts of the letter?
3) Next, try the "V." Does it balance? As with the "L," snip off the number "1" sections then try balancing again. Then snip off the "2" sections. Does it finally balance? Where do you think the center of mass was before you snipped off any sections?
4) Now for the letter "O." Can you balance it? Why not? Where is the center of gravity for the "O"? Now snip off section "1." Can you balance this letter "C"? Snip off the "2" sections. Does it balance now? Where would you estimate the center of mass is? Finally, snip off the "3" sections. Where is the center of mass?
5) Lastly, try the lower case "d." Before you balance it, try to predict where the center of mass will be. Will it balance? After you find, out, predict what will happen if you snip of the "1" section. How will the center of mass change? Then go ahead and try it.

We've discovered that sometimes the center of mass is actually outside an object. Strange, but true. Let's do some more experimenting with objects that have their center of mass outside of themselves. This concept is a key to performing amazing balancing tricks.

In this next activity, we will be balancing a cardboard V not on your balance stand, but on a tightrope made of thread. You will discover for yourself another basic principle of balancing.

ACTIVITY 1.5: Balancing the letter V

You will need: scissors, a piece of thread or thin string (20 cm is plenty), and a cardboard letter V with very long "arms."

1) Cut a cardboard V that has very long "arms." We will be trimming these arms bit by bit, so make sure they are pretty long to start with. The exact length or width is not important. Just cut something similar to what you see in this drawing.
2) Have an assistant hold the ends of the string and pull it tight. Then put the V upside down on the string, so that it balances. Rock it a bit and see how well it maintains its balance. Remember what you discovered about the letter V in the previous activity. Where would the center of mass be for this V?
3) Now trim a bit off each "arm" of the V. Make sure you trim the same amount off each side. Try balancing it again. Does it still balance?
4) Trim the V again, the same amount off both sides. Balance it again. Rock it back and forth a bit. Is it still stable?
5) Keep trimming off a bit from both sides and checking how it balances each time. What happens when the arms get very short? Keep trimming until you reach a point where it does not balance anymore.
6) Take this unstable V and balance it on your balance stand. Does it balance? Is the center of mass inside or outside this V?

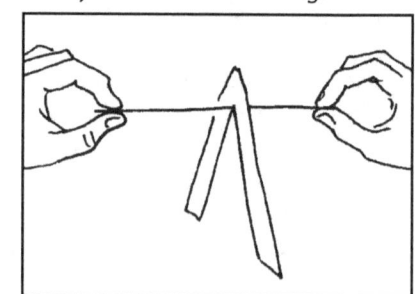

7) Where do you think the center of mass was for your original long-armed V? What happens as the center of mass gets closer and closer to the point of the V?

A circus act for letters!

So now we've discovered that the way to achieve stable balance is to keep the center of mass below the balance point. (The string was the balance point in our last activity). Science museums in big cities sometimes have an exhibit where you can ride a bicycle on a tightrope, often very high up in the air. This can be done safely by making sure that there is a huge amount of weight well below the wire— so much weight that the bicycle can't possibly become unbalanced. This picture shows a famous "daredevil" act where a man got his wife to act as the center of mass below the rope. It's the same principle as the bicycles at science museums.

If you watch gymnasts do routines on the balance beam, you will notice that if they think they are starting to fall, they will quickly lower their bodies. They need to get their center of mass closer to the balance point in order to gain stability.

Now it's time for some tricks. Here are some balancing acts that use the principle of keeping the center of mass below the balance point.

ACTIVITY 1.6: Balance two forks on a glass

You will need: two forks, a toothpick, and a glass
  (Since these objects are often on tables at restaurants, this is a good trick to do while you are waiting for a meal!)

1) Put the two forks together so that their tines interlock.
2) Wedge the toothpick into a crack where the forks intersect.
3) Rest the toothpick on the edge of the glass.

If you need a video demo, there is one on the playlist for this curriculum. YouTube.com/TheBasementWorkshop, click on the "Discovering Motion" playlist.

You have probably seen pictures of the Leaning Tower of Pisa. It didn't start out as a leaning tower, of course. It was built to be the bell tower of the cathedral in Pisa, Italy. Over the course of hundreds of years, one side has gradually sunk into the ground. What the builders didn't know was that the site they chose to build on had different types of rocks and soil underneath it. So basically, one part of the building was on harder rock and one part on softer rock. The softer rock started to compress and sink under the weight of the building, causing the tower to lean.

In the 1990s, it was determined that the Leaning Tower was leaning so much that is was no longer safe for people to be inside of it. Architects measured the rate at which the lean increased each year and calculated that eventually the tower would fall over. It would still be quite a few years until that happened, but they decided to go ahead and fix it now instead of waiting until the tower got any more dangerous. They figured out a way to adjust the foundation underneath the tower so that the higher side was brought down a bit. In 2004, the tower was declared to be safe. Tourists can now climb the stairs to the top and look out over the city.

How can you predict when an object is leaning enough that it will fall over? It's a matter of calculating where the center of mass is, then determining whether that center of mass is located over the base of the object. In this series of drawings, the dotted line shows where the center of mass is located over the base. In the first drawing, the center of mass is directly over the center of the base. In the third drawing, the building has leaned so much that the center of mass is no longer over any part of the base, but over the grass. When the center of mass of an object shifts so that it is no longer over any part of the base, the object falls over.

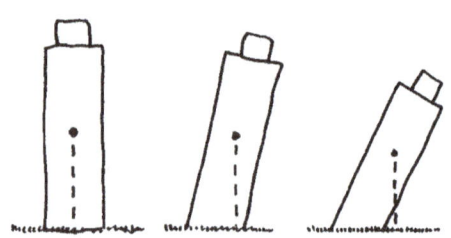

In these two drawings of the Leaning Tower, we've proposed a somewhat ridiculous scenario with a crowd of tourists grouped together on the lower side of the tower. So far, we've assumed that the weight of the tower is distributed equally, with no part of it being much lighter or heavier than any other part. However, when calculating center of mass, you have to take into account any areas that heavier (denser) than others. In our cartoon scenario, the weight of the tourists has greatly increased the weight of that side of the tower, causing a shift of the center of mass. With the additional weight of the tourists, the center of mass has moved closer to the edge of the base, and, therefore, closer to possible tragedy.

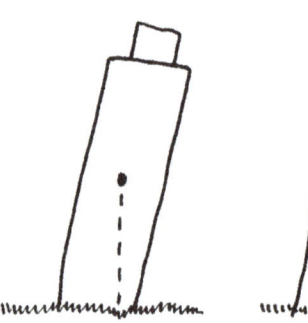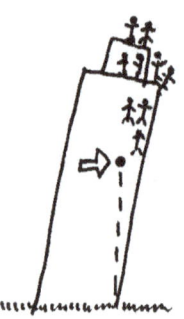

ACTIVITY 1.7: The leaning beverage can trick

You will need: *an empty can of the type shown in this picture. The can must have a bottom that looks like this one, with an angular section between the side and the flat bottom. You'll also need some water that you can gradually pour into the can.*

1) Start by trying to balance the empty can. (If you happen to have a full can, try to balance that as well. Prove to yourself that neither will balance.)
2) Pour water into the can until it is about 1/3 of the way full. Try to balance it again. If it does not balance, add or subtract a little bit of water until the can balances.
3) If you have balanced the can well, you will be able to push the can gently and get it to roll around in a circle!

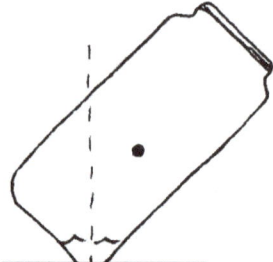 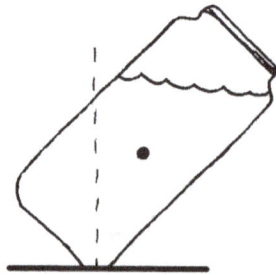 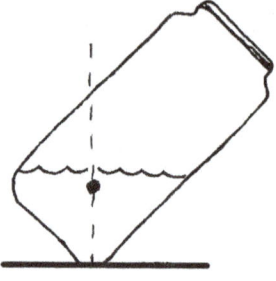

If the can has too little water, the weight of the can itself will be the most significant factor in the center of mass, causing the center of mass to be far from the base. (The "base" is that very small edge touching the table.)

If the can has too much water, the center of mass will again be to one side of the balancing point, causing it to topple over.

If the can has just the right amount of water in it, the center of mass will be directly over the base.

Center of mass is important in many branches of science and engineering. Engineers who design vehicles are very concerned about center of mass. Cars that have their center of mass closer to the ground are less likely to flip over (a special concern for racing cars that go around turns at very high speeds). If an airplane has its center of mass too far forward, it will tend to be less maneuverable and be difficult to control during take-off and landing. A plane with a center of mass too far back will be very maneuverable but will be less stable during flight. A helicopter must be designed so that the center of mass can shift backward when the pilot wants to go forward. This is why a helicopter looks like it is tilting nose-down when it is flying forward—the center of mass moves behind the rotor.

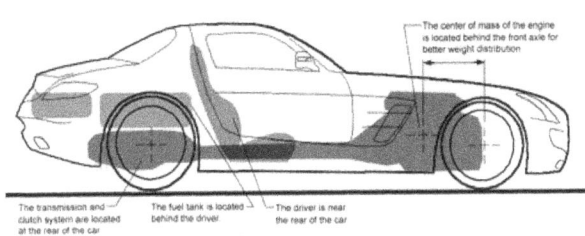

In sports, high-jumpers learn to bend in such a way that their bodies go over the bar even though their center of mass does not. Long jumpers shift their center of mass as efficiently as possible in order to maximize the power of their jump. Athletes shift their center of mass all the time as they run and jump, though they don't think about what they are doing in scientific terms. Scientists called **kinesiologists** *(kin-EE-see-OL-o-gists)* study body movement and develop equipment or therapies based on the science and math of body movement.

In astronomy, center of mass is very important. When a moon orbits a planet, or a planet orbits a star, they are actually both moving around a central point called the "barycenter." This point can be inside one of the bodies, or at a point in space. The Earth and the moon orbit each other at a point that is about 1,710 kilometers (1062 miles) below the surface of the Earth. It is this point, the barycenter, that orbits around the Sun. It's as if all the mass of both the earth and sun is concentrated into this one point.

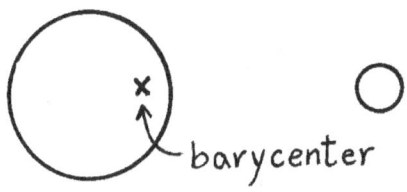

ACTIVITY 1.8: Experiment with center of mass and airplane flight

You will need: a piece of paper, a large paper clip, and instructions for making a paper airplane if you don't know how

1) *Make your airplane.*
2) *Put the paper clip on the nose. See how this flies.*
3) *Move the clip back a bit and try again. Better or worse?*
4) *Move the clip back again. Keep moving the clip back and see how it flies with the clip at the back. Is there an optimum place for the center of mass?*
Extension: What happens if you put more than one clip on the optimum center of mass position?

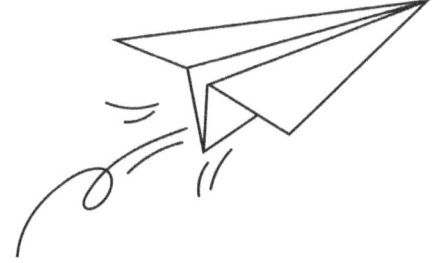

ACTIVITY 1.9: "The Impossible Hop"

You won't need anything special for this experiment, but some people like to place a dollar bill on the ground and make a bet with someone that they can't jump over it. You'll win the bet and keep the dollar bill most of the time.

1) *Bend over and hold on to the tips of your shoes (or your toes if you are wearing just socks).*
2) *Try to jump forward without letting go of your shoes/toes. (In the dollar bill version, you place the dollar bill in front of their feet and tell them to jump over the dollar without letting go of their shoes/toes.)*
3) *Most people find this trick impossible because when you jump you shift your center of mass forward. In order to compensate for this sudden shift forward, you instinctive counterbalance using your arms. If you can't counterbalance with your arms because you are holding on to your toes, you will fall over.*

ACTIVITY 1.10: Supplemental information about Archimedes

Archimedes was probably born in the year 287 BC. We are more sure of the year of his death than of his birth because his death occurred during part of a Roman war campaign that took place in 212 BC. As we learned in this chapter, he was born in the city of Syracuse, on the island of Sicily (shown in red on the map). His father, Phidias, was an astronomer, but that's all we know about his birth family. One of Archimedes' friends wrote a book about his life, but unfortunately, it has been lost or destroyed. We know the book existed because other ancient writers mentioned the book, but we don't have the book itself. We also don't have any portraits of Archimedes, so we don't know what he looked like. Paintings of Archimedes, like the one shown below, are completely from the imagination of the painter.

"Archimedes Thoughtful" by Domenico Fetti, painted in 1620

It is possible that as a young man, Archimedes traveled to Alexandria, Egypt, to complete his education. Alexandria had the best library in the world at that time and thus attracted scholars from all over the world. Wealthy families would often send their children to Alexandria in the same way that we send students off to universities. We know that Archimedes knew the head librarian, **Eratosthenes** *(air-uh-TOS-thuh-neez),* the man famous for estimating the circumference of the earth. (Yes, people back then knew that the earth was round, not flat.) One reason to think that Archimedes received part of his education in Alexandria is that he would have been about the right age to have been a student of the famous Greek mathematician **Euclid** *(YEW-klid)* who lived in Alexandria at that time. This would explain how Archimedes became such a mathematical genius at a relatively young age. Archimedes would also have studied philosophy in addition to mathematics. Socrates, Plato, and Aristotle pre-dated Archimedes, so he undoubtedly read their books (as handwritten scrolls, not as printed books).

Mathematics in the ancient world consisted mostly of geometry. By Archimedes' day, mathematicians had already discovered everything we learn today in high school geometry, and much more. After learning everything known about geometry at that time, Archimedes went on to make many discoveries of his own. He was fascinated with the relationship of a circle's diameter to its circumference—the value that we call "pi" (3.1415...) He was able to approximate the value of pi (π) to be 3.141 by using straight-sided polygons both inside and outside of a circle. It is fairly easy to calculate the perimeter (distance around the outside) of a figure with square sides if you assume all the sides are the same length and you know the length of one side. In this figure, you can see that the measurement around the outside of the circle (its circumference) must be more than the perimeter of the smaller polygon, but less than the perimeter of the larger polygon. The correct value lies somewhere between the perimeters of these polygons. If you keep increasing the number of sides of the polygons, you will get closer and closer to the value for the circumference of the circle.

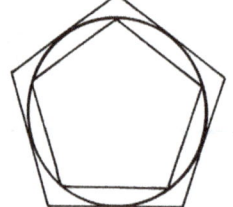 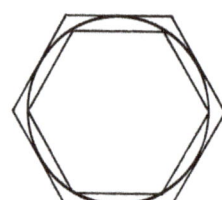 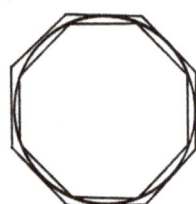

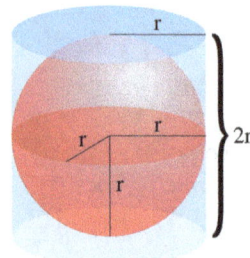

Archimedes considered that one of his greatest achievements was figuring out the relationship between the volume of a sphere and the volume of a cylinder that exactly contains the sphere, as shown in this diagram. The letter "r" represents the word "radius" which is the measurement from the center of a circle to its edge. The red sphere has exactly 2/3 the volume of the cylinder. It also has 2/3 the amount of surface area as the cylinder. The volume of the sphere is $4.3\pi r^3$. Why he considered this to be his greateest achievement is unknown. Calculating the value of "pi" would prove to be much more useful to future mathematicians and scientists.

Archimedes also studied parabolas. A **parabola** is the shape that a ball makes when you toss it in the air. The shape of the parabola will depend on the angle at which you toss it. Two parabolas are shown here. Archimedes used known rules of geometry to show that the area of the purple shape in the top drawing is equal to 4/3 of the area in the blue triangle in the bottom drawing. Not impressed? That's okay. Archimedes didn't need anyone to be impressed. He simply loved math. To him, it was a fun activity, not a chore.

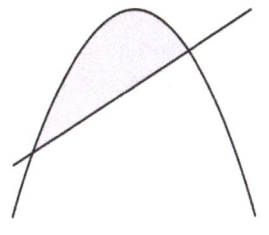

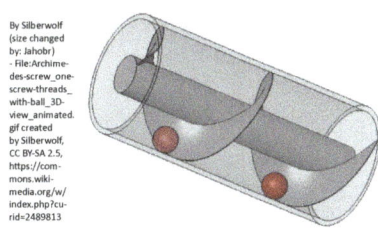

Sometimes his geometry studies had obvious practial applications, and this allowed him to also become an inventor. His study of spirals inspired him to invent a tool we now call the "Archimedes screw," which consists of a long spiral shape inside a tube. (You can see a moving animation of this image by going to the Wikipedia page on Archimedes.)

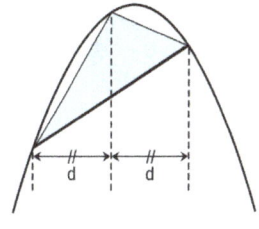

This device was a great help to farmers who needed to water their crops. They could put one end of the tube into an irrigation canal and then crank the handle on the other end to bring that water up to higher ground. In modern times, you can find Archimedes screws in some types of machines, including grain elevators and combine harvesters used in agriculture.

By the time Archimedes was a young adult, his genius had caught the attention of the king of Syracuse. The most famous story about Archimedes comes from this period in his life. It is said that the king had commissioned the making of a gold crown and had given the goldsmith the amount of gold necessary. When he received the finished crown, the king thought there was something not quite right about it. He wondered whether the goldsmith had cheated him by using a cheaper metal to make the crown and then covered it with a thin layer of gold (keeping the rest of the gold for himself). He asked Archimedes to find a way to test the crown to see if it was make of pure gold, but without damaging the crown in any way. The story has several variations at this point. One version says that Archimedes was so obsessed with this problem that he forgot to do basic things like change his clothes or take a bath. When he began to stink, his friends took him to the public bath and tossed him in. Other versions simply have him stepping into a bathrub. All versions agree that the tub was so full of water that as he stepped in, some of the water sloshed out. He immediately realized that his body was now occupying the space that used to be taken up by that sloshed water. The volume of his body in the tub was equal to the volume of water displaced out of the tub. This principle (buoyancy) could be used to find the volume of the gold crown. When combined with information about the weights of silver and gold, he would be able to work a way to determine whether the crown was pure gold. He was so excited about his realization that he jumped out of the tub and ran down the street yelling **"Eureka!"** which was Greek for "I found it!" Being a bit absent minded, Archimedes was unaware that he was running down the street totally naked. We don't know how accurate this story is, but because it is so entertaining, there's no doubt it will continue to be told for many years to come.

The king of Syracuse would continue to rely on Archimedes for help, particularly with his military. Syracuse was constantly under threat of invasion by superior armies such as the Romans. Archimedes found ways to improve the design of his city's catapults so they could throw heavier weights and be more accurate. He also designed a device called "the claw" which was said to be able to lift enemy ships out of the water, dumping all the sailors and their weapons into the sea. To defend their harbor, Archimedes found a way to use bronze shields as mirrors, angling them so they could direct beams of intense sunlight at ships, blinding the sailors or causing the sails to catch fire.

Archimedes wrote many short books, mostly on geometry or mathematical puzzles. In his book titled *The Sand Reckoner*, he tried to figure out how many sand grains would fill the universe. The most important thing we learn in this book is that Archimedes believed that the sun was the center of the solar system, not the earth. This idea somehow got lost for hundreds of years until it was rediscovered by Nicolaus Copernicus in the 1500s.

Some of Archimedes' books were copied by scribes and preserved until modern times when they could be printed into bound books. Other books have been lost or destroyed over the centuries. In 1906, an amazing discovery was made in the city of Istanbul (known in ancient times as Constantinople). A Danish professor was there to study a very old book from the 1300s which had been kept in a monastery library for hundreds of years. The pages were made of goat skin parchment, a common material for books at that time. Parchment was very expensive and was never wasted; scribes often recycled old parchment by carefully scraping the ink letters. The professor noticed some very light lettering behind the darker lettering of the Medieval prayers and wondered if this parchment had been recycled. You can see some of this light lettering, plus some circles, on the pages shown here. After further examination, he realized that this light lettering was a copy of text by Archimedes. When experts were able to examine the entire book, they found three of Archimedes' books that no one had ever seen. What a discovery!

Recycled parchment books are called "palimpsests."

ACTIVITY 1.11: Supplemental Videos

There is a YouTube playlist for this book located at YouTube.com/TheBasementWorkshop. Click around until you find the playlist called "Discovering Motion." (You might have to click on "Created Playlists.")

I've already watched all these videos and made sure they are don't contain anything you shouldn't be watching, but even so, make sure your parents or guardians know you are accessing YouTube.

I used to be able to put labels on the videos marking which ones go with which chapter, but I can't do that anymore, so I just try to make sure they come in approximately the right order as you read through the book. For this chapter, watch the first few videos that are about center of mass/gravity and Archimedes.

ACTIVITY 1.12: Review questions

See if you can remember the answers to these questions. If not, go back into the text and find the answer. (There is an answer key in the teacher's guide.)

1) Archimedes was born in the city of _____ on the island of _____.

2) TRUE or FALSE? The balance point of a wooden rod can also be called the center of gravity.

3) TRUE or FALSE? A trapezoid is a special type of parallelogram.

4) To find the center of a triangle, you find the _____ of each side and then connect that to the angle directly opposite to it. The place where these three lines _____ will be the center of gravity.

5) TRUE or FALSE? "Center of gravity" and "center of mass" are the same as long as we are on Earth.

6) TRUE or FALSE? Some shapes have their center of mass at a point that is outside the shape.

7) TRUE or FALSE? When doing calculations, you can assume that the entire mass of the object exists at a single point, the point we call center of mass. Mathematically, the rest of the object weighs nothing.

8) TRUE or FALSE? A balanced object becomes less stable if its center of mass is lowered.

9) TRUE or FALSE? The Leaning Tower of Pisa was designed to lean, in order to attract tourists.

10) What was the original function of the Leaning Tower? _____

11-13) Name three practical applications for center of mass in industry or sports:

_____

_____

_____

14) TRUE or FALSE? A barycenter is the point around which two objects orbit.

15) Who was Eratosthenes? _____

16) What scientific achievement is Eratosthenes famous for? _____

17) Was Archimedes older than Euclid? _____

18) Was Plato older than Archimedes? ____

19) The measurement of the outside (perimeter) of a circle is called its _____.

20) The distance from the center of a circle to its edge is called the _____.

21) TRUE or FALSE? The Archimedes screw can only be used to pump water.

22) TRUE or FALSE? No one knew that the earth goes around the sun until Copernicus in the 1500s.

23) TRUE or FALSE? Archimedes was against war and refused to use his genius for anything related to military.

24) Archimedes wrote a short book in which he tried to figure out how many _____ it would take to fill the universe.

25) Why were parchment books sometimes recycled? _____

For those of you working from a paperback copy of this book (not digital), patterns are downloadable at www.ellenjmchenry.com/discovering-motion-printables

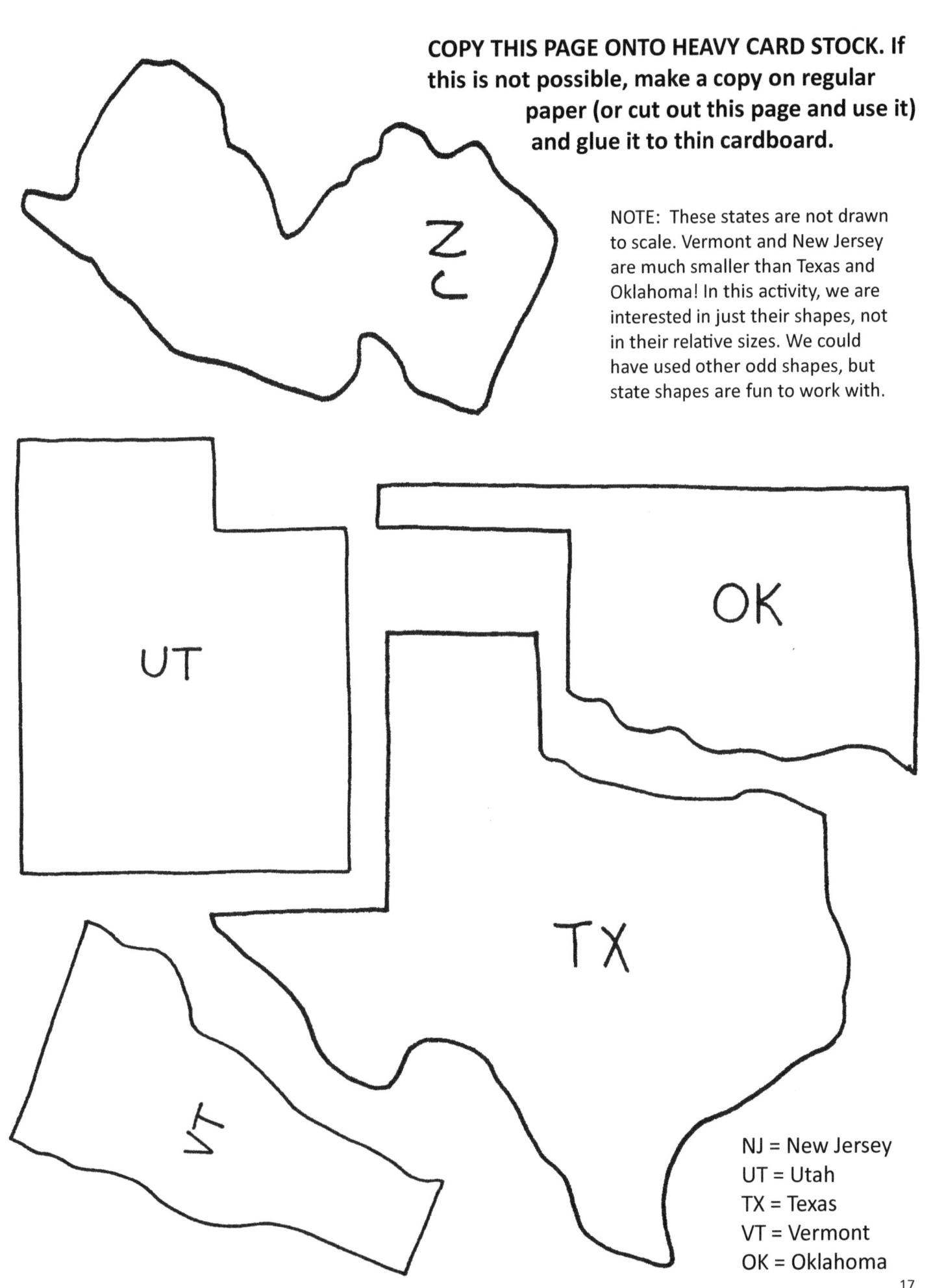

For those of you working from a paperback copy of this book (not digital), patterns are downloadable at www.ellenjmchenry.com/discovering-motion-printables

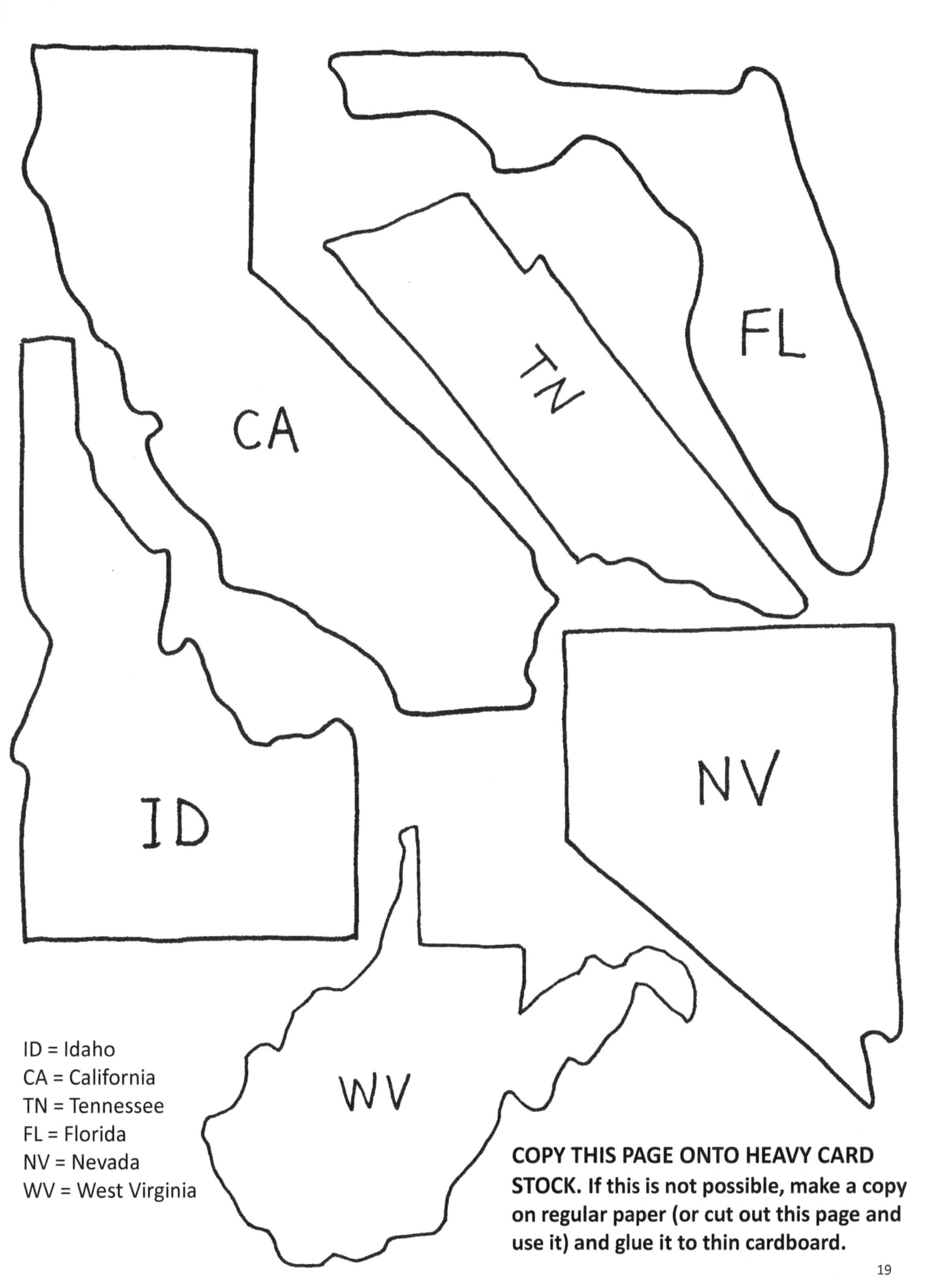

For those of you working from a paperback copy of this book (not digital), patterns are downloadable at www.ellenjmchenry.com/discovering-motion-printables

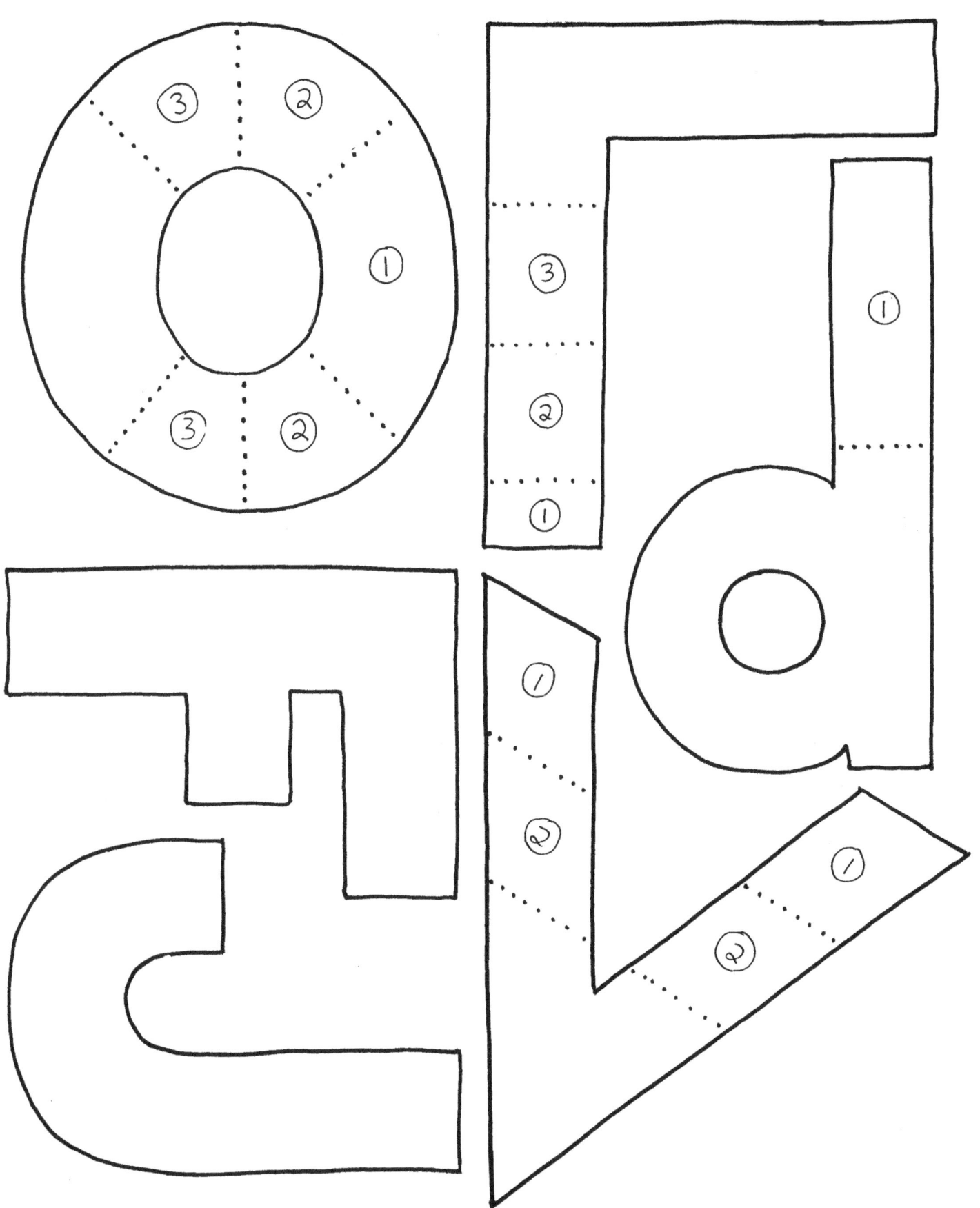

Copy this page onto heavy card stock paper. If this is not possible, you can copy it onto regular paper (or cut this page out of the book) and glue to thin cardboard.)

Adventure 2: Balancing, Lifting and Turning

Our adventure with balancing continues, as we send our volunteers to spy on Archimedes again.

Stop! We're going to interrupt our interview with Archimedes so that we can give you the opportunity to make essentially the same discovery that he is about to make. However, to make it a bit easier for you, we're going to work with paper clips instead of the flat rocks that Archimedes is using.

ACTIVITY 2.1: Balancing numbers

You will need: a few dozen paper clips that are all the same size, a pin, tape, a glue stick, scissors, a pencil that has still has its eraser (or an "eraser cap"), a can of soup (or any other fairly heavy can that can be a firm base), and a copy of the following pattern page printed onto regular paper

1) Push the pin into the pencil's eraser. My pencil had a nice "cap" eraser, but you can make it work with the pencil's eraser if it is big enough and if you don't poke it too many times.
2) Tape the pencil to the side of the can. This completes your balancing base. I like to use masking or duct tape, but any tape will do.
3) Fold the paper in half lengthwise, keeping the numbers on the edge. Fold in half again. Then once again. Be very exact with your folds! Run some glue stick between the folds. Close all the folds again, press down firmly. You should now have something that looks like a paper ruler. (See photo on following page.)
4) Cut out the V notch (striped area).
5) Place paper clips right under the numbers, as shown in here in photo.

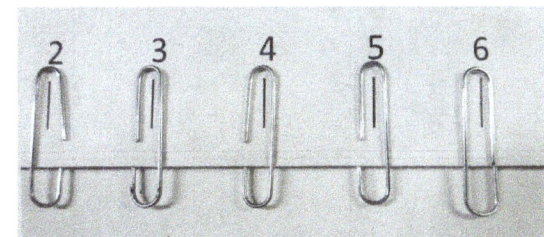

6) Balance the paper strip by placing the V notch right over the pin. If it doesn't balance, adjust the paper clips a tiny bit until it balances. If the strip wants to fall forward or backward, you can bend the top edge of the strip just slightly. Experiment a bit until your strip is stable on the pin, as shown in photo.

It's not the prettiest contraption, but it will work.

7) The rest of the paper clips will be used as weights. Open them just a bit so you can easily hook them onto the metal loops under the numbers.

8) Try hanging a clip on both number 6's. Does it balance? If it doesn't balance, adjust the end clips until it does.

9) Now let's try what Archimedes did and balance unequal numbers of clips. Hang one clip on the 7 on one side, and 7 clips on the 1 or the other side. It should balance. This means that the position at which you put a clip affects how much downward "pull" it will have. Clips placed at high numbers (which are far away from the fulcrum) will seem to have more "weight" than those at lower numbers.

10) Try some simple math. You can show that 3+3 = 6. Put 2 clips on the number 3 on one side, and 1 clip on the number 6 on the other side. It should balance. This is shown in the photo. Two 3's equals one 6.

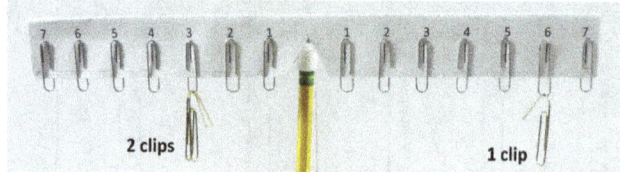

11) Now try your own combinations. Here are some suggestions: 4+3= 7   5+2 = 4+3   1+2+3= 6   2x3= 6

12) You can even do equations with parentheses. (1x3) means three clips hanging on the 1. (3x1) means one clip on the 3. Try to set up these equations on your balance bar:
   (2x2) + 1 = 5   (1x3) + (1x2) = 5   (2x3) = (3x2)   (2x7) = (2x2) + (2x5)   1+2+3+4+5+6+7 = 7x4

CHALLENGE: Set up an equation on your balance bar and then have a friend try to figure it out and write it down.

24

Archimedes had discovered what we call "the principle of the lever." Archimedes' most famous quote is "Eureka!" but his second-most famous quote is, "Give me a lever long enough, and I will move the world." Some variations of this quote add the phrase, "and a place to put a fulcrum" or "and a firm place on which to stand." Archimedes had hit upon a basic principle of physics that works at any scale, though actually moving the earth is impossible, of course.

Not only did Archimedes discover that a lever can help move very heavy objects, he also discovered that levers are governed by mathematics. You don't need to actually run an experiment with 10-ton objects to know what force is needed to lift them. You can easily calculate this with pencil and paper. The limiting factors in any real-life test of the principle of the lever are the strength of the lever (to not break when force is applied) and the strength of the fulcrum (to not crack under all the pressure being applied downward onto it). So in real life, this principle could appear to fail simply because of the limitations of the materials you are working with.

Levers give us *mechanical advantage*. The amount of mechanical advantage a lever gives us can be calculated by measuring the distance of the lever on each side of the fulcrum. The side of the lever that has the load resting on it is called the *resistance arm*. It is called the resistance arm, and not the load arm, because the resistance can be any kind of resisting force, not just a weight (a load) pressing down. The side of the lever that is applying the downward force to lift the object is called the *force arm*.

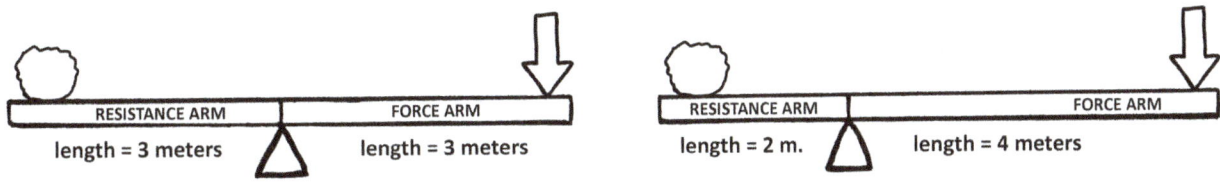

The amount of mechanical advantage is calculated by **dividing the length of the force arm by the length of the resistance arm** (FA/RA). In the diagram on the left, this gives us 3 divided by 3 (3/3) which is equal to 1. By definition, to get any mechanical advantage, we need a number higher than 1. In the diagram on the right, if we divide the length of the force arm by the length of the resistance arm, we get 4/2 which is equal to 2. This lever gives us a mechanical advantage of 2. The higher the number, the easier it will be to lift the load. The mechanical advantage does not have to be a whole number. You can have a mechanical advantage of 1.5 or 7.3 0.25. However, a mechanical advantage less than 1 means it isn't an advantage at all—it means the work will be harder for you!

ACTIVITY 2.2: Calculate the mechanical advantage of these levers using (Force arm/Resistance arm)

The answers are printed on the bottom of the reverse side of this page. (Don't check your answers until you have completed all of them. This will prevent you from accidentally seeing right answers before you do the problems.)
NOTE: The measurements (meters, centimeters, inches) won't be in your answer. Your answer will be just a number.

1) 2 cm / 12 cm

Mechanical advantage = _____

2) 6 m / 3 m

Mechanical advantage = _____

3) 5 m / 1 m

Mechanical advantage = _____

4) 8 m / 2 m

Mechanical advantage = _____

5) 8 in. / 24 in.

Mechanical advantage = _____

6) 200 cm / 300 cm

Mechanical advantage = _____

ACTIVITY 2.3: Solve these balancing puzzles

Find the missing numbers in each lever puzzle below. Remember, these levers are balanced, so that means that the weight times the distance on one side equals the weight times the distance on the other side.

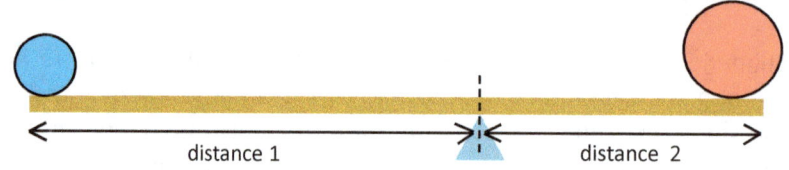

(weight of blue circle) x (distance 1) = (weight of red circle) x (distance 2)

The answers are printed on the back of the next page. Don't check your answers until you are finished with all of the puzzles.
NOTE: Your answers will have units attached to them this time, such as kilograms, meters, feet, or pounds

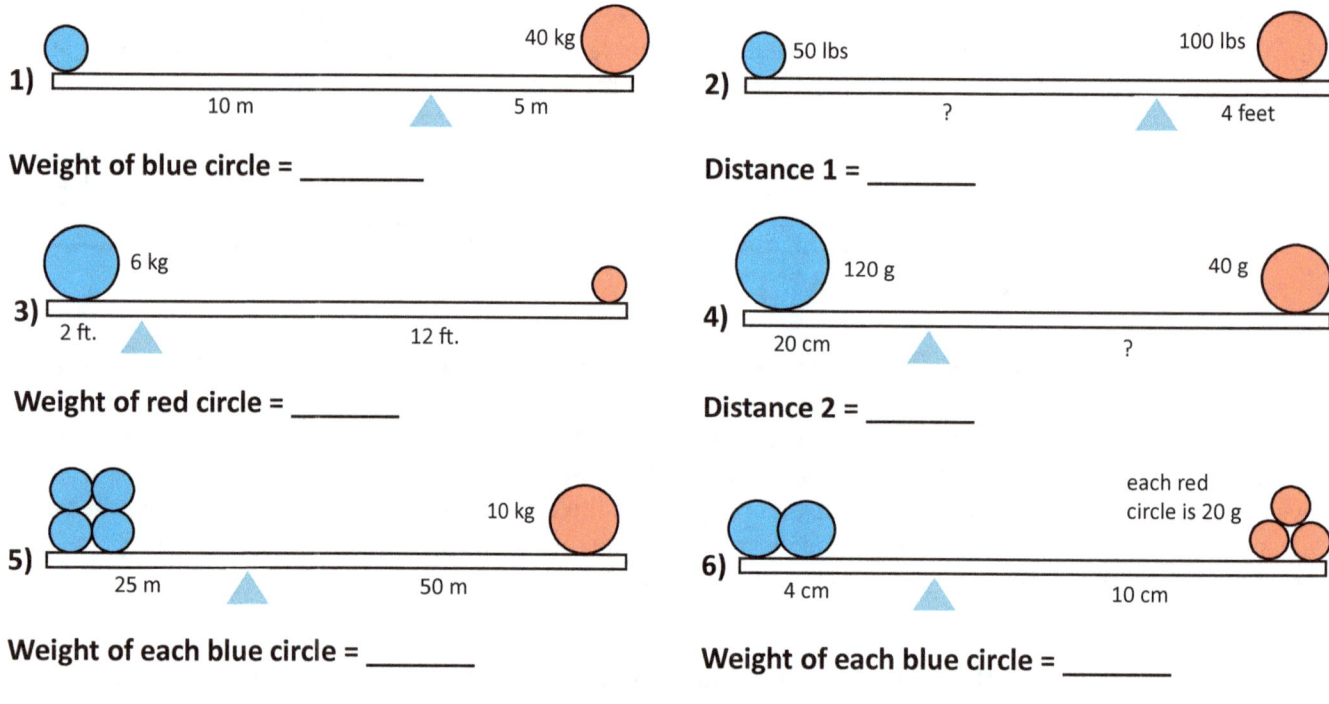

1) Weight of blue circle = _____

2) Distance 1 = _____

3) Weight of red circle = _____

4) Distance 2 = _____

5) Weight of each blue circle = _____

6) Weight of each blue circle = _____

Archimedes made another discovery about levers. He found that although a really long lever gives you excellent mechanical advantage, the distance that the lever has to move is greatly increased. An elephant and a mouse can operate a teeter totter (seesaw) together, but the mouse will have a wild ride, going high up in the air. The elephant's ride will be boring by comparison. It will hardly go up and down much at all. But even though the mouse and the elephant are having such different experiences, the amount of "work" each one is doing is the same.

ANSWERS TO ACTIVITY 2.2   1) 6   2) 1/2 or .5   3) 5   4) 1/4 or .25   5) 3   6) 3/2 or 1.5

Believe it or not, there is a lot more we can say about levers! Engineers came up with a way to classify different types of levers. There are three ways you can arrange the force, load and fulcrum.

FIRST CLASS LEVERS have the fulcrum in the middle, the load on one end and the force on the other end. The classic playground seesaw (which is hard to find nowadays) is a first class lever. Other examples include scissors and light switches.

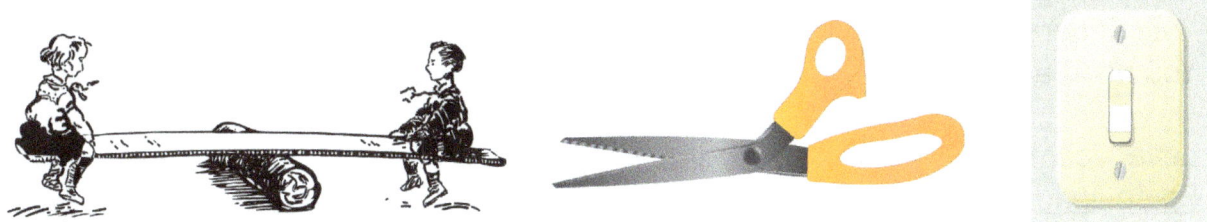

SECOND CLASS LEVERS have the load in the middle and the force and fulcrum at the ends. Examples of second class levers include wheelbarrows, doors and staplers. In the wheelbarrow, the wheel functions as the fulcrum. In the door and the stapler, a hinge functions as the fulcrum.

THIRD CLASS LEVERS have the force in the middle and the load and fulcrum at the ends. Examples of third class levers include baseball bats, brooms, and chopsticks. Any time you are using a stick to swat or bat or push, you are using a third class lever. Other examples include tennis rackets, hammers, fishing rods, or even your arm.

Even if we understand the differences between the three types of levers, it can be hard to remember which one is first or second or third class. You might be saying, "Why do we even care? When we will ever need to know which is which?" Fair enough question, but if you end up going into any type of engineering, physical therapy, kinesiology, or sports medicine, you will run into these classifications in the first year of your studies and you will be expected to know the answers on a test. If you don't plan on going into a profession that uses levers, you never know where this information might show up in every day life. If nothing else, you can impress your relatives at the next family gathering, identifying levers being used to serve and eat food. This next activity will give you a song that makes learning these classifications very easy.

ACTIVITY 2.4:   "The Lever Rap" Song

*Let's have some fun with this not-funny physics stuff. On the reverse side of this page you will find the lyrics to a rap song about the three types of levers. You can access the audio files for this song by going to **www.ellenjmchenry.com/music**. Scroll down until you see the Lever Rap song. There is a version with the lyrics and also a "karaoke" version with just the drum beat.*

The Lever Rap
by Ellen J. McHenry

If you have a job that's hard to do,
Like pulling a nail or turning a screw;
Your fingers won't do it, so be very clever,
You need a machine that's called a lever.

Levers come in classes: first, second, third.
<u>Force</u>, <u>load</u> and <u>fulcrum</u> are the key words.
In a <u>first class</u> lever, the fulcrum's in the middle.
If I seesaw with you, you seem to weigh little!

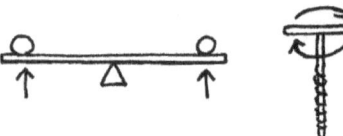

A <u>second class</u> lever has the fulcrum at one end;
The load's in the middle—give a ride to a friend
In a big wheelbarrow, where the force is your muscle,
If you call your friend "The Load," you'll get into a tussle!

If you want some speed to swat or bat,
The <u>third class</u> lever is where it's at.
The fulcrum's at one end, the load's at the other,
The force is in the middle. But don't hit your brother!

When you use any lever, you'll find, of course,
You can always trade distance for a gain in force.
If you make the lever long, moving loads is a snap.
And this is the end of the lever rap.

"Give me a lever long enough, and a fulcrum on which to place it, and I shall move the world!"
                                                            Archimedes (about 200 BC)

www.ellenjmchenry.com

ANSWERS to ACTIVITY 2.3: 1) 20 kg   2) 8 feet   3) 1 kg   4) 60 cm   5) 5 kg   6) 75 g

ACTIVITY 2.5: Experiment showing mechanical advantage in a second class lever (wheelbarrow)

If you are not able to do this experiment right now, but you have Internet access, you can watch a video version of this lab by going to the YouTube playlist for this curriculum and finding the video titled "Simple Machines: The Lever" by funsciencedemos.

You will need: *a meter stick (or yard stick), three large binder clips, a pencil, some books, a spring scale that measures in grams, a thick rubber band, a paper clip, and a weight (a can of soup will work)*

Use your materials make this set up:

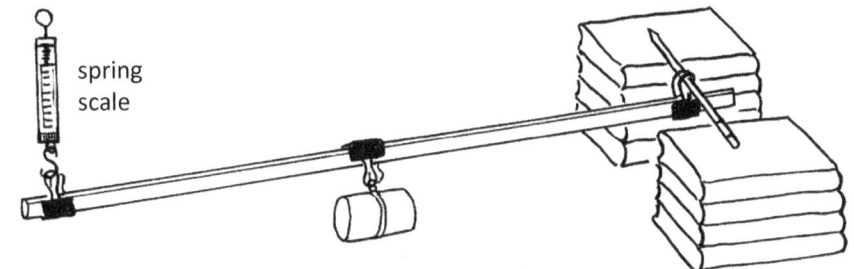

You've set up a second class lever, with the fulcrum being the pencil. The load will hang from the clip in the middle and the force will be a lifting force at the end with the spring scale attached (so you will be able to measure the force). In a wheelbarrow, you'd have a wheel instead of the pencil and there would be a metal basket in the middle instead of a clip with a weight hanging from it.

Place the thick rubber band around the soup can (or whatever you are using for a weight) and use a paper clips to hang it from the middle clip. Slide the binder clip all the way to the end where the spring scale is. The clips can be touching. Lift the stick up while holding the top of the spring scale so that the scale will be measure the force needed to lift the weight. If the weight was directly under the scale, the scale would show you exactly how heavy the weight is. However, since the weight is not directly under the scale, the scale will read just a bit less than the actual amount of weight. But for our purposes here, it will be close enough.

Now move the clip with the weight to the middle of the stick. Lift the stick again and look at what the reading is on the spring scale. Does it read less? Then move the weight closer to the pencil. Read the scale again. Finally, move the weight as close to the pencil as you can get it. Lift the stick and check the scale. How much force is required to lift the can when it is almost at the fulcrum?

Thinking about what you just saw in this experiment, would longer handles on a wheelbarrow make it easier to lift loads? Why is the wheel of a wheelbarrow usually right underneath the metal basket?

In the Lever Rap, the first sentence says, "If you have a job that's hard to do, like pulling a nail or turning a screw..." According to this song, a screwdriver is a lever? But a screwdriver doesn't lift—sit twists. A screwdriver looks more like a wheel and axle than it does a lever.

You've probably studied **simple machines** when you were a little younger. If so, you might remember that usually we are told that there are six different kinds of simple machines: levers, wedges, pulleys, screws, inclined planes, and the wheel and axle. Actually, this list can be boiled down to just two: **the lever and the inclined plane**. All the rest are variations of these two. It's easy to see how a wedge is very much like an inclined plane. Just turn an inclined plane up so its tip is facing down and you have a wedge. It's a bit harder to see how a screw is derived from an inclined plane, but if you cut a long triangle out of paper then wrap it around a pencil, you can see the connection.

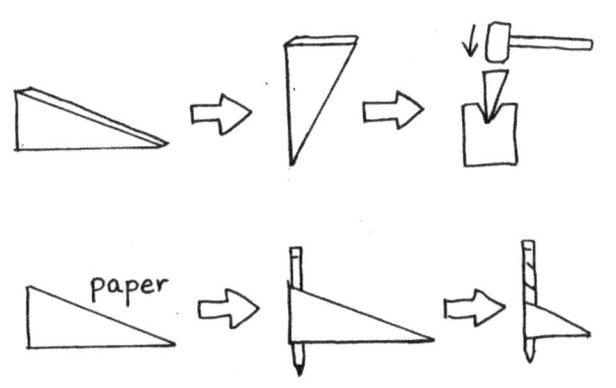 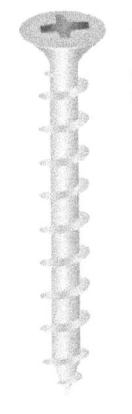   

Yes, a screw is a combination of several simple machines. Something that has more than one simple machine is a *complex (or compound) machine*. A screw certainly does not seem complex but technically it does combine several simple machines. It even has a hidden lever.

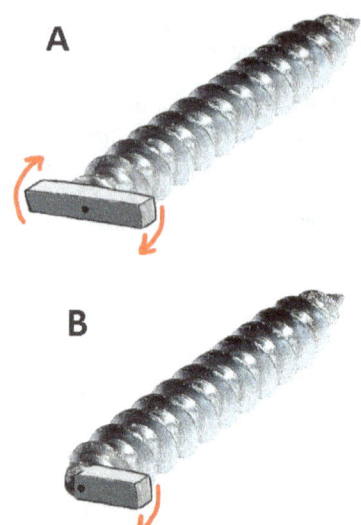

A screw would still work if we modified the head and made it look like a bar instead of a circle. (Diagram A) If your fingers were strong enough, you could twist the bar to make the screw turn. Now that it looks like a bar, it reminds us of a seesaw, a first class lever, with a fulcrum in the middle. One end goes down while the other end goes up.

Technically, we could still make the screw turn if we cut off one half of the bar. (B) It would be much harder to turn the screw, but if you were putting the screw into something soft, like Styrofoam, it would work. Now it looks like the fulcrum is at one end and the force is at the other end (the red arrow) reminding us of a second class lever. If we wanted to make this lever easier to turn, we could increase its length. In real life, we would not want a long bar sticking out from a screw. Instead, we rely on a screwdriver to give us mechanical advantage.

Wheels (including the head of this screw before we chopped it up) are a variation on the principle of the lever, but it does get complicated. A wheel is like a infinite number of levers all combined into a circle shape.

The windlass was already in use by Archimedes' time.

A wheel and axle are like two cylinders stuck together. The larger cylinder is designated as the wheel and the smaller cylinder is the axle. In this drawing of a "windlass," the axle is the cylinder around which the rope is wound, and the wheel really does look like a wheel. In other situations, the wheel might be harder to identify, but in the windlass it is very easy.

The circle with the larger diameter (the wheel) gives you mechanical advantage. The way you calculate the mechanical advantage is to measure the radius of both wheel and axle. The radius is the distance from the center of a circle to its edge.

MECHANICAL ADVANTAGE = $\frac{\text{radius of wheel}}{\text{radius of axle}}$

The line means "divided by." (When we write fractions we use this "divided by" line, though often we forget what it means.)

If the radius of the wheel in this picture is 60 centimeters and the radius of the axle is 20 centimeters, the wheel will give us a mechanical advantage of 3 (60/20=3). We'd have three times our normal lifting power. If we wanted even more lifting power, we could make the axle even smaller. An axle of only 10 centimeters would give us a mechanical advantage of 6 (60/10=6). However, as we discovered with our levers, when you achieve greater mechanical advantage (a gain in force), you always have an increase in distance. Here, the increase in distance will mean that you have to turn the wheel twice as many times.

The screwdriver that we might have used to turn our screw (instead of reshaping its head into a bar!) is also a wheel and axle. The long metal shaft is the axle and the handle is the wheel. According to our formula for mechanical advantage, to make a powerful screwdriver the handle should be substantially larger than the diameter of the long shaft. We are limited, however, by the size of our hands; we have to be able to comfortably wrap our fingers around the handle. Another limiting factor for handle size is our preference for having tools that don't take up too much space in our toolbox.

A ***gear*** is an example of a wheel that does not have an axle. It has a pivot point in the center, but it does not have another cylinder coming out of it. Gears can be smooth, or they can have ***teeth***. Gears can be used to gain mechanical advantage. Similar to the wheels and axles on the previous page, the diameter of the gears can be used to determine mechanical advantage.

One gear will be the ***driver***. This is the gear that will be either connected to a motor or will be turned by hand. The other gear is called the ***follower*** (or the ***driven***).

MECHANICAL ADVANTAGE = $\frac{\text{radius of follower}}{\text{radius of driver}}$

The red gear will be our driver. It has a radius of 20 centimeters. The blue follower has a radius of 40 centimeters. Our mechanical advantage is 40/20 = 2. This means that the "work" you put into turning the red gear will be doubled. Good deal!

You can also calculate the mechanical advantage by using the number of teeth:

MECHANICAL ADVANTAGE = $\frac{\text{number of teeth on follower}}{\text{number of teeth of driver}}$

This would be 26/13 = 2, the same answer we got using the radius. We'd also get the same answer if we used the circumference (distance around the outside) of the circles.

Now let's make the blue gear our driver. For every full turn of the blue gear, the red follower goes around twice. The result will be that the smaller gear will be turning twice as fast as the large one. This situation is called a ***speed multiplier.*** Sometimes this is exactly what you want in a complex machine. You want to have a gear that speeds up motion. However, sometimes you want more power, not speed. When we had the smaller red gear as our driver, we had a mechanical advantage of 2, making our turning motion twice as powerful (even though we had to do more turning). When we have a mechanical advantage of more then 1, the gears are acting as a ***force multiplier.***

A classic example of a force multiplier is a 10-speed bicycle. The lowest ("first") gear is very small. You use the smallest gear when you are going up a steep hill and you don't mind sacrificing speed in order to achieve a gain in force. If you try using the lowest gear while riding on a flat surface, you realize how incredibly fast you must pedal. But when you were riding up that hill, the speed at which you were pedaling seemed about right.

If you watch gears in action, you'll notice that the driver makes the follower turn in the opposite direction. In cases where you need your follower to go the same direction as your driver, you can put a gear between them. This middle gear is called an ***idler*** (IDE-ler) because it doesn't affect the mechanical advantage at all. (To be "idle" means to do nothing.) The size of the idler isn't important; any size will work. This diagram shows a tiny idler but you could also have a large idler and tiny drivers and followers. You can also have more than one idler. An odd number of idlers will result in the driver and follower turning the same direction, but an even number will make them turn in opposite directions.

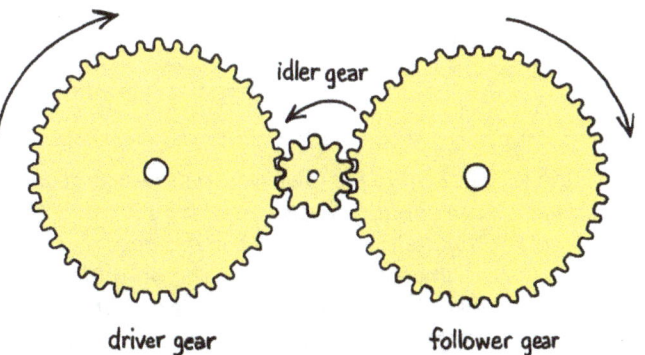

How about this mnemonic *(nem-ON-ik)*? The word "driver" is similar to the word "diver," (You won't forget that the real word is driver, not diver.) A diver is down below the surface, so this helps you remember that the "driver" gear radius is below the line. The two letter O's in the word "follower" will then remind you of round air bubbles being created by the diver. Air bubbles float up, so the word with the O's (follower) is on the top, above the diver.

ACTIVITY 2.6: Do some thinking and a little math (Answers at the bottom of page 45.)

Answer these questions for the gears shown below.
1) If gear A is the driver, will this result in a speed multiplier or a force multiplier? _____
2) If gear B is the driver, what will be the mechanical advantage if gear A has a radius of 24 cm and gear B has a radius of 12 cm? _____

Answer these questions for the gears shown below. (Notice the arrow on the yellow gear.)
3) Which direction will the green gear turn, clockwise or counterclockwise (anticlockwise)? _____
4) Would the green gear still turn the same way if you removed one of the purple gears? _____

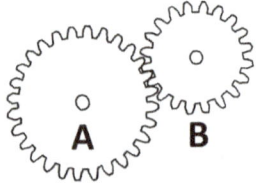

5) If a driver gear has 18 teeth, and its follower has 36 teeth, will this result in multiplication of speed or force? _____

6) What is the mechanical advantage of gears where the follower has a radius of 12 cm and the driver's radius is 60 cm? _____

7) Look at the colored gears again. Will the green gear be experiencing more, or fewer, complete turns per minute than the yellow gear is? (What did we learn about idlers? Do they contribute to force or speed?) _____

8) An employee in the Fabulo Tool Company had a great idea. Why not make a super powerful screwdriver that gives the user a mechanical advantage of 20? Most screwdrivers only give an advantage of 2-5. The CEO and the board of the company weren't as enthusiastic as the employee about this wonderful idea. All Fabulo screwdrivers have a standard shaft size of half (.5) a centimeter. Why the lack of enthusiasm for a screwdriver with a mechanical advantage of 20? _____

_____

9) Eratosthenes was sent to the town well to draw a large bucket of water. Without any mechanical advantage, he can only lift a bucket weighing 150 mina. (One mina equals about 450 grams, or one pound.) Fortunately, Archimedes had designed a windlass for the well. The wheel of the windlass had a diameter of 45 daktyloi. (The word "daktyloi" comes from the Greek word for "fingers." One daktylos was about 2 cm.) The axle of the windlass was about 15 daktyloi. By using the windlass, will Eratosthenes be able to lift a bucket weighing 300 mina? _____

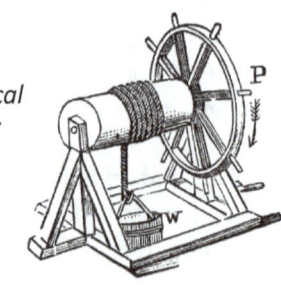

Meanwhile, back in ancient Greece, our volunteers have moved on to another episode in the life of Archimedes. We find him on a road in his home town of Syracuse. Archimedes has used gears to make a machine that will measure out long distances. Greeks did not measure in miles or kilometers, of course. Archimedes' friends might be measuring in "stadia" (about 185 meters or 200 yards).

The most famous example of the use of gears in the ancient world is the Antikythera mechanism. It was found in a shipwreck off the coast of Greece in 1901. No one knew for sure what this device did until 2008 when a team of researchers from Cardiff University used x-rays to decipher tiny bits of writing and to get a better look at some of the inner gears. It became clear that this ingenious device was a "clock" for the motion of the sun, moon, and planets, and could predict solar and lunar eclipses. Historians are fairly sure that the device was made during the second century BC (the 100s). Since Archimedes lived during the third century BC (the 200s), he can't have made the device himself. However, the device is so clever that historians tend to think that it must have been based on similar devices designed by Archimedes. We know that Archimedes was the greatest inventor in ancient Greece and was one of the few people with a good enough understanding of gears to be able to invent something like this.

CC BY 2.5, https://commons.wikimedia.org/w/index.php?curid=469865

Archimedes was not only an expert in the use of gears, but he also seems to have been the first person in history to realize the incredible potential of pulleys. Pulleys consist of a group of wheels that work together to give mechanical advantage. The stimulus that got him thinking about the power of pulleys was the threat of invasion by the Roman and Carthaginian armies. Sicily was, unfortunately, located between these two empires. The king of Syracuse asked Archimedes to please save their city if he could. So, (as we learned in our first adventure), Archimedes set to work designing both offensive and defensive devices. He showed the craftsmen how to improve their catapults, making them able to hurl much heavier objects with increased accuracy. His most famous defensive device was called "the claw." Archimedes put a huge metal grappling hook on the end of a very strong rope that was suspended from a huge lever hanging over the city's wall. A system of pulleys allowed soldiers to have such an incredible mechanical advantage that they could pull an enemy ship up out of the water after the hook had grabbed it. As the end of ship went up into the air, people and gear fell out. Then the soldiers manning the pulleys could suddenly release the ship, causing it to go crashing down into the water, hopefully breaking the hull.

The painting shown here is from the year 1600. The artist choose to show the claw as an iron hand. It is unlikely that it really looked like this.

ACTIVITY 2.7: Watch some videos right now, if you can

The very best way to learn about pulleys is to see them in action. If you can possibly do so, take time right now to access a few videos. You can go to the YouTube playlist for this curriculum, or you can search for some yourself. We highly recommend starting with "Simple Machines: The Pulley" by funsciencedemos, and episode 228 on the "Smarter Every Day" channel. (This curriculum doesn't have any business connections with those channels.)

Here is a recap of what you learned in the videos.

In diagram A, the pulley is changing the direction of the rope but it is not giving any mechanical advantage. In diagram B, the pulley is giving a mechanical advantage of 2, so it is half as hard to lift the weight. Why does simply changing the location of the pulley make such a difference? In diagram A, all the weight is pulling down on the rope that runs from the weight to the pulley. In diagram B, you have two ropes holding up the weight, therefore, each rope only has to bear half of the weight. The other half of the weight is being supported by the hook on the ceiling (not shown in the diagram).

A quick and easy way to calculate the mechanical advantage of a pulley system is to count the number of ropes holding up the weight.

The diagrams below (C-F) show some pulley systems that use both fixed and movable pulleys. The red W's show you how much of the weight each rope is holding. In diagram C, each rope is holding half of the weight, indicated by W divided by 2, (W/2). In diagram D, each rope is holding one third of the weight, (W/3), because the force arrow is pulling up, not down. In diagram E, we don't have any additional mechanical advantage because now the force arrow is pulling down, so we still have three ropes bearing the weight. In diagram F, we have four ropes holding the weight, so each rope only has to bear one fourth of the weight, (W/4), and we have a mechanical advantage of 4.

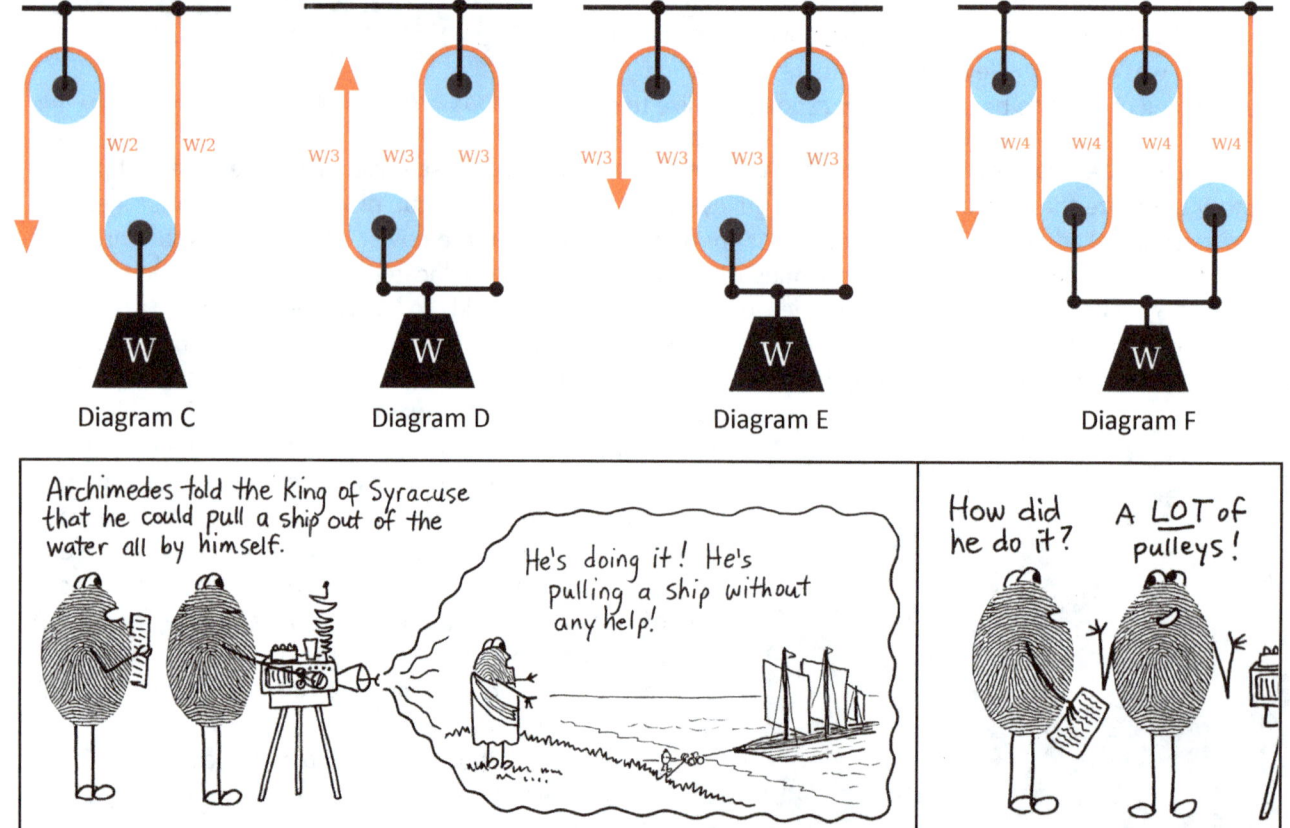

36

ACTIVITY 2.8: Online Gear Games

Gears are hard to make, which is why we're not going to suggest doing a hands-on lab where you have to make your own gears. If you happen to have a 3D printer and want to make some gears, it is easy to find patterns online. Instead, we're going to recommend a few websites where you can experiment with gears.

https://javalab.org/en/gear_en/

This is an easy way to play with gears! You can place as many gears as you want to on the virtual board. See if you can get the last gear in line to spin super fast. Then, see if you can make the last gear go so slowly that it doesn't even look like it is moving. Make several lines of gears going out from a central gear. Cover the board with gears or make them form a square. What else can you do with them?

(If this web address doesn't seem to work, try searching for "Javalab gears" using a search engine.)

https://www.engineering.com/gamespuzzles/connectit.aspx

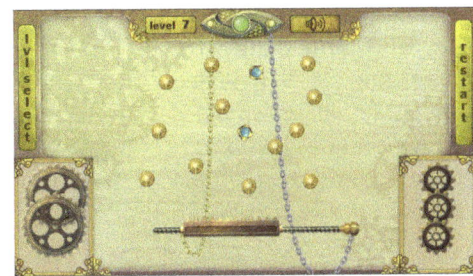

This is actually a game where you progress through levels of difficulty. The goal is to arrange your gears so that the last one in the line matches up with a toothed bar that slides over to finish the puzzle. The first few levels are easy so that you can get the idea. By the time you hit level 8 or so, you really have to think! (HINT: You will eventually have to use a 2-step strategy, moving the bar halfway, then reorganizing your gears to move it the rest of the way.)

ACTIVITY 2.9: How much of the lever rap can you remember?

You might want to listen to the song at least one more time before doing this activity. Then, see if you can fill in these blanks without looking at the lyrics. (If you really get stuck, go ahead and peek.)

If you have a _____ that's hard to do, like _____ a _____, or turning a _____,
Your _____ won't do it, so be very _____; you need a machine that's called a _____.

Levers come in classes: _____, _____, _____,
Force, _____ and _____ are the key words.
In a first class lever, the fulcrum's in the _____,
If I _____ with you, you seem to weigh _____.

A _____ class lever, has the _____ at one end,
The _____'s in the middle, give a _____ to a friend
In a big _____ where the force is your _____.
Don't call your friend "The _____" or you'll get into a tussle!

If you want some speed to _____ or _____,
The _____ class lever is where it's at.
The fulcrum's at one end, the _____'s at the other.
The _____ is in the middle, but don't hit your _____!

When you use any _____, you'll find, of course,
You can always trade _____ for a gain in _____.
If you make the lever _____, moving loads is a snap.
And this is the _____ of the _____ _____!

ALSO...
watch epdisodes 12-15
of the *Eureka!* series.

ACTIVITY 2.10: A review word puzzle (answer key in teacher's section)

ACROSS:
2) The island where he lived.
3) Archimedes studied this arched shape-- the path traced out by any projectile.
5) In a ____ class lever the fulcrum is in the middle.
6) The driver gear turns the ____.
7) The parts of a lever are the force, load and _____.
8) Archimedes used a giant lever and some pulleys to make "the ____" (a device that could pick up ships).
12) Archimedes loved to study circles. He tried to calculate the value of "pi" which is the _____ of a circle divided by its diameter.
16) Archimedes certainly would have discovered that this type of gear can be used to make the driver and the follower turn in the same direction (not to multiply force or speed).
18) Archimedes did a lot of work with center of _____ though he would not have called it by this name.
19) The _____ device might have been based on earlier inventions by Archimedes.
21) This city in Egypt was famous for its library.
22) This means "I found it!" in Greek.

DOWN:
1) Archimedes considered that one of his greatest achievements was figuring out the relationship between the _____ of a sphere and the _____ of a cylinder that exactly contains the sphere. (same word in both blanks)
2) The city where Archimedes lived
4) The principle of _____ was what Archimedes discovered while in this bathtub.
9) He was the librarian of the best library in the ancient world. He also calculated the circumference of the earth.
10) The most famous mathematician before Archimedes. He might have been Archimedes' teacher.
11) The "Archimedes _____" is a simple machine that ancient people used to move large volumes of water.
13) In a first class lever (like a seesaw) one side is the force arm and the other side is the _____ arm.
14) The ancient Greeks were experts at this type of math. It deals with lines and shapes.
15) Archimedes used a series of this type of simple machine to make it possible for him to pull a ship by himself.
17) This wheel and axle structure was already in use before Archimedes was born.
20) The language Archimedes spoke

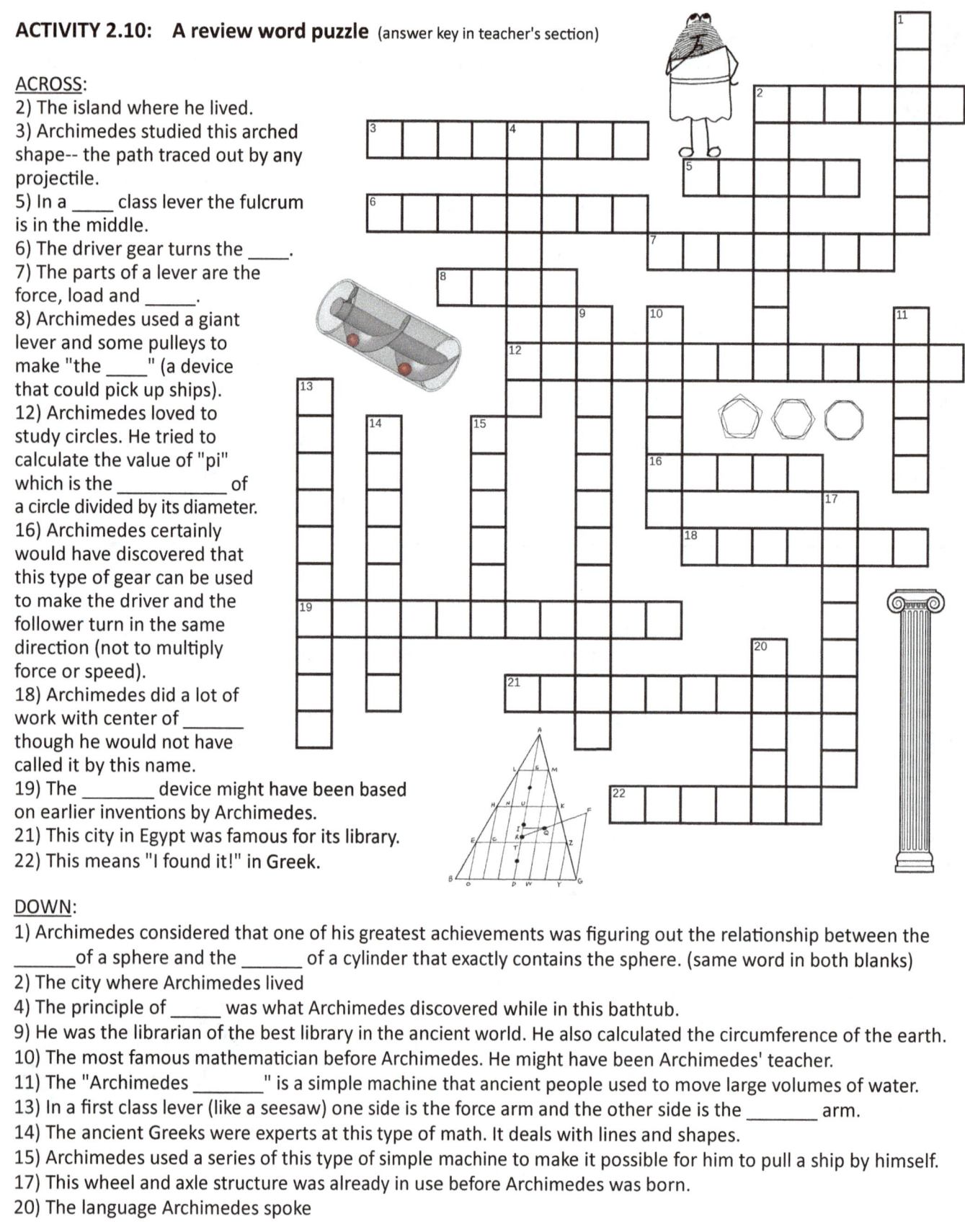

ANSWERS for 2.6
1) speed 2) 2 3) clockwise 4) yes 5) force (When the driver is smaller than the followers, this results in force multiplication.)
6) 1/5 (one fifth) Mechanical advantage of less than 1 means are not gaining force, but gaining speed instead. 7) fewer turns per minute
8) A mechanical advantage of 20 would mean that the screwdriver handle would have to be 10 cm in diameter, which is too large for an ordinary person's hand. 10/.5=20 9) yes (The mechanical advantage given by the windlass is 3. Therefore, with the windlass, he would be able to lift three times what he can lift without the windlass. The windlass will increase his lifting power to 450 mina.)

38

Adventure 3: Swinging with Galileo

Our time travel adventure is going to fast forward now, from about 250 BC to about 1600 AD. We are going to send our volunteers to spy on one of the most famous scientists of all time: Galileo Galilei. He was born in the city of Pisa, in what we now call the country of Italy; but back in his time, each Italian city was like a tiny, independent country. Sometimes these city-states even fought wars against each other.

The city of Pisa (which you will remember from our center of gravity adventure) belonged to the larger city of Florence. Florence was, and still is, a very cultural city, filled with artists and musicians. Since Galileo's father was a musician, he decided to move his family to Florence when Galileo was 8 years old. Thus, Galileo grew up surrounded by beautiful Renaissance buildings and sculptures, reading Latin and Greek literature, and listening to the finest music available at that time. His father was very interested in music theory—the mathematical structure of music. He taught Galileo much about both the mathematical basis of music. Galileo was successful at everything he studied: Greek, Latin, literature, poetry, music, and art. Though none of his paintings have survived till this day, we know from the writings of people who knew him that he was a very talented artist. It was also said that he could play the lute (a guitar-like instrument) as well as the professionals of his day.

Galileo's father knew all too well how hard it was to support a family on a musician's or artist's income, so he told Galileo that he had to study medicine and become a doctor. Galileo would have preferred to become a priest or a monk at that point in this life, or maybe an artist, but he had to conform to the wishes of his father and he was packed off to study medicine at a school in Pisa.

While studying in Pisa, Galileo often went to church services in the cathedral.

What surprised Galileo? You can find out for youself by doing the experiment on the next page.

ACTIVITY 3.1: Make a swinging chandelier

You will need: *a copy of the chandelier page (opposite) printed onto card stock, scissors, a pin, a stopwatch or clock (For those with paperbacks, patterns are downloadable at www.ellenjmchenry.com/discovering-motion-printables)*

1) Print the following pattern page onto card stock. Trim off the chandelier piece, (trim right on the straight line—don't actually cut out the chandlier) but make sure to keep the other two stripes, as you will be using them later. If you are working from a paperback copy of this book (instead of a digital book), you can cut out the chandelier page and glue the chandelier strip to thin cardboard or thick paper. (You can also download patterns here: www.ellenjmchenry.com/discovering-motion-printables)

2) Push a pin through the paper right at the top of the chandelier chain, about a centimeter from the top. Wiggle the pin a bit so the hole enlarges just slightly. The strip of paper should be able to swing freely from the pin. Make sure the paper is hanging straight and is not curled at all. Also, make sure you are in a place where the air is very still. If you are near a heater, the paper can be blown about by warm currents of air. You can either hold the pin with your fingers (if you can hold it very still), or you can tape one end of the pin to the edge of a table. Or, if you have a bulletin board handy, you can stick the pin into it.

3) First, we will do an experiment without a timer. You will pull the chandelier to the side and let it go and then watch it until it stops. (Don't pull it too far to the side. If you imagine that the chandelier is the hour hand of a clock, pointing at 6 o'clock, pull it to no more than 4 or 8 o'clock, maybe a little less.) As you watch it swing, the important thing to notice is the "beat" of the back-and-forth motion. Keep time with it in your head— perhaps sing a song in your mind, with the chandelier providing the "beat." Is the beat steady?

4) Now we will use the timer. Get the stop watch or clock ready. You will count how many times the chandelier swings in 10 seconds. Pull the chandelier to one side and then let it go and start it swinging. (Again, don't pull it too far to the side, no more than the 4 or 8 o'clock position.

5) Now try it again, but this time pull it to only the 5 or 7 o'clock position. Time for another 10 seconds.

6) Try it again at least one more time.

7) What did you discover about the time of the swings? Did the little swings seem to take as long as the big ones? If you were accurate with your timing, you should have found the results to be very close to identical, if not identical. This is what Galileo discovered. Those tiny swings (as it is almost stopping) take just as much time as the bigger ones!

It appears that the distance the pendulum travels back and forth does not affect how long it takes to complete a full swing. You came to this conclusion and so did Galileo. This is entirely true if you don't start the pendulum swinging too high. As long as the swings are relatively small (less than about 10 degrees from the vertical starting point) the timing is so accurate that a pendulum can be used as a time-keeping device inside a clock. However, for larger swings, pendulums are not perfect time-keepers. It turns out that if you release a pendulum from a very high point, those first couple of swings take more time than the smaller ones. Mathematicians can calculate exactly how much difference there is between large swings and small swings. This is what their equation looks like:

This kind of equation is called an "infinite series."

$$T = 2\pi\sqrt{\frac{L}{g}}\left(1 + \frac{1}{16}\theta_o^2 + \frac{11}{3072}\theta^4 + \frac{173}{737,280}\theta^6 + \ldots\right)$$

Hmm... let's not delve into that one! Let's just say there is some difference between big swings and little swings, and big swings are not useful for time keeping. In the next activity, we will see this difference.

ACTIVITY 3.2: Wide swings compared to narrow swings

You will need: *a long object such as a meter stick (yard stick) or two rulers taped together, or any long skinny object you have at hand such as a golf club or ball bat, a piece of string, scissors, tape (duct tape works well), and your timer*

1) Tape a loop of string to the end of your long object. Hold it so that it can swing freely. Make sure to not to move while holding the object; you don't want any body motion to influence your results.

(continued on next page)

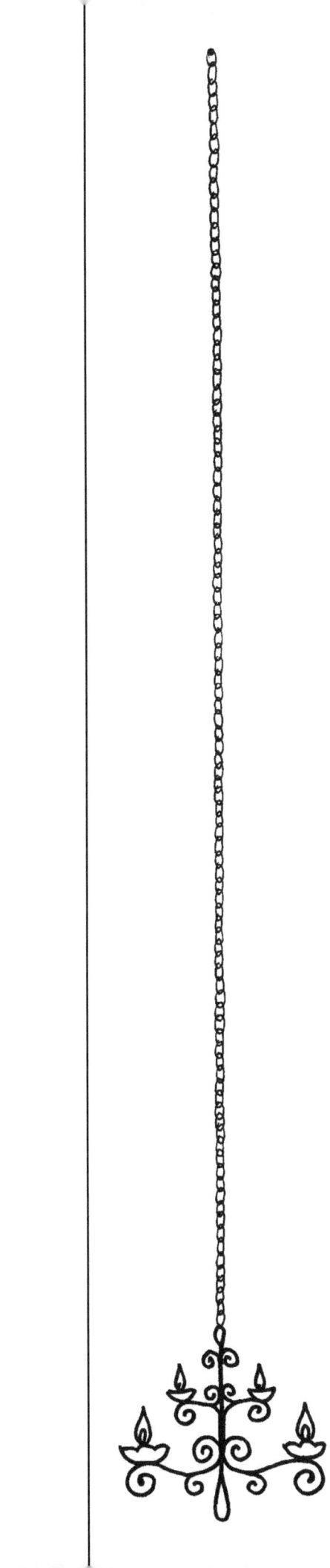

2) Pull the object up to a very high starting point, almost straight up. Get your timer ready. Let go and time how many swings it does in 20 seconds.
3) Now try it again, but only pull back the object a tiny bit so the swings will be small. Time another 20 seconds.
4) Did the very large swings take a bit longer than the small ones?

You'd think that after making this interesting discovery, Galileo would have realized that a pendulum might be used as the timing mechanims for a clock. But he didn't. Instead, perhaps because he was in medical school at the time, he thought about reversing the situation and instead of using a pulse rate to time the pendulum, perhaps a pendulum could be used to time pulse rate. He designed a simple pendulum device for doctors that he called a "pulsilogium." With this device, doctors could determine an exact measurement of how fast a patient's heart was beating. Galileo took great delight in designing and building practical tools. Years later, he would finally realize the potential of the pendulum for time-keeping and would sketch a design for a pendulum clock. Sadly, this was right before he died and he never got to build the clock. But this is getting ahead of our story.

Galileo was very busy trying to keep up with his university classes. He wanted to take a math class in addition to this medical classes, but his father had made it clear that he should not waste time studying math while he was trying to finish his medical degree. But then, one day, when Galileo was at home during a school break, a visitor came to the house. Ostilio Ricci *(ree-chee)* was a friend of Galileo's father, and just happened to be a professional mathematician. Ricci struck up a conversation with Galileo and the topic eventually turned to math. Ricci could see that Galileo was extremely interested in math, and told his father that he should let Galileo come to a geometry class he was teaching. Galileo's father said, "Absolutely not!" but it was too late—Galileo now knew about the math class. He began sneaking into the back of the classroom and hiding. Ricci was teaching from Euclid's *Elements*. (Yes, Euclid's geometry textbooks were still being used even though they were 1,800 years old!) Galileo eventually found a copy of the *Elements* in a book shop in Pisa, and began studying it on his own. He took the book home during school breaks but kept it hidden under a pile of medical books. When his father wasn't looking, he'd pull out the geometry book, perhaps even hiding it inside the cover of a medical book.

Because Ricci was a friend of his father, Galileo had opportunities to talk with Ricci from time to time, outside of the classroom, of course. Galileo's questions about geometry so impressed Ricci that he asked Galileo who his math teacher was. Galileo was silent for a while, then confessed that he had been sneaking into Ricci's classes. Ricci laughed and told Galileo that he was welcome to attend any of his lectures, and to come right in and sit anywhere he wanted to. He also began speaking to Galileo's father, telling of the boy's incredible aptitude for math. Ricci could tell that Galileo was never going to be a doctor—he was a born mathematician. At first, Galileo's father was furious. But after months of coaxing, he finally gave in and allowed Galileo to switch his major from medicine to math. Galileo promised his father that he would that he would find a way to make a living doing math.

Ricci then introduced Galileo to the writings of Archimedes. Now Galileo had a role model who suited him very well. Like Archimedes, Galileo would be not only a mathematician but an inventor, as well.

Our time machine is now going to skip to a little bit later in Galileo's life, when he has already been a professor of mathematics for several years, and has become somewhat famous in northern Italy. At this point in his life, he was able to spend some time doing experiments. He remembered how interested he'd been in pendulums and decided to pick up this thoughts where he'd left off years ago.

Once again, you can discover these things at the same time that Galileo is discovering them (in our story), but first, here are some helpful vocabulary words.

The weight on the end of a pendulum is called the **bob**. The place where it hangs is usually referred to as the **pivot** and the string is called the **rod**. When the pendulum is hanging straight down and not moving at all, it is said to be at its **equilibrium point**.

Once the pendulum starts to move, there are names for the aspects of its movement. The size of a swing is called the **amplitude**. The amplitude is measured in degrees—the same degrees that you use to measure angles in geometry. One complete swing back and forth is called a **cycle**. The time it takes for a pendulum to complete one cycle (a complete swing) is called its **period** (meaning the period of time it takes). The number of cycles (swings) per second (or per minute) is called the **frequency**.

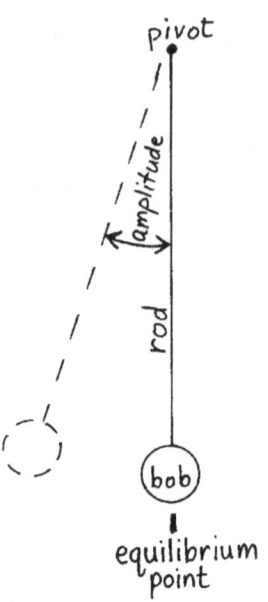

ACTIVITY 3.3: What happens when you change the weight of the bob?

You will need: a piece of thread about 40 cm (16 in.) long, 10 paper clips, your timer

PART 1:
1) Tie a paper clip onto the end of the thread. Open the clip a bit to make a "hook." Put another paper clip onto the hook. Tie a tiny knot near the top of the thread so you can hold it at the same place every time.
2) Use the timer to record how many swings your pendulum makes in 20 seconds.
3) Add a few more paper clips to the hook and then time it again. (Don't make a chain—put them all on the hook.)
4) Add a few more paper clips and time it again.
5) Load the last of your clips onto the hook and time it again.
6) What can you conclude about how weight affects the frequency of a pendulum?

NOTE: You may have some slight differences in the numbers. In all science experiments, there is something called the "margin of error." Since you are human, you are not perfect. Even though it seemed to you that you released the pendulum at the exact moment you heard the word "go," the truth is that you released it a fraction of a second differently each time. You were also off by a fraction of a second each time you called out, "Stop." There may also have been small air currents that were not noticeable to you but nevertheless affected your outcomes each time. Life's little irregularities left their mark on each of your experiments. Therefore, we must expect small errors. Counts that were only one number off can be considered as identical results.

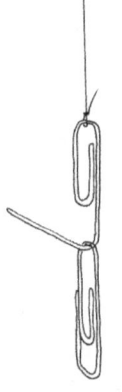

PART 2:
We will repeat this experiment, but instead of counting how many cycles it completes in 20 seconds, you will count how many cycles it completes from the time it starts swinging until it stops.

1) Load one clip onto the hook. Pull it back just a bit (low amplitude) and let go. Count how many cycles it makes until it comes to a complete stop.
2) Add all the rest of the clips and repeat. Make sure you pull back the pendulum the same amount. Count how many cycles it makes until it comes to a complete stop. (To speed up this experiment you can do both 1) and 2) at the same time using different pendulums.)
How does the weight of the bob affect the motion of a pendulum? Does it swing longer?

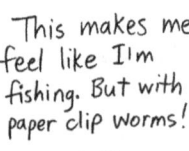

BONUS EXPERIMENT: The importance of center of mass

1) Take 10 paper clips and attach them end to end, so that they hang down in a long chain.
2) Loop a piece of thread through the top paper clip and tape the thread to the edge of a table so that the top clip in the chain is right below the table, almost touching it. Make sure the chain can swing freely.
3) Hang your pendulum from PART 1, putting 9 clips on the hook. Each of our pendulums now has total 10 clips. Make sure the thread is taped so that this pendulum is exactly the same length as the chain of clips.
4) Pull back both pendulums the same amount, and notice any differences in their swing. Is their period the same?
5) Watch them until they stop. Does one swing longer? Can you adjust the length of the thread chain so its period matches the chain? This experiment demonstrates the importance of center of mass. It makes a difference if the mass is distributed evenly throughout the rod or occurs mostly at the bottom of the rod.

So **the weight of the bob has no effect on the period** of a pendulum—it only affects how long it will keep swinging. Many people find this surprising. They assume that adding more weight will slow it down. They also think that gravity pulling on the bob is what eventually brings the pendulum to a stop. Not so. **Gravity does not slow down a pendulum.** What slows down a pendulum is what slows down any moving object: **friction**. Where can friction be found in a pendulum? There aren't many possibilities, are there? In our chandelier pendulum, the pin was rubbing on the inside of the tiny hole in the paper. Doesn't seem like that would be enough friction to stop the pendulum, but it eventually does. There is another source of friction that is not obvious at all: the friction of air molecules rubbing against the pendulum as it swings. Air molecules don't seem like they would cause much friction, but they do contribute enough friction to be a factor we can't overlook. A friction-free pendulum would go on swinging forever!

ACTIVITY 3.4: Find out what speeds up or slows down a pendulum

You will need: scissors, your paper chandelier, and the rest of the page it was cut from.

PART 1: Does the width of a pendulum affect how fast it swings?

1) Cut along the remaining line on the sheet from which you cut the chandelier strip. Now you should have three strips of paper that are the same length but different widths. (The thinnest strip will be the chandelier.)
2) Place the medium and large strips under the chandelier strip. Line them up so that the tops and bottoms match up perfectly, and so that they are also centered perfectly. Put the pin through the hole at the top of the chandelier and poke straight through all three strips. (Or one at at time if the pin has trouble going through.) Wiggle the pin so that the hole is large enough that all three strips swing freely. Let the strips hang straight down. Spread the tops apart just slightly on the pin so that when they swing they will not be touching each other.
3) Pinch them in the middle (so that they won't slip apart) and pull them back. Then let go. They should all start swinging at precisely the same moment.
4) What happens? Do they basically swing at the same rate? *(After they swing for a while, there will probably be more difference in their swings. This may be due to the greater effect of air resistance on the wider pendulums, or to some other cause.)*

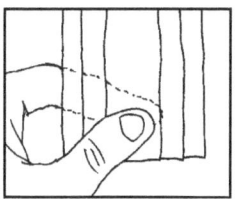

PART 2: Let's try varying the length of the pendulums

1) Use those wide strips to cut three more narrow strips the exact width of the chandelier strip.
2) Leave the chandelier strip full length. Cut the other three strips so that you have one that is 3/4 the length of the chandelier, one that is 1/4 the length of the chandelier, and one that is 1/4 of the length.
3) Line these strips up so that their tops match exactly. Push the pin through that hole in the top of the chandelier chain one last time, then poke it through the other three strips (or pierce them one at a time if you have trouble getting the pin through). You should now have found strips of different lengths hanging from your pin.
4) Spread the strips apart so that they can swing freely without touching each other. (If the strips are curled a bit lengthwise, run your fingers down them a few times and they should straighten out.)
5) Use your finger to pull back all four at once, then let go. What happens? Does the length of a pendulum affect how fast it swings back and forth?

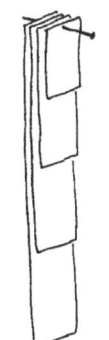

Now you know how to make pendulums go faster or slower: adjust the length. You know how to make a pendulum swing longer: add weight to the bob. You are now ready to make a reliable time-keeping pendulum. Can you make a pendulum that ticks off seconds, like a clock?

Galileo eventually invented a pendulum clock, but, sadly, it was after he went blind. His son drew sketches of his clock idea, but both he and Galileo died before the clock could be built. A Dutch scientist name Christian Huygens picked up the idea and was the first to build a pendulum clock.

ACTIVITY 3.5: Can you make a pendulum that ticks off seconds?

You will need: your thread and paper clip hook from activity 3.3, another paper clip, and a way to count off seconds (you might try www.metronomeonline.com)

1) Measure the thread. If it is shorter than about 35 cm, attach a new piece of thread that is more than 35 cm.
2) Wind the top of the thread around the other paper clip, as shown in diagram on right. This will give you a way to quickly and easily adjust the length of the thread.
3) Hold the clip and the wound thread so that it won't slip while the pendulum is swinging. Start the pendulum swinging and compare a complete cycle (back and forth) to the seconds ticking on the metronome or stopwatch.
4) Adjust the height of the bob, up or down, to better match the ticking seconds. Keep adjusting until you get your pendulum to tick off accurate seconds. The longer you let the pendulum run, the more information you will get about its accuracy. It might seem very accurate for the first few seconds, but after several dozen seconds it might be out of sync.

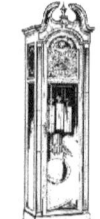

Here is an interesting question: Is it possible to make a working pendulum clock for a doll house? (How long would the pendulum have to be if you wanted the clock to tick off seconds?)

BONUS ACTIVITY: Turn on some music and try to make your adjustable pendulum keep time to the music.

As Galileo worked with pendulums, he began to sense that their behavior was based on mathematical principles. Perhaps there was a mathematical relationship between the period of a pendulum and the weight of the bob or the length of the rod? He didn't know what he was looking for—he just started experimenting and writing down measurements.

It is thought that Galileo also used the stars as time keepers. It is hard to imagine someone keeping a pendulum going for 24 hours, until a star reappeared in the same place the following night, but some science historians claim that this was, indeed, done. You can perform Galileo's experiment (minus the water clock and the 24-hour vigil), and discover some pendulum math for yourself. However, we've streamlined the process so that you don't have to start from scratch like Galileo did; your experiments will lead straight to a conclusion. Galileo's experiments probably took several months to complete; yours will only take a few minutes.

ACTIVITY 3.6: Discover some pendulum math

You will need: *a long piece of thread (90 cm or so), a coin (a penny is fine), a ruler that can measure centimeters, tape, and a stopwatch*

1) Tape the coin to the end of the thread.
2) Pinch the thread at <u>precisely</u> 5 cm above the coin. (You can make a small dot of ink on the thread if you find it helpful.)
3) Count this pendulum's cycles for 15 seconds. (One cycle is "out and back.") You might want to do this several times so that you can be sure of your number. Record your answer below, where it says 5 cm _____.
4) Now pinch the thread at exactly 20 cm above the coin. This pendulum is **4 times** as long as your first one. Count number of cycles for 15 seconds and record your answer below.
5) Pinch the thread at exactly 45 cm above the coin. This pendulum is **9 times** as long as the original. Count the cycles for 15 seconds and record the results below.
6) Finally, pinch the thread at exactly 80 centimeters above the coin. This pendulum is **16 times** as long as the original. Again, count the cycles for 15 seconds.

Your results:

5 cm _____   20 cm _____   45 cm _____   80 cm _____
              (4 times as long) (9 times as long) (16 times as long)

7) Can you find a pattern in this number sequence? What do you think the frequency would be if the length of the pendulum was extended to 125 cm (25 times as long)? _____ What about 180 cm (36 times as long)? _____
(Notice that 4, 9, 16, 25, and 36 are "square" numbers. 4=2x2, 9=3x3, 16=4x4, 36=6x6, 25=5x5)
Even if you can't find the pattern, continue on to the next section.

8) Now plot your results on this graph. The points will make a distinct shape. Make a smooth "best fit" curve along the path of these points. After you make the curve, you will see that the curve can be used to estimate the frequency for any length. Put your pencil on any length number, then move the pencil to the right until it bumps into the curve. Then look up and see what frequency number is right above your pencil point.

What would the frequency be for:
50 cm? _____   10 cm? _____

Extend the ends of the curve on both ends, following the shape of the curve. (This is called **extrapolation**.)

Now estimate the frequency for:
100 cm _____   2.5 cm _____

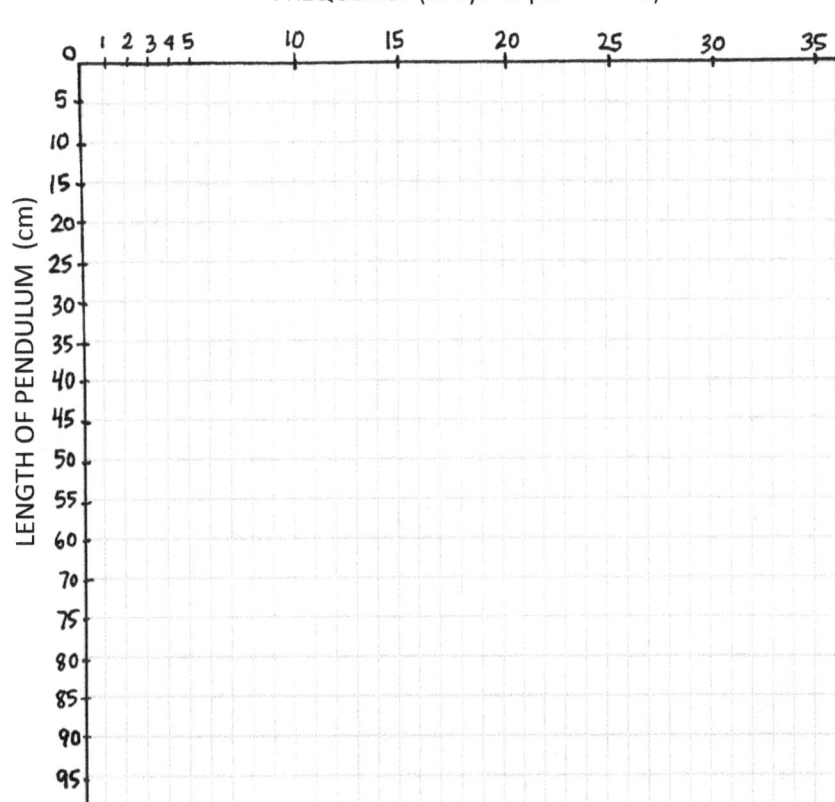

WHAT YOU SHOULD NOTICE:

When the length is increased to be 4 times longer, the frequency slows to 1/2 the rate. When the length is increased to be 9 times longer, the frequency slows to 1/3 the rate. When the length is increased to be 16 times longer, the frequency slows to 1/4 the rate. On the next page, we'll discuss this pattern.

This pattern is called the "**Inverse Square Law**." When the length is increased N times, the frequency decreases by 1/√N (1 divided by the square root of N). If you extended the pendulum to 125 cm, that would mean that you had made it 25 times as long. The square root of 25 is 5, so the frequency would be 1/5 of the original rate. If you shortened the pendulum to 50 cm, you would have multiplied the length by 10, which is not a perfect square number. The math is a little messier, but you can still calculate the new frequency. Find the square root of 10, (about 3.16). The frequency would be 1/3.16 times slower. Just divide your original frequency by 3.16

Use the inverse square law to find the frequency if the pendulum was 30 cm: _____ Then go back to your graph on the previous page and find the place where the 30 cm line hits your curve. Look straight up from that point to see what frequency number it corresponds to. Does this match the calculation you just did?

We now turn some dials on our S.N.O.O.P. machine and go forward in time. We stop in Paris, France, in 1851, where a man named **Léon Foucault** *(fu-ko)* has attached a very long pendulum to the ceiling of the Panthéon. But before we turn on the S.N.O.O.P. machine, here is some information about the building.

You may have heard of the original Pantheon in Rome, which was built around the year 25 BC and dedicated to many of the gods that the Romans worshiped at that time. (It is now a Christian church.) The Panthéon in Paris was commissioned by King Louis XV in 1758 and was completed in 1790, about the time that the French revolution began. The Paris Panthéon was built as an upgrade to the Church of St. Genevieve (a female patron saint of Paris). The columns in the front are modeled after ancient Greek temples.

Pantheon in Rome

Panthéon in Paris (a 1791 painting)

The Panthéon in Paris has a very high dome. Here is what you'd see if you stood inside and looked up at the ceiling. There is a beautiful painting at the very top of the dome. You can see the blue colors of the sky in the painting.

Somehow or other, Foucault managed to convince the people in charge of the Panthéon to let him hang a pendulum from the top of that dome. (He had done this experiment in his lab and already knew it would work.)

As we learned in previous activities, the length of a pendulum is what controls how fast it swings. The rope hanging from the dome was very long, so the pendulum swung very, very slowly. Foucault's reason for hanging such a long pendulum was to demonstrate that the earth rotates, spinning on its north-south axis. Foucault made marks on the floor in a large circle around the swinging pendulum. The marks would keep track of the direction in which the pendulum was swinging. The bob on the pendulum was extremely heavy, and this helped the pendulum to keep going for many hours. Over the course of the day, it looked like the pendulum was changing direction with respect to the marks on the floor. Foucault asserted that it was not the pendulum that was moving, but the earth.

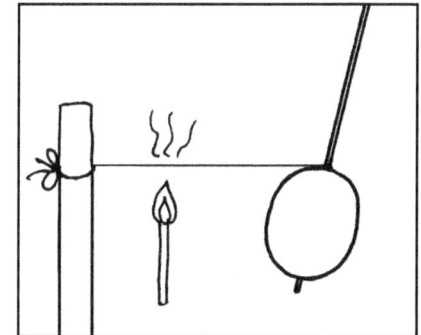

This demonstration would have been easier to understand if the pendulum had been hanging over the North Pole. You can imagine the earth spinning on its axis, rotating underneath the pendulum. Because the earth is turning counterclockwise with respect to the North Pole, it would look like the pendulum was going clockwise. But Foucault's pendulum was not at a pole; it was in Europe. Why did it still trace out a circle? As long as the pendulum is not on the equator, it will trace out a circle, though more slowly than at the poles. At the poles, it would look like the pendulum traced a circle in 24 hours. Away from the poles, it takes longer than 24 hours. At the equator, it does not go in a circle at all, just back and forth.

One problem Foucault had in setting up the demonstration was that he needed the pendulum to go back and forth in a straight line, with no movement side to side. Even the slightest movement in a sideways direction would start the pendulum going in a long oval instead of a straight line. If he let the bob go with this hands, he would always accidentally introduce some sideways motion. So he devised an ingenious starting mechanism that used a match to cut a string that was pulling the pendulum back. When the match burned through the string, the pendulum was released. (When they tried this experiment at the South Pole, they were not allowed to use any fire, so no match for starting the pendulum. This made the experiment harder to do.)

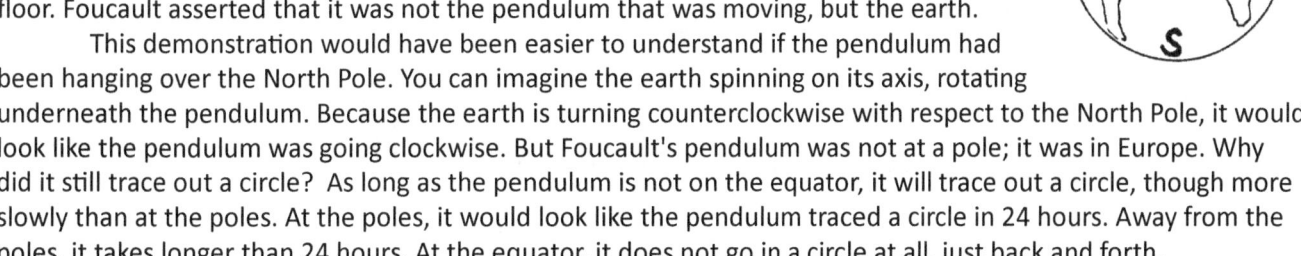

Foucault pendulums on display in public buildings use electromagnets to help them keep going. Even pendulums with very heavy bobs will eventually lose energy, slow down, and stop. Pendulums in public spaces need to keep going continuously, so their swings are given a tiny extra boost by the clever use of magnetism generated by electricity. These devices can also counteract any tendency of the pendulums to start swinging in an oval pattern instead of straight back and forth.

ACTIVITY 3.7: Watch videos on the Foucault pendulum

You can access videos about Foucault pendulums by using the YouTube playlist for this curriculum.

So far, we have been working with single pendulums. What would happen if we joined two or more pendulums together? Each pendulum would have its own distinct motion, but its motion would also be affected by the motion of any other pendulum to which it was attached. What kind of motion would result? Anything interesting, or just a mess?

In this next activity, you will experiment with a compound pendulum. A compound pendulum is made of two or more pendulums attached together in some way. Compound pendulums come in many shapes. Mobiles (hanging sculptures) qualify as compound pendulums, although they usually aren't designed for motion.

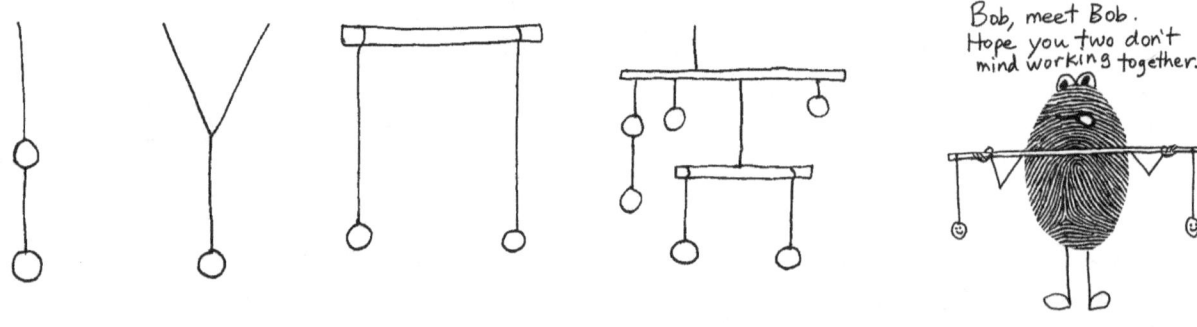

ACTIVITY 3.8:   Pendulums that take turns

You will need: some thread, two coins, tape, a drinking straw, scissors

1) Cut two pieces of thread about 40 centimeters (16 inches) long.
2) Tape a coin to the end of each thread.
3) Tape the free ends of the threads to the edge of a table. The distance between the threads should be just a little less than the length of the straw.
4) Make snips in both ends of the straw. (If it has a flexible part, trim that off.) Don't cut a "V," just a straight slit.
5) Slip the notches of the straw onto the thread to make a bar that goes between the threads, as shown. The straw should at least be several inches (8-10 cm or so) above the coins. If the straw is too close to the coins, it will be too difficult to observe the phenomen we want to observe.
6) Start one coin swinging but don't touch the other one.
   Watch for a while and see what happens. It should look like the pendulums are passing their energy back and forth. When one pendulum reaches the height of its energy, the other one will be at rest, and for a few seconds it won't move at all.
7) You can adjust the height of the straw by sliding the threads through the notches. Try the experiment again with the straw up high, then down low.

CAN YOU "COMMUNICATE" WITH YOUR COINS?

1) Re-tape your pendulums so that one is substantially longer than the other. Keep the straw positioned between the threads, and straight across (parallel to the floor).
2) Tap or lightly push the straw in a way that makes only one coin swing. You'll have to discover how to do this on your own. Just do a little experimenting.
3) Now see if you can do the reverse, getting the other coin to swing while the first one stays still.

<u>Here is the explanation of what is going on:</u>

To get one of the coins to start swinging, you have to make your taps match the natural period of the pendulum. It's almost if as each pendulum has a narrow range of "hearing" and can only "hear" taps that match the frequency at which it would swing if it was in motion. The pendulums "ignore" taps that are not at that frequency.

A word often used to describe this phenonemon is "**resonance.**" When something "resonates," it is responding to its natural frequency. The classic example of this is an opera singer shattering a glass. The note sung by the singer just happens to be at the natural frequency of the glass, so the glass "hears" the note and starts to vibrate. Then it vibrates too much and shatters. The word resonate has a broader meaning, also. If someone says, "That resonates with me," they mean that what someone else just said matches the way they, themselves, are feeling or thinking. The other person's thoughts lined up with their own thoughts.

It was fun to experiment with resonance in our compound pendulum activity. The principle of resonance has an interesting practical application in architecture. Skyscrapers can sway back and forth during wind storms, acting like upside-down pendulums. This can be very scary if you are in the top floor when the swaying starts! Theoretcially, if the winds are strong enough (such as in a hurricane) they could get a building moving back and forth fast enough that the streel structure begins ripping apart. However, architects have developed a way to prevent too much swaying.

In the world's tallest buildings, the top floors have a vertical shaft in the center where a large pendulum hangs. As you might guess, the bob of this pendulum weighs many tons. Architects use mathematical formulas to determine how much weight to use, how long to made the rod, and exactly where to place it in the building. These formulas take into account how tall the building is, how wide it is, and the strength and flexibility of the materials from which it is built. They "tune" the pendulum to resonate with the building. When the building starts to sway, the motion of the building begins to be transfered to the pendulum. As the motion is transfered to the pendulum, the building sways less. Even if some of the energy of the

The tuned mass damper in Taipei 101, in Taiwan

pendulum gets transfered back into the building, it will be far less than the original amount of energy. Part of the design, however, is to prevent energy from going back into the building. It's amazing that something so much smaller than the building can stop it from swaying.

The pendulums in skyscrapers are called **tuned mass dampers**. To dampen something is to decrease its energy or activity level. Tuned mass dampers don't have to look like pendulums, however, and they are found in other places, not just tall buildings. Sometimes they look like cylinders or blocks that shift back and forth. Here are tuned mass dampers on power lines, helping the wires not to sway in the wind.

Bridges also use tuned mass dampers to reduce their movement. Dampers on bridges can be designed to move against the motion of the bridge, thus preventing resonance. Bridge dampers are usually some type of spring mechanism attached to a large weight. It is said that when armies march across bridges they are told to stop marching and just walk, to prevent any chance that the bridge might resonate to the frequency of the footsteps.

ACTIVITY 3.9: Watch a few short videos about tuned mass dampers

There are several videos about tuned mass dampers in the YouTube playlist. Now would be a great time to watch them!

You're probably almost ready to leave the topic of pendulums and move on to something else, but we need to cover one more major concept. Pendulums belong to a group of mechanisms called ***oscillators***. To oscillate is to "vary in magnitude or position in a regular manner about a central point." This definition is broad enough that it can be applied to many kinds of motion, as well as to situations where something shrinks and grows. The central point doesn't have to be literal, it can be a point on a graph or a point in time. For example, some people think that the universe might go through cycles of expansion and contraction. This idea is called the oscillating universe. Sound waves oscillate, and the machine that measures these oscillations is called an ***oscilloscope*** (shown here).

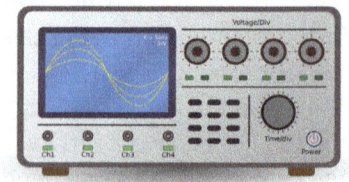

Pendulums don't have to swing back and forth; they can also go up and down. A heavy ball bouncing up and down on the end of a spring can be called a "vertical pendulum." Vertical pendulums have many of the same characteristics as swinging pendulums. But instead of reading about them, why don't you discover their characteristics for yourself?

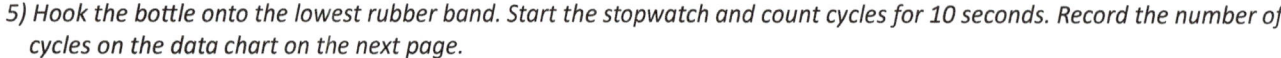

ACTIVITY 3.10:   A pendulum that bounces instead of swings

You will need: *at least a dozen rubber bands (thin ones, if possible, and all exactly the same size, and as new as possible since rubber bands loose elasticity as they age), your paper clip hook, a bottle of water (at least 335 ml, but no more that 500 ml), a stopwatch, a pencil, and a way to hang your vertical pendulum from someplace near the ceiling (or perhaps from the top of an open door)*

PREPARATION:

1) Join about ten rubber bands, end to end, as shown in diagram on right. Thin one are better than thick ones, but size is less important than making sure all the bands are identical in size and stretchiness. (Avoid old rubber bands, as age makes them brittle.)

2) Find a way to hang this rubber band chain from a very high place so that the bottom of the chain is several feet off the floor. The chain should be able to stretch down to the floor.

3) Put a rubber band around the neck of the water bottle. Wrap it several times so it is tight. Then put your paper clip hook onto it and position the hook so that you can hang the bottle from one of the rubber bands.

4) Hang the bottle of the bottom band. If the hanging bottle touches the floor, you'll need to either increase the height of the chain, or decrease the number of rubber bands in the chain, or decrease the amount of water in the bottle. The bottle needs to be able to go up and down.

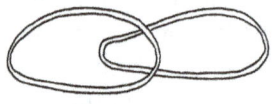
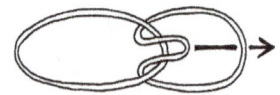
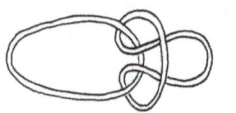
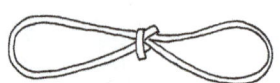

NOTES ABOUT DOING THE EXPERIMENTS:

These general directions might have to be adjusted to your situation. Your rubber bands might be smaller or larger than average, for example. If you need to use more or less than 10 rubber bands, that's fine—you don't need to use all the squares on the data chart. Also, you may have trouble counting the number of cycles (up and down) towards the end of the 10 seconds. Just to the best you can. Keep the rhythm in your head as the bottle slows down. You also might might need to use fractions in your answers. For example, a bob might complete 6.5 cycles in 10 seconds.

TIP: It helps if you have two people to do these experiments, one to watch the clock and one to watch the bottle. If you are working on your own, hold the stopwatch right near the bottle so you can see both.

5) Hook the bottle onto the lowest rubber band. Start the stopwatch and count cycles for 10 seconds. Record the number of cycles on the data chart on the next page.

6) Move up the chain to the next highest rubber band. Count cycles for another 10 seconds. Record data.

7) Continue up the rubber band chain. You might have to stop at about 3 from the top. It can be very difficult or impossible to time just one rubber band.

8) Now adjust the amount of water in the bottle by pouring out half of it. If your rubber bands no longer stretch to the floor, add a few more to the chain.

9) Start back down at your lowest rubber band and count cycles for 10 seconds. Record data. Continue up the chain.

11) After you have recorded all your data in the chart, plot each point on the graph.

*12) Look at the general trend of your dots. Are they making a line? A curve? Draw a line (whether straight or curved) that shows the general trend of your data, but doesn't necessarily go through every point. This is called making a "**best fit**" curve. If some of your data points are a little high or low, don't worry about it. Make the line smooth—no zigzags.*

DATA CHART:

bounces in 10 sec.

# of rubber bands	full bottle	half-full bottle
1		
2		
3		
4		
5		
6		
7		
8		
9		
10		
11		
12		

GRAPH YOUR DATA:

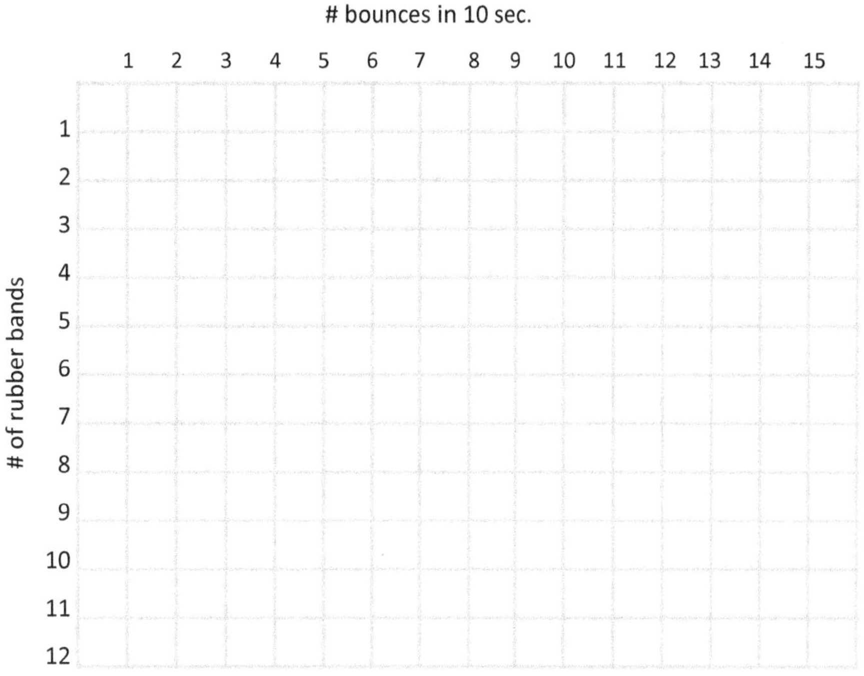

bounces in 10 sec.

(Graph grid: x-axis 1–15 # bounces in 10 sec.; y-axis 1–12 # of rubber bands)

☐ = full bottle   ☐ = half-full bottle

Follow-up questions:

1) Which bounced longer, the heavier bottle or the lighter bottle? _____

2) In your thread pendulums, which swung for a longer time, 1 clip or 10 clips? _____

3) Does the weight of the bob control how long both swinging and vertical pendulums are in motion? _____

4) Did the bottles seem to come to rest more quickly than your thread pendulums? _____

Can you suggest a reason why? _____

5) A longer rubber band chain produced (greater/fewer) number of cycles per 10 seconds. (cirlce one)

6) Does the length of the rod control frequency in both swinging and vertical pendulums? _____

7) Do these curves look familiar? Where did you see this shape previously? _____

Extend the lines/curves you drew on your graph so that they are a little bit longer but still follow the same curved shape. This will let you do something called ***extrapolation***. Extrapolation is a very important tool in scientific research. It lets scientists make reasonable guesses about what would have happened at points above and below their actual data points. For example, if you were only able to use 4 rubber bands (which was the case for the heavier bottle in the sample data), what would have happened if you had been able to use 6 or 7? If you extend the ends of your line a bit, you can make reasonable guesses for what would have happened in actual experiments if you had been able to do them.

Use extrapolation to answer these questions:

8) For the lighter bottle, estimate the number of cycles per 10 seconds for:
   1 rubber band more than your highest value:_____   2 rubber bands more than your highest value: _____

9) For the heavier bottle, use extrapolation to estimate the number of cycles per 10 seconds for:
   1 rubber band more than your highest value:_____   1 rubber band less than your smallest value: _____

Your patience with watching vertical pendulums is probably wearing thin, so we'll just do a thought experiment to conclude our work on this topic. Imagine a bob hanging from a very stiff spring that is hard to pull. Give that imaginary bob a pull downwards and let it go. It snaps back almost instantly. The bob didn't really go up and down at all. Now imagine a spring that is very easy to stretch. Perhaps it is very long. Put a bob on the end of that "soft" spring and give it a pull downwards. The soft spring allows the bob to pull it down a considerable distance. It then slowly shrinks, taking the bob on a very gentle ride upwards. Then all the way down again. The bob gets a very long ride as the soft spring stretches out again and again. So we can see that with vertical pendulums, other factors are important, such as the size and the stiffness of the spring. Our thread pendulums didn't have any stiffness issues; we only had to consider the length of the thread.

stiff spring

soft spring

Now let's put a pen in the bob of a vertical pendulum and a roll of paper behind it. We'll pull out the paper at a steady rate and let the pen in the bob touch the paper. The shape traced onto the paper will be a gently rolling wave shape. (There is a video of this on the playlist. You might want to watch it right now.) As the vertical pendulum experiences dampening due to friction and to heat generated in the spring, and begins to run out of energy, the height of the waves will decrease, but the length of the waves will remain the same. (The length of the wave is the frequency of the cycles, which does not change.)

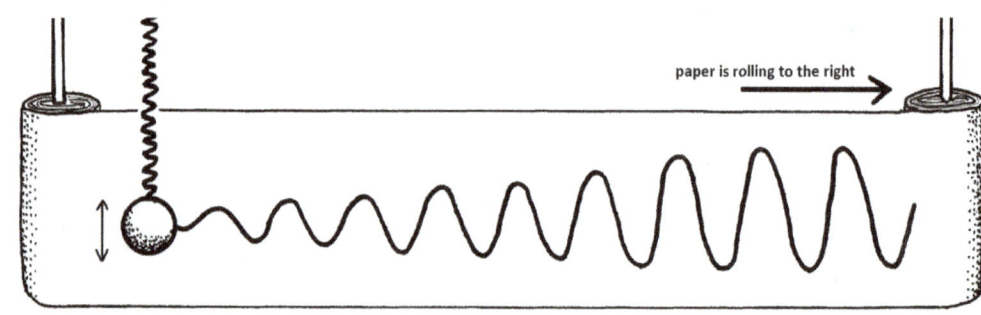

The pen is on the side of the pendulum that you can't see.

paper is rolling to the right

Physicists are very aware of the connection between up-and-down oscillations and this wave pattern. The wave is called a **sinusoidal** *(sine-yu-soy-dul)* wave, or just **sine** wave. Sine waves are a mathematical way to record oscillating motion. There are many equations associated with sine waves, such as the one shown here.

$$x(t) = x_0 \cos\left(\sqrt{\frac{k}{m}}t\right) + \frac{v_0}{\sqrt{\frac{k}{m}}} \sin\left(\sqrt{\frac{k}{m}}t\right)$$

We need one more vocabulary word to finish this discussion. Physicists often add the word "harmonic" to this type of oscillation, and call it **harmonic oscillation**. Pendulums are called **simple harmonic oscillators.** You might recognize the word "harmonic" (and the related word "harmony") as musical terms. Sound waves are oscillations of air molecules. The pitch (high or low) is related to the length of the wave (distance from peak to peak).

Our last stop with the S.N.O.O.P. machine for this chapter will be Scotland in 1844. We will meet a mathematician named Hugh Blackburn who is inventing a machine that can record the harmonic oscillations of two pendulums that are at right angles.

The readers would also like to see what Hugh is holding. We don't have any of the images that his machine drew, nor any details about what his machine actually looked like, so we'll have to use modern images that are probably very similar. Hugh Blackburn's work inspired other mathematicians and physicists to do more investigations, and as a result, several types of **harmonographs** have been developed. Hugh's machine likely only had two pendulums, but the most popular harmonograph today (judging by the number of YouTube videos about it) is the "three-pendulum rotary harmonograph." The word "rotary" refers to something that goes around in circles. One of the three pendulums swivels around a pivot point, instead of going back and forth.

The author of this book built her own three-pendulum rotary harmonograph, shown here in the photo. You can see the pen against the blue background. The rotary table looks like it has a sheet of black paper lying on it. Perhaps the machine is set up for making a white-on-black pattern like the one shown below. (A video of this machine in operation is on the playlist.) The paper table sitting on top of the rotary pendulum goes around in a circle while the top "arms" that are attached to the side pendulums go back and forth. The combined motion of all three pendulums creates the pattern.

The operator doesn't have much control over which design the machine will make—it seems to have a mind of its own. Although it makes similar patterns, it never does exactly the same thing twice.

All of these patterns were drawn by the author's harmonograph. Visitors will often compare these designs to Spirograph®. The main difference is that with a harmonograph, you have a dampening effect as the pendulum loses energy and slows down. This means that the drawing spirals inward and never retraces the same path.
With Spirograph, once the pattern is established, the pen traces over the same lines again and again and again.

ACTIVITY 3.11: Review/quiz (Answers are in the teacher's section.)

TRUE or FALSE?
1) ____ The weight of the bob is what controls the period of the pendulum.
2) ____ Pendulum cycles for very large swings are a tiny bit faster than for very small swings.
3) ____ Galileo's father was very supportive of his son's interest in math.
4) ____ Foucault hung a pendulum from the dome of the Pantheon in Rome.
5) ____ Foucault wanted to demonstrate the rotation of the earth.
6) ____ Euclid's geometry book was only used in ancient Greece.
7) ____ Foucault discovered the Inverse Square Law.
8) ____ The Inverse Square Law lets you predict what will happen to frequency if you increase length.
9) ____ Air molecules have no effect on swinging pendulums.
10) ____ Pendulums can bounce as well as swing.

Fill in the blanks.
11-12) Name two places you might find a "tuned mass damper." _____ and _____
13) Galileo watched a chandelier in the cathedral of this city: _____.
14) A _____ wave is how mathematicians represent oscillating motion.
15) If you want to speed up your pendulum you should _____.
16) This is when you extend data lines above and below your experimental results: _____
17) When something vibrates at its natural frequency we say that it _____.
18) To vary in magnitude or position in a regular manner about a central point: _____
19) The _____ of a pendulum is the number of cycles per second.
20) Foucault pendulums in modern public buildings use _____ to keep them going.

Questions from previous chapters:
21) Which one of these is NOT a way to calculate mechanical advantage?
   a) radius of wheel/radius of axle   b) radius of follower/radius of driver
   c) length of lever/length of fulcrum   d) length of force arm/length of resistance arm
22) TRUE or FALSE? ____ The balance point of an object can be outside of the object.
23) TRUE or FALSE? ____ An idler gear has no effect on mechanical advantage.
24) What shape does a ball trace out in the air when you give it a toss?
   a) oval   b) semi-circle   b) ellilpse   d) parabola   e) parallelogram   f) parable
25) Which one of these is a second class lever?
   a) seesaw   b) scissors   c) wheelbarrow   d) baseball bat   e) broom

Name the simple machines:
26) _____
27) _____
28) _____
29) _____
30) _____

And don't forget to check the playlist for additional pendulum videos!

Adventure 4: Inertia with Newton
(Newton's 1st Law)

Oddly enough, this fourth adventure about motion starts out with people thinking about NOT moving. If you are not familiar with the word *inertia*, it is pronounced like this: *in-ER-shah*. The word inertia comes from the Latin word "inert," which means "lazy." Scientists all the way back to ancient Greece had noticed this tendency of stationary (non-moving) objects to stay put. They came up with various theories, but in the end, Aristotle's ideas won the day and were handed down for well over a thousand years. Aristotle believed that moving objects slowed down because they wanted to go back to being lazy and not moving.

In the 14th century (1300s), a scientist named Jean Buridan challenged Aristotle's idea and proposed that objects didn't slow down and stop because they wanted to, but because something was affecting them, causing them to slow down whether they wanted to or not. He also proposed that air was one of the things that made objects slow down. Buridan brought us one step closer to understanding intertia, though he didn't call it that. Buridan's research was continued by several of his students who refined his ideas and helped science make a few more steps in the right direction.

Aristotle

In the 1500s, an Italian scientist named Benedetti began studying motion and came to the conclusion that Aristotle had been wrong about many things and that although Buridan was on the right track, he had not discovered one important truth about motion—that the forces that pushed on objects and caused them to move only did so in straight lines. He believed that "pushing forces" could only push in one direction. Thus, when we see what looks like curved motion, such a ball following a curved path through the air, it is the result of more than one force affecting the object. This was quite a revolutionary thought!

Also in the 1500s, Copernicus and Galileo added their insights to the study of motion. Copernicus, who was an astronomer, pointed out that objects at rest (as we see them sitting on a table) are not really at rest at all because the table is sitting on planet Earth which is traveling through space at a very high speed, orbiting the Sun. This point did not turn out to be as helpful as Galileo's conclusion that "an object moving on a level surface will continue in the same direction at a constant speed unless disturbed." Galileo later stated that he thought that motion is "relative," meaning that whether or not something appears to be in motion depends on your viewpoint. (If you are riding in a car, everything inside the car—the steering wheel, the seats, your travel mug, your dog—looks stationary (not moving), while the landscape outside the car appears to be zooming past. However, to someone standing in that landscape, it is your car that appears to be moving.)

Finally, along came Isaac Newton, in the 1600s. Many people don't realize that Newton wasn't the first person to think about objects moving or being at rest. He knew about all of these previous scientists and their theories. He admitted that he was building on the work of scientists that had gone before him, and he once said, "If I have seen further, it is by standing on the shoulders of giants."

Newton was born in December of 1642. Galileo had died in January of that year.

Isaac was born prematurely and was so small that no one thought he would survive. His mother wrote that newborn Isaac could have fit into a quart-size mug. Three years later, Isaac's mother decided to remarry.

Although Isaac's mother had brought him home again (after the stepfather's death when Isaac was 11) this school was a distance from home, and therefore his mother arranged for him to live under the care of a gentleman who ran an apothecary shop. (Today we'd call him a pharmacist.) Isaac was fascinated as he watched the apothecary mix potions. When the apothecary saw Isaac's interest in science, he gave him permission to experiment with the chemicals. He also loaned Isaac a book called "The Mysteries of Nature and Art." This book had information about simple chemistry experiments, as well as instructions for building projects like kites and model windmills. The apothecary must have supplied young Isaac with building materials as well as chemicals, because we know that it was during this time that Isaac began building many kinds of models and simple machines. Perhaps he even tried out the book's instructions for making "fire drakes"—kites with firecrackers tied to their tails. Without any video games or social media to distract him, he had plenty of time for making inventions.

When Isaac graduated from this school at age 17, his mother brought him back home to help run the family farm. Isaac was much less successful as a farmer than as a student or an inventor. Fortunately, the head master at the school saw that Isaac was a very gifted learner and was able to persuade his mother to let him go to college. His uncle helped him get into Trinity College, Cambridge.

Many famous scientists have attended Trinity College, Cambridge, including the author's favorite scientist, James Clerk Maxwell.

Newton had to work his way through college, unlike most of his peers who were from rich families. He did not have time to go to college social events, but this did not bother him because he really preferred being alone with his books. One of his teachers introduced him to the writings of Galileo, Copernicus, and the famous French mathematician/philosopher René Descartes. *(Ren-ay Day-cart)* While reading Descartes's book, "La Geometrie," Newton fell in love with mathematics. By borrowing books from the college library, Newton taught himself (in only one year) almost everything that was known about math at that point in time. Upon graduation, he was all set to continue on as a graduate student, but in that year, 1665, the university had to close due to a plague epidemic. It remained closed for 18 months. Newton spent this time thinking and experimenting.

During this time at home, Newton made key discoveries about the nature of light and about gravity and motion. He also began inventing calculus. (Not *taking* calculus— *inventing* it!) It was also during this time that his famous observation of the falling apple supposedly took place, although some of his later writings, and the timing of them with respect to correspondence with famed scientist Robert Hooke, reveal that he probabaly did not completely realize the implications of gravity until about a decade after the apple incident.

Let's S.N.O.O.P. around in Newton's private workshop on a day when he is experimenting with moving and stationary objects, and doing some deep thinking about the concept of inertia.

You can do similar experiments to see the principles of inertia that Newton observed.

ACTIVITY 4.1:   The coin and card trick

You will need: *a playing card (or index card) and a large coin*

1) Put the coin at the center of the card.
2) Balance this on the tip of one finger.
3) With your other hand, give the card a quick flick.

The coin should stay on your finger while the card goes flying off. The coin stays on your finger because of its inertia. It is at rest and wants to stay at rest. (Large coins work better than small ones because they have more inertia. Playing cards work best because they are very smooth.) The motion of the card underneath it happens so fast that the energy overcomes any friction and is not transferred to the coin. This is a small version of the classic "pulling the tablecloth out from under the dishes" trick. If you want to see this classic tablecloth trick, there is a video on the playlist.

ACTIVITY 4.2:   The "drop the coin into the bottle" trick

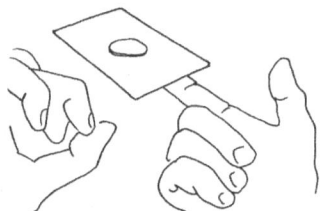

You will need: *a small coin (such as a US dime), a bottle that has a neck just slightly larger than the coin, and a round hoop of some kind (hoop ideas: an embroidery hoop, a roll of masking tape, a circle cut from a large plastic bottle, or a strip of thin cardboard taped into a circle)*
NOTE: The larger the hoop, the more dramatic this trick looks!

1) Balance the coin on top of the hoop, then balance the hoop on top of the bottle, as shown.
2) Put your finger inside the hoop, then pull the hoop out of the way, moving your hand very quickly.
3) The coin should drop right through the neck of the bottle. If the neck of the bottle is just slightly bigger than the coin, this make the trick more amazing.
4) Try the trick again, but this time put your hand outside the hoop instead of inside. Does it still work? (If your hoop is very stiff, putting your finger outside the hoop might not make any difference.)

The fact that you had to balance the coin on top of the hoop ensured that the coin was directly over the neck of the bottle. Gravity is a great help in getting things vertically straight ("plumb"). When you pulled quickly on the inside of the hoop, the energy from the moving hoop did not have time to get transferred to the coin, so the coin was unaffected by the motion of the hoop. Thus, the coin stayed directly over the neck of the bottle and gravity pulled it straight down into the bottle.

If you push the hoop from the outside, your finger bends the hoop inward (assuming you have a flexible hoop) and ruins the perfect circle shape. It was this perfect circular shape that helped to keep the coin lined up. If the shape of the hoop is distorted, the coin loses its perfect alignment and therefore drops slightly off center.

The playlist has a version of this trick using an egg instead of a coin, and a glass of water instead of a bottle.

ACTIVITY 4.3: "Hanging by a Thread"

You will need: *two pieces of thread (each about a meter [yard] long) and a heavy book.*

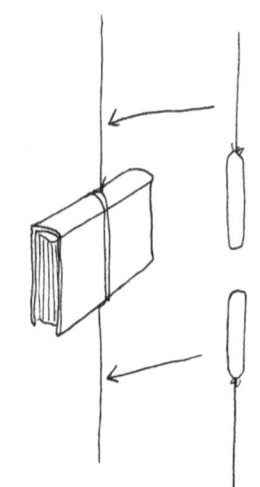

1) *Tie the end of one piece of thread around the book. Adjust the loop of thread so the book hangs balanced.*
2) *Tie the end of the other piece of thread around the book, but so that it hangs off on the other side.*
3) *Hold one of the threads so that the book hangs in the air with the other thread dangling beneath it.*
4) *Predict which thread will break if you yank very hard on the bottom thread.*
5) *Yank! Which thread broke?*

You'd think the that top thread would break because it already has a lot of weight pulling down on it. It might not be able to take any more weight, right? But the book has a lot of inertia and doesn't want to be pulled down. Just like we witnessed in our two previous demonstrations, a quick motion isn't enough to overcome the object's inertia.

Inertia also applies to objects that are in motion, not just stationary objects. The principle of inertia says that objects want to do what they are already doing. If they are sitting still, they will keep sitting still until something forces them to move. However, if an object is already moving, Newton concluded that it will keep on moving unless a force stops it. Often, this force is friction. Friction is when two things rub together, slowing motion.

ACTIVITY 4.4: An object in motion keeps on going

You will need: *a cylindrical object (such as a battery), a book (or block), a length of floor space*

1) *Set the battery on top of the book. (or whatever objects you are using)*
2) *Start sliding the book very slowly so that the battery goes along with it and does not roll off.*
3) *Once the book is moving quickly, suddenly stop pushing it. What happens?*

The battery should go rolling off the front of the book, continuing in motion. You didn't push the battery off the book. The inertia of the battery overcame the friction between it and the book and made it keep on moving.

ACTIVITY 4.5: A thought experiment

Imagine yourself riding in a car. Hopefully you have your seatbelt on, because the car is going to stop very quickly. Something darts across the car's path and the driver slams on the breaks. Wham! What happens to your body? Do you jolt forward suddenly? That's your inertia in action. As you were riding along, you were going the same speed as the car. It felt like you were sitting still because everything inside the car looked stationary to you. But when the car slowed down, your body kept on going until it hit the seatbelt. Your body was in motion and would have continued forward had it not been for the seatbelt. Bodies in motion (you, in this case) will continue in motion unless acted upon by an outside force (the restraining power of the seatbelt).

ACTIVITY 4.6: Watch Episode 1 in the "Eureka" series (on inertia)

Use the playlist, or simply search for Eureka episode 1.

Newton did not write about his discoveries right away. In fact, that was an issue that would create problems for him later in life. He was interested in science mostly to satisfy his own curiosity and cared very little if anyone else knew what he did. (Little did he know, but someone else was also working on inventing calculus at the same time as he was. The two men would eventually get into an argument that lasted for years. If Newton had published his ideas right after he had thought of them, this dispute would probably never have happened.) So Newton's laws of motion did not appear in writing until 1687, when, at the insistence of a friend, he finally published his work in a book known as **Principia**. (Newton said *Prin-chip-ia*, Latin scholars say *Prin-kip-ia* and most scientists say *Prin-sip-ia*. Take your pick!) This book has been called the most influential science book ever written.

Principia Mathematica contains not only Newton's laws of motion and gravity, but also how they can be applied to astronomy. Newton was able to explain the motion of the planets in a way that solved many of the problems that astronomers were having with their calculations. But more on that later. Right now we want to highlight the very first law of motion that appeared in *Principia*.

Portrait of Newton from 1689

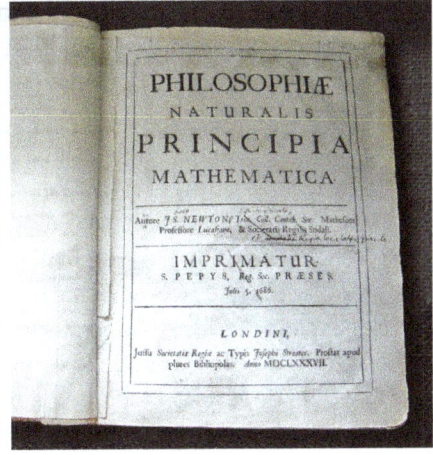
Newton's own copy of his book, with all his hand-written corrections.

The book was written in Latin, which was the language of scholars in that day. Latin was used because at that time, scientists in every country could read Latin. The Latin word for law is "lex," so Newton's first law was written as LEX I. The Latin says: "Every body perseveres in its state of rest, or of uniform motion in a straight line, unless it is compelled to change that state by forces impressed thereon." In simpler English:

This is how Newton's first law looks in the first edition of Principia.

A body at rest will stay at rest, and a body in motion will stay in motion, until an external force acts upon it.

The idea that motion is essentially always in a straight line did not originate with Newton. Robert Hooke had already proposed that motion which appears to be curved (like the path of a ball thrown into the air) is actually a combination of two or more straight-line motions. (We saw this happening in the harmongraph patterns in the last chapter. In a two-pendulum harmongraph, you have only two back-and-forth (straight) motions happening. Yet if you connect the two motions with rods that have a pen attached, the pen will draw circular and oval patterns.) Hooke had written a letter to Newton asking his opinion about this idea, particularly as it related to the path of the moon around the earth.

After the publication of *Principia*, Newton instantly became famous. He went from being a recluse (loner) surrounded by books to attending dinners hosted by kings and queens. He gradually became used to being around people and eventually learned to enjoy being the center of attention. But enough of Newton's social life, let's get back to the science...

Are all objects equally "lazy"? Do they all have the same amount of inertia? In the "Eureka!" video we saw the cartoon character try to move a large boulder. It had too much inertia and he could not make it budge. Is there a way we can measure inertia?

The amount of inertia something has can be measured by finding its **mass**. The word "mass" comes from Greek and means "a barley cake or lump of dough." Mass is usually measured using a **scale**, or **balance**. The earliest scales we know of are seen in ancient Egyptian paintings. In this painting we see the god Anubis weighing a human soul with a "pan balance." The soul sits in one pan (inside the little jar) and an ostrich feather is placed on the other pan. The ostrich feather was a symbol of Ma'at, the goddess of truth and justice. If the pan with the soul in it went down, the person went to an unpleasant afterlife. If the pan with the soul stayed even or went up, this would indicate that the soul was not weighed down with bad deeds, but had the lightness of truth and justice, and therefore the soul would be rewarded with a pleasant afterlife.

Many of us have a flat scale in our bathrooms or bedrooms. This type of scale has a mechanism that can detect and measure the amount of pressure that is pressing down on top of it. Gravity pulls us down onto the scale. The bigger you are, the more "stuff" there is for gravity to pull on. More stuff registers as "heavier." These gravity-based scales are fine as long as everything that you weigh is on planet earth. Since most of us rarely leave the earth, this isn't a problem. However, if you were to take your bathroom scale to the moon, you would find that your scale will tell you that you weigh about 1/6 as much as you do on earth. This is because the gravity of the moon is about 1/6 that of earth. Gravity is directly related to how much mass something has, and the moon has much less mass than the earth. (Oddly enough, various parts of the moon have varying amounts of mass, so your weight on the moon would depend on where you put your scale. For maximum weight, you'd put your scale on one of those dark gray splotches. Those areas have more mass because they've been hit with giant meteors. We think that perhaps at least some of that extra meteoritic mass still lies below the moon's crust.)

This brings us to an important difference between the words "mass" and "weight." Your mass did not change when you went to the moon; your body remained the same. The scale on the moon is telling you how much you weigh *while on the moon*. For someone born and raised in a moon colony, their moon weight would be their normal weight, and they would be fascinated by the heavier readings on their scale when they visited the earth. So which scale is correct? The moon colonists would think their scales are right. We'll have to think about this problem and see if there is a way to measure mass without relying on gravity. But first, let's meet one more type of scale.

A scale you often meet in physics classes is the "spring scale." (We met this scale briefly in activity 2.5.) The spring inside the scale is just stretchy enough to be able to measure the weight of small to medium-sized objects. For small (or lightweight) objects, you need a scale that measures up to 100 grams. For objects that weigh as much as a soup can or a pineapple, you can use scales that measure up to 1,000- 5,000 grams.

The spring in a 1,000 gram scale will be much stiffer than the spring in a 100 gram scale. This means that the larger scales are much less sensitive to tiny amounts of weight. You would not want to measure something very light using a larger scale. You must choose the correct spring scale for the object you want to weigh.

Spring scales can do something that regular scales can't do—they can also measure force. Often, they will be labeled not only in grams (g) but also in units called newtons (N). **One newton is equal to about 100 grams.** But first, how much is a gram, anyway?

ACTIVITY 4.7: Measuring grams using your spring scale

This activity assumes that you have been able to buy or borrow a spring scale. If you don't have one, you can watch Spring Scale Lab 1 on the playlist.

You will need: *your spring scale and various objects you want to weigh*

1) Your spring scale probably has a hook on the end. This is because very often physics labs use weights that have rings on their tops. To use the spring scale with something that does not have a ring, you can either wrap a rubberband around the ojbect and slip the hook under the rubberband, or you can make a "sling" out of a scrap of cloth. Sew or tie a piece of string to each corner of the cloth and the tie the strings together.
2) Weigh various objects and record their weight in both grams and kilograms. 1 kg = 1,000 g

object	g	kg	object	g	kg

Here are some approximations to help you remember how much a gram is. Some of you might already be very familiar with grams, but others might not. The metric system is used by all scientists around the world. Since it is based on the number 10, calculations are much easier than using a system based on the number 12.

A gram is about the weight of one paper clip.

A pencil weighs about 10 grams

An apple weighs about 100 grams

A kilogram is about the weight of a pineapple.

Let's take time right now to watch two more episodes in the "Eureka!" series.

In episode 2, you will be reminded that it is just as hard to stop a moving object as it was to get it moving. A memorable example of this is when an airplane lands. To slow the speed of the plane after landing on the runway, the engines have to work just as hard in reverse as they did in forward mode during take-off. In this second episode, you will also be introduced to one of the "standards" that scientists use to define their units of measurement. There is one special platinum block, (held in a very secure place), that is the "official" kilogram weight. All other weights are compared to this one. The study of standards is its own branch of science. Also, governments of large countries usually have an entire department devoted to standard. They monitor weights and measures of all kinds, making sure that scales and other measuring devices are accurate. Next time you are at a gas (petrol) station, look for a stamp that shows how recently someone has been there to make sure the pump is dispensing accurate gallons or liters.

In episode 7, they will review the difference between mass and weight. You will see and hear them say that gravity is pulling down at 10 meters per second, per second. Don't worry about this for now —we will get to this eventually. You don't need to understand this in order to understand the episode. You will see the cartoon character go to the moon to decrease his weight as shown on a spring scale that measures in newtons. You will also see an old-fashioned balance (sort of like the Egyptian one) showing that this type of scale works the same on the moon as it does on earth, proving that mass does not change, only weight as measured by gravity.

ACTIVITY 4.8:   Watch episodes 2 and 7 from the "Eureka!" series

The videos helped us to understand that your weight is determined by the gravitational field you are in. Your weight varies from planet to planet. On Jupiter you'd weigh so much that your bones and muscles would barely be able to keep you standing up. On Mars, you'd weigh only about one third of your earth weight. But the mass of an object does not change as it goes from planet to planet, only its weight does. So, if you need gravity to measure how much something weighs, how would an astronaut in a weightless environment measure an object's mass? Would they ever need to measure mass? Is there another way to measure mass?

Mars colonists would experience less gravity.

ACTIVITY 4.9:   Measure mass _without_ using gravity (using an *inertial balance*)

You will need: *a hanger, a small plastic cup (or a sturdy paper one), several dozen coins (all the same kind, pennies are perfect), duct tape, and your stopwatch*

(Note: An alternate way to make the arm of this balance is to use a hacksaw blade. Clamp the end of the blade to a table leg. The wide, flat shape of the blade will prevent the arm from bending down as much as the coat hanger will. You might also try taping two hacksaw blades together (overlap the ends) to give a larger oscillation.)

TIP: It will help if you have someone to operate the stopwatch for you. They can call out "Go!" when the start the watch, and then they can stop it when you call out "10!"

1) Unwrap the coiled part of the hanger near the hook. Straighten it out, then bend a long loop at then end opposite the hanger hook. Make the loop at a 90 degree angle from the hanger hook.
2) Tape the small cup into the hanger hook.
3) Duct tape (or clamp) this contraption to the edge of a table so that the cup end is dangling way out from the edge. Make sure it is securely fastenend. You should be able to "twang" the cup back and forth without the base coming off the table.

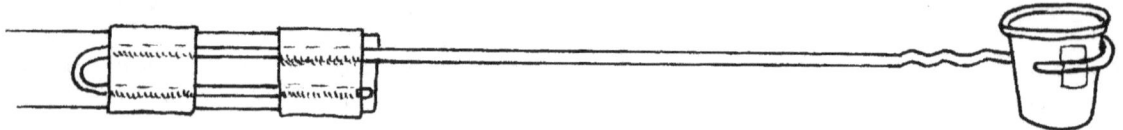

What you have made is called an <u>inertial balance</u>. Yours isn't nearly as accurate as a real one, but it will be good enough for this experiment.

5) Put 5 coins into the cup. Then pull the cup back about 10 cm (5 inches) or so and then let go. Count how many seconds it takes for the balance to go back and forth 10 times. (Count how many times the cup comes back to you.)
6) Now put 5 more coins into the cup, for a total of 10. Pull the wire back the same distance you did before and let it go. Again, count how many seconds it takes for the cup to come back to you 10 times. (If you have trouble estimating how far back to pull the spring each time, you can put the back of a chair right at that spot. Then you just pull the spring back until it touches the chair. This will ensure that you pull it the same distance every time.)
7) Add 5 more coins and count again. Then 5 more. Then 5 more.
8) Keep adding coins until your wire bends down too much.
9) Make a graph of this experiment, with one axis being the number of coins and the other being the number of seconds.
10) When your results are all recorded, draw a "best -fit" straight line showing the general trend of the data.

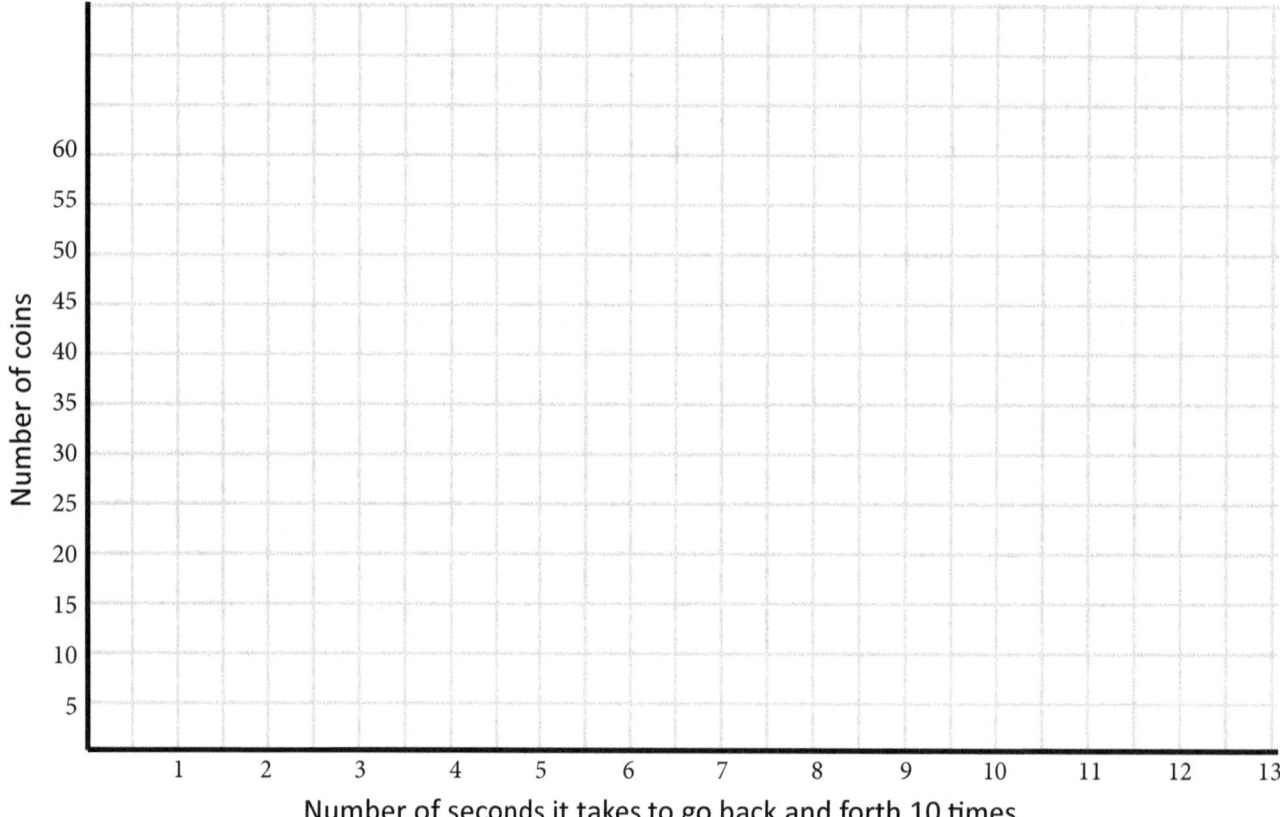

11) Do some extrapolation.
    How many seconds for a number 10 greater than the highest number of coins on your graph? _____
    What about for 50 more coins, instead of 10 more? _____
    Which number are you more sure of: 10 greater, or 50 greater? _____

This lab allowed us to discover the true definition of "mass." **Mass is the measure of an object's resistance to a change in velocity**. This definition makes no sense at all unless you have worked with an interial balance. Now that you have used one, you can see how this definition is possible.

The force you applied to the wire each time you pulled it back was always (approximately) the same. When the cup contained very little mass, it was happy to change directions many times per second. The more mass we put into the cup, the more interita it had, making it better at resisting a change in direction, and thus, giving us slower swings back and forth.

This method of measuring mass would work in space, as long as you securely tied your weights to the end of the spring. This is how things are weighed on the International Space Station. Astronauts have to keep track of their weight (their mass) if they are in a weightless environment for a long time because being weightless can lead to loss of muscle mass. Also, some experiments that are done on the space station require finding mass. For example, they might want to find out how fast an experimental plant was growing.

ACTIVITY 4.10: Watch astronauts using inertial balances on the ISS

If you go to the playlist, you can watch an astronaut weighing himself in the space station using an inertial balance that bobs up and down. There is also a short video clip of a researcher putting a sample into a digital inertial balance. (If the videos are not there, just search for "astronuts weigh things in space.")

ACTIVITY 4.11: One last demo about inertia   *(There is a video of this on the playlist.)*

NOTE: If you don't have a long rod, you can do a smaller scale demo with a long kitchen knife.

You will need: an apple, a hammer, a dowel rod with one end sharped to a point (does not have to be sharp)

1) Push the sharpened end of the dowel rod through the apple, going straight through the core.
2) Position the apple at the bottom of the rod while holding it near the top.
3) Bang on the top of the dowel rod with the hammer. Does the apple start creeping up the rod? (If not, loosen the apple just a bit and try again.)
4) Try hitting the bottom of the rod on the floor. Does the apple go back down again?

The apple, like all objects, has inertia and wants to keep on doing what it is doing. When you hit the top of the rod with the hammer, you forced the rod to go down suddenly. The apple's inertia caused it to want to remain in place. As the rod jerked down bit by bit, it looked like the apple was going up, but it was simply staying in place while the rod moved down. When you banged the rod on the floor, you caused the apple to be in motion, headed down towards the floor. When the rod hit the floor, it stopped, but the apple tried to obey Newton's first law and stayed in motion. The friction between the apple and the rod soon overcame the downward force of the moving apple, however, and caused it to stop moving.

Turn the page and you can find out what is going on here.

ACTIVITY 4.12: More about Newton's life

Newton had been fascinated with chemistry since childhood. You might remember that he lived with an apothecary while attending school. After the publication of *Principia* in 1687 (when he was 45), he achieved enough financial success that he had money he could spend on leisure activities. He decided to spend more of his time and money studying "alchemy," which was a mix of superstition and actual chemistry, and something he had dabbled with since early adulthood. Alchemists of previous centuries tried to find a magic elixir of life that could give immortality. They also were famous for trying to turn things into gold. We don't know exactly what Newton's goals were, but we do know from his notes that many of his experiments involved tasting the chemicals. We also know that his lab was stocked with many substances that we now know are toxic, including lead, arsenic, antimony and mercury. In his notes, he wrote that mercury tasted "strong, sourish, and ungrateful."

1771 painting of an alchemist discovering the element phosphorus.

All this tasting of chemicals took its toll, and by 1693 (when he was about 50 years old) he became ill. He had many neurological symptoms consistent with the symptoms of heavy metal poisoning. He almost destroyed his genius brain. For a year, he suffered from insomina, confusion, memory problems and depression. No one at that time connected his illness to his chemicals. They just chalked it up to the fact that he was an odd person to begin with. Amazingly, he managed to recover from this terrible bout of poisoning, and became functional again, though he never regained the brilliance of his youth.

In that day, it was customary for someone who had achieved academic greatness to be given prestigious appointments at a universtity or in the government. Newton had received a few promotions, but some of his friends thought he should have received more. One of his supporters was able to have him appointed as Warden of the Royal Mint in 1696. The Mint was where all the English coins were manufactured. Newton took his new job more seriously than anyone had anticipated. He knew that counterfeiting had become a serious problem. It was all too easy for someone to make coins that looked just like the real ones. It is estimated that by the early 1690s, 10 percent of all coins in the realm were counterfeit. Newton vowed to end this scandal. His knowledge of chemistry and math helped him to find solutions to the coinage problems. The result was the Great Recoinage of 1696. All old coins were recalled and replaced by new ones. Newton also decided to go after counterfeiters with more intensity than others before him had done. Technically, the punishment for making fake coins was death, but catching and convicting someone was very difficult. Newton decided to play detective and personally gathered evidence against 28 counterfeiters. One of these counterfeiters was particularly notorious and had devised massive schemes to fool even people at the Mint. Newton saw through his schemes and was able to get him arrested. When the counterfeiter's friends were able to get him released from prison on the grounds that the evidence had not been conclusive enough, Newton was able to gather even more evidence and have him arrested a second time. This time the conviction stuck, and the man was publicly executed. The Mint was then put under Newton's complete control, and he remained in charge of the Mint until his death in 1727.

Newton's portraits usually show him wearing a wig. That was the fashion of his day. It is believed that perhaps this fashion came about because of head lice. Men would keep their heads shaved so that lice had no place to live, then they would powder the wigs to prevent lice from living in the wig. Both of these portraits are from later in Newton's life. The one on the right definitely shows him in a wig. Do you think the portrait on the left shows his natural hair or a wig?

ACTIVITY 4.13: Review word puzzle

Fill in the blanks with the correct letters, then use the number code to spell out two famous quotes by Newton.

1) The title of Newton's book: ___ ___ ___ ___ ___ ___ ___ ___ ___
   31 89 22 34 43 108 49 109 33

2) A baseball bat is a ___ ___ ___ ___ ___ ___ ___ ___ ___ lever.
   16 29 91 111 11 44 15 25

3) He said, "I think, therefore I am." ___ ___ ___ ___ ___ ___ ___ ___
   107 8 95 73 94 21 18 37

4) The technical word for going up and down, or back and forth: ___ ___ ___ ___ ___ ___ ___ ___ ___ ___
   23 84 13 32 50 83 28 41 47 24

5) When something vibrates at its natural frequency: ___ ___ ___ ___ ___ ___ ___ ___
   86 30 56 55 45 33 99 35

6) A machine that makes circular patterns using pendulums: ___ ___ ___ ___ ___ ___ ___ ___ ___ ___
   54 73 2 19 74 75 1 94 12 72 59

7) The tendency of something to keep on doing what it is already doing: ___ ___ ___ ___ ___ ___
   52 71 48 93 6 90

8) The measure of an object's resistance to a change in velocity: ___ ___ ___ ___
   69 12 76 110

9) "You can always trade distance for a gain in ___ ___ ___ ___ ___."
   27 70 57

10) The wave shape that records harmonic motion: ___ ___ ___ ___ wave
    77 67 78 106

11) The number of pendulum swings per minute (or second) is called its ___ ___ ___ Q ___ ___ ___ ___
    97 60 101 98 14 7

12) Newton spent the last part of his life as head of this governmental office: ___ ___ ___ ___ ___ ___ ___ ___
    20 80 44 11 102 5 46 36

13) These were used to decipher writing on the Antikythera mechanism (page 35): ___ - ___ ___ ___
    9 104 81 66

14) The pivot point in a lever: ___ ___ ___ ___ ___ ___
    85 87 62 39 96

15) The weight at the end of a pendulum: ___ ___ ___
    38 70 88

16) Foucault set up his famous pendulum at the Paris ___ ___ ___ ___ ___ ___ ___
    10 51 71 40 17 92 55 68

17) To decrease something's energy or activity level: ___ ___ ___ ___ ___
    100 3 79 61 105 53

18) If you wrap an inclined plane around an axle you get a ___ ___ ___ ___
    58 93 112

19) Simple machines give you mechanical ___ ___ ___ ___ ___ ___ ___
    63 4 3 64 42 103 65

20) The curved shape a ball makes after you throw it into the air. (pg. 13) ___ ___ ___ ___ ___ ___
    61 82 26

Famous quotes by Isaac Newton:

1) ___ ___ ___ ___ ___ ___ ___ ___ ___ ___ ___ ___ ___ ___ ___ ___ ___ ___ ___ ___ ___ ___ ___ ___ ___
   1 2 3 4 5 6 7 8 9 10 11 12 13 14 15 16 17 18 19 20 21 22 23 24 25

   ___ ___ ___ ___ ___ ___ ___ ___ ___ ___ ___ ___ ___ ___ ___ ___ ___ ___ ___ ___ ___ ___
   26 27 28 29 30 31 32 33 34 35 36 37 38 39 40 41 42 43 44 45 46 47 36

   ___ ___ ___ ___ ___ ___ ___ ___ ___ ___ ___ ___ ___ ___ ___ ___ ___ ___ ___ ___ ___ ___
   48 9 49 50 51 52 53 112 54 55 56 57 16 58 21 59 60 61 62 63 64 65 42 66

   ___ ___ ___ ___ ___ ___ ___ .
   67 68 69 70 36 13 26 71

2) ___ ___ ___ ___ ___ ___ ___ ___ ___ ___ ___ ___ ___ ___ ___ ___ ___ ___ ___ ___ ___ ___ ___ ___ ___ ___
   72 11 73 40 74 5 76 79 80 85 89 90 92 78 111 82 94 91 77 6 75 21 11 92 67 95

   ___ ___ ___ ___ ___ ___ ___ ___ . ___ ___ ___ ___ ___ ___ ___ ___ ___ ___ ___ ___ ___ ___ ___ ___ ___
   96 80 97 93 13 98 99 100 88 87 16 102 81 103 104 105 83 6 106 84 36 27 86 109 8 46 107

   ___ ___ ___ ___ ___ ___ ___ . (Written in his notebook while still a teenager)
   108 110 40 2 101 42 29

67

ACTIVITY 4.14: Just for fun: "Catch-the-coin" inertia trick

You will need: a coin

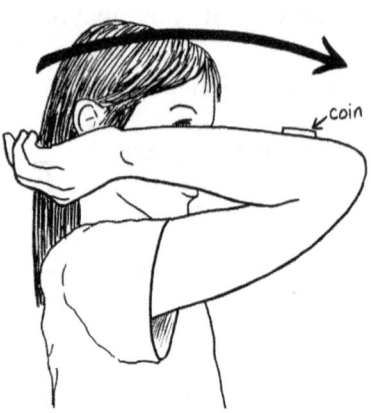

1) Place the coin on your elbow as shown in the pictture.
2) Drop your elbow very quickly and catch the coin in your hand.
   (as shown by arrow)
3) This trick works because the inertia of the coin causes it to stay in place just long enough that you can catch it.

ACTIVITY 4.15: Optional: Watch a video documentary about Newton's life

If you like documentaries, check out the 45-minute video on the "Biography" channel on YouTube.
"Sir Isaac Newton: Unhappy Scientific Genius | Full Documentary | Biography"

Adventure 5: Slowing and stopping with Da Vinci

Not only do objects at rest stay at rest because of inertia, they also stay at rest because of **friction**. You are already very familiar with friction. It gives traction to things like shoes and tires that must not slip on the surfaces on which they traverse. It allows you to keep a tight grip on things you are carrying. It causes your hands to heat up when you rub them together. Too much friction, however, can cause your skin to blister or burn, or your tires to leave black skid marks on the road. Friction has upsides and downsides.

Friction is a force that always opposes motion. It is the "anti-motion" force in the universe. There are two kinds of friction: *static* friction and *kinetic* (kin-ET-ick) friction. You feel the force of static friction as you push on a very heavy object trying to get it to move. The force you feel right before it finally budges is the maximum force of static friction for that object. After the heavy object starts moving, you feel the force of **kinetic friction** as you push it across the floor with great difficulty.

One of the first people to study friction was Leonardo Da Vinci. We know from his notebooks that he began doing friction experiments in 1493.

Here is one of the many tiny sketches from Leonardo's notebooks. He drew as he was thinking. Sometimes he wrote notes about the drawings but other times he did not write anything, or just a few words, leaving us to guess what he was thinking. This sketch leads us to believe that

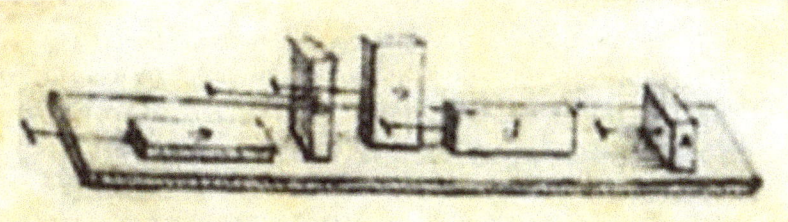

he was experimenting with surface area. These blocks look very similar, almost as if it was the same block drawn in different positions. Depending on how it was tipped, the block would have different amounts of surface area in contact with the table. You can see what looks like handles on the blocks, indicating that they would have been pulled across the table. Leonardo probably measured force by using counterweights hanging off the edge of the table, as seen here on the right. This was his version of a spring scale. (But he didn't measure in newtons!)

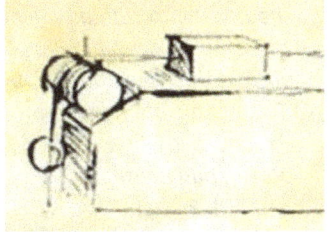

You can discover for yourself what Leonardo was discovering.

ACTIVITY 5.1: The relationship of a block's surface area to its friction

You will need: *a full food box (cereal, crackers, etc.), a meter (yard) of string, your spring scale*

1) Lay the box down flat. Make a loop of string around it so the spring scale will be able to pull it. Slowly and gently pull on the box using the spring scale. Watch the reading on the scale as the box slides across the table. Write down that reading.
2) Now stand the box upright. Again, loop the string about the box (near the bottom so you don't topple it over when you pull). Pull the box, note the reading on the scale, and write it down.
3) Now lay the box on its long, narrow side. Again, loop the string around and pull the box using the spring scale. Note the reading and write it down.
4) How did these readings compare?

      1   2   3

What's going on here?

In the first trial, the box's largest side was against the table, giving the largest surface area affected by friction. However, the weight of the box was distributed evenly across that whole surface area so that each square centimeter of the surface area only felt a small portion of the weight. In the second trial, the smallest side was against the table. The weight of the box was the same, so the weight of the box was concentrated into that smaller surface area. In the third trial the surface area was between the first two trials. In each trial, the mass of the box was the same, just distributed across the surface differently. If the surface are was large, each part of it had less pressure. If the surface area was small, each part of it had more pressure. **But overall, the pressure remained the same.** This might remind us of the principle of the lever. A small amount of large force is equal to a large amount of small force. The outcome is the same. (Arithmetic also shows us this principle: Two 10's is the same as ten 2's.)

The next activity is designed to show that sometimes many small frictional forces can add up to be as strong as one very large force. This demonstration was always done with phone books back in the day when phone companies still sent out printed books to their customers. In big cities, these phone books could be quite thick. Phone books were ideal for this demonstration partly because they had so many pages, but also because the pages were quite thin and flimsy and easy to tear, making the final result even more amazing.

ACTIVITY 5.2: The power of surface area

You will need: *two books (or thick magazines, or thick notebooks) Make sure not to use good books that can't be accidentally ripped! We won't try to rip the books, but accidents can happen.*

1) Open both books to the last page. Then overlap the books so that one of those last pages is on top of the other last page.
2) Keeping the books overlapped like this, begin turning over one page of each book, alternating between books. Make sure most of the page overlaps—don't just overlap the edges, use the whole page. You should end up with all the pages of both books overlapping in an alternating pattern.
3) If the covers keep flapping open, use a large rubber band or a piece of string to keep them in place. Put the band or string around the whole bundle, but don't tie it tightly. Tie it just tight enough to keep the covers closed.
4) Two people are needed for this next step. Each person holds onto the spine of one of the books/magazines. Pull in opposite directions, trying to pull the books/magazines apart. Assuming your books/magazines have enough pages, you shouldn't be able to pull them apart.
    THE PLAYLIST HAS VIDEOS OF THIS DEMONSTRATION BEING DONE WITH PHONE BOOKS.

ACTIVITY 5.3: The importance of surface texture

You will need: half a dozen blocks of identical size and weight, tape, scissors, a large wooden board or cardboard panel (such as the tri-fold cardboard panels sold for school projects), various textures to wrap around the blocks (such as aluminum foil, waxed paper, regular paper, very fine sandpaper, rough sandpaper, plastic wrap, a balloon, etc.)

NOTE #1: Ideally, use wooden building blocks or small pieces of 2x4 lumber. If you don't have any wooden blocks, you might be able to construct blocks using Legos® as long as they are identical in size. If neither of these ideas works for you, perhaps you could use granola bars (still in their wrappers). You will need these blocks for activity 4.13, so don't unwrap them when done.
NOTE #2: Ideally, you want to test all of these friction wrappers all together, in one go. But if you only have a few blocks, you'll have to do this experiment in several steps, testing just a few wrappings at a time.

PRE-TEST: Put all the blocks at one end of the panel, lined up and ready to "race." Begin to lift the end of the panel with the blocks very slowly. Continue lifting until all the blocks begin to slid down the slope. If the blocks are identical, they would all begin to slide at approximately the same time. Try it several times to confirm that the blocks all slide at about the same rate.

1) Cut pieces of materials to a size that will wrap around a block once. Wrap each block in one material. Apply tape on the top side to secure the wrappings.
2) If you are using a balloon, you might be able to cut off the slender "neck" and then stretch the opening wide enough to get the block inside the balloon.
3) Line all the blocks up at one end of the panel, as you did in the "pre-test." Predict which block will slide down the slope first, which will be last.
4) Slowly lift the edge of the panel where the blocks are sitting. Lift very, very slowly! Keep lifting slowly. Eventually, one of the blocks will begin sliding down the slope. Keep lifting at the same slow rate. Another block will begin sliding. Keep lifting until all the blocks have slid down to the bottom.
5) Repeat the experiment to see if you get the same results. And maybe a third time?
6) Were your predictions correct? Did any of the blocks surprise you? How important is the texture of the surface area?

Here are my blocks lined up and ready to go. (The blue one is wrapped in a balloon.)

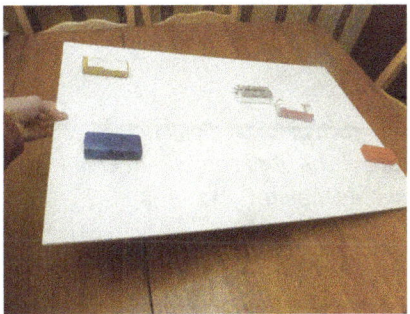

Blocks begin to slide.
(I used a sheet of foamcore for my slide surface.)

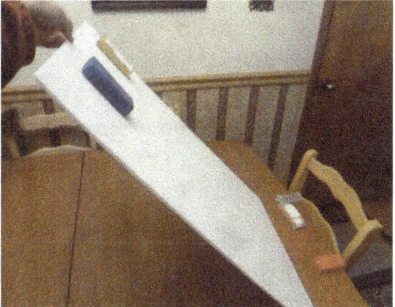

You might be surprised how long it takes for certain blocks to slide!

Mechanical engineers who design things like roads, sidewalks, and racetracks must find a way to quantify how much friction there is between various types of surfaces. In their equations, the number that represents how much friction is contributed by the surface texture is called the "**coefficient** *(co-eh-FISH-ent)* *of friction*." This number often sits in an equation that looks like this: $F_f = \mu F_N$ where the Greek letter "mu" (μ) is the coefficient of friction. The way to read this equation is: "The Force of friction is equal to the coefficient of friction times the Normal Force." (We'll learn what the "normal" force is very soon.) In algebra, a "coefficient" is a number that sits in front of a variable. In the expression "3x" the number 3 is the coefficient. The coefficient of friction almost always turns out to be a number between 0 and 2. Drag racing tires rubbing against a concrete surface give one of the highest coefficients of friction, at almost 4.

But what does this 4 mean? Is it 4 newtons? 4 kilograms? 4 kilometer per second? 4 what?

Drag racing tires give a very high coefficent of friction.

The coefficient of friction is what we call a "dimensionless" number. It isn't 4 of anything. It's just plain 4. Engineers who work with coefficients of friction all the time get very used to dealing with these numbers so the coefficient number has meaning for them. If they are designing something where steel will rub against aluminum, they can look up this combination of materials in their engineering charts and find that someone has already determined that the coefficient of friction between these two materials is about 0.61. They will know if this number is acceptable for their project.

In the next activity you will find the coefficient of friction for each of your blocks. This test will be for the coefficient of **static friction** because we will be measuring maximum force right <u>before</u> the block moves.

ACTIVITY 5.4: Calculate the coefficients of (static) friction for your blocks

You will need: your blocks and ramp from previous activity, plus two meter sticks ("yard sticks" marked with centimeters)

1) Lay one meter stick along the edge of your ramp, as shown. Make sure the zero end is even with the end of the ramp. This will be your horizontal measure.
2) The other meter stick will be held straight up so it can measure how high the ramp is when the block finally begins to slide. This will be your vertical measurement. TIP: It might make your math easier if you put this vertical stick right at a nice even number on the bottom stick, such as the 30, 40 or 50 centimeter mark.
3) Put one of the blocks at the starting point, hold the vertical meter stick in place, and begin slowing raising the ramp.
4) When the block slips, stop raising the ramp and read on the meter stick how high the ramp is. Record this number on the chart below. (Best practice is to take this reading 3 times and average the results.) Under "Material 1" write what your block is wrapped in (paper, foil, rubber, etc). Under "Material 2" write what your ramp is made of (metal, wood, cardboard).
5) Do this for each block.

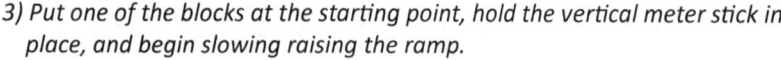
(1)

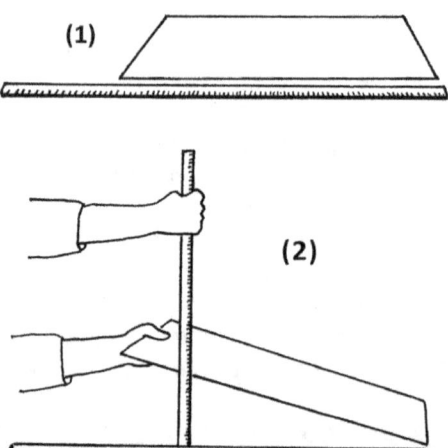
(2)

6) **Calculate the coefficient of static friction (μ_s) for each block by dividing the vertical measure by the horizontal measure. (v/h)** When finished, your chart will look very much like the data charts that engineers use! These might be just numbers to you, but to a mechanical engineer they carry a lot of meaning.

Material 1	Material 2	vert. measure	horiz. measure	μ_s

Here are some actual coefficients of static friction from an engineering handbook:
    diamond—diamond: 0.1
    brick—wood: 0.6
    wood—wood: 0.54
    steel—steel: 0.78
    aluminum—steel: 0.6
    glass—glass: 0.1

ACTIVITY 5.5: Comparing STATIC FRICTION to KINETIC FRICTION

You will need: your spring scale, your blocks from the previous experiments, a piece of string long enough to tie around one of your blocks, and possibly a small weight (a small book?) to put on top of your blocks

NOTE: For this experiment to work well, you may need to add weight to your blocks. You want the weight you are pulling with your spring scale to register somewhere in the middle of the scale. For example, if you are using a 500 g scale, then you want to pull something that will register 200-400 g. Try setting a book on top of your block to make it register higher on the scale. If you are using lightweight blocks, a 100 g scale would be best.

1) Tie the string around one of your blocks. Slide the spring scale hook under the string. Set the block on the table.
2) Begin to pull the block using the spring scale. (Add weight if the block moves immediately.) You will need to watch the scale carefully. Increase your pulling force very gradually. (If you use a weight, make sure to use the same weight on top of all the blocks.)
2) Notice how the reading on the spring scale increases even though the block does not move. The scale is showing you the force of **STATIC friction**. (The word "static" means "not moving.") The static friction will increase as you pull harder.
3) The block will suddenly begin to move. Get a reading on the scale right as the object moves. You will probably need to do this several times to make sure of the reading. Notice that the reading jumps up quite high just as the object begins to move, then the reading decreases. The highest reading you see will be the maximum STATIC friction force for this object. Record this number on the chart below.
4) Now keep dragging the block at a constant speed and see what the scale reads. This reading, as the block is in motion, is called **KINETIC friction**. ("Kinetic" means "moving.") Record this number in the chart below.
5) Repeat the experiment for at least three of your other blocks and record results. (If you added weight to the first block, add the same weight to all the other blocks.)
6) Look at your data table. Is maximum static friction always larger than kinetic friction? _____
   Does the coefficient of friction play a role in static friction? _____ In kinetic friction? _____

BONUS STEP:
7) Go back to the first block you used in steps 1-3. Try three different amounts of weight on top of the block. Compare these results with your very first results in line 1.
   Does mass affect the force of static friction? _____ Of kinetic friction? _____

which block?	F_{sf} (force of STATIC friction)	F_{kf} (force of KINETIC friction)

It is important to emphasize that in this experiment we learned that the force of static friction isn't something we can see with our eyes. The only way we know the force is there is to measure the push or pull acting on the object. Our spring scale told us that pulling force was being applied to the block even though it looked like nothing was happening. As long as the block was not moving, the applied force and the force of static friction were equal. The force of static friction starts at zero when no pulling force is being applied, and then steadily increases as the pulling force increases. When the pulling force is equal to the coefficient of friction, that's the maximum amount of static friction possible. As soon as the pulling force becomes greater than the maximum static friction possible, then the block jerks into motion and we no longer have static friction, but kinetic friction.

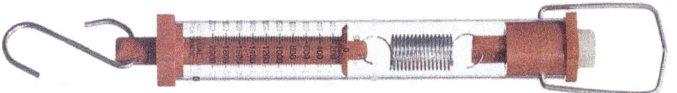

Spring scales can show hidden forces.

Researchers at Cambridge were among the first to study friction at the microscopic level. They saw that even flat surfaces look rough if you look close enough. Even smooth surfaces looked like tiny mountain ranges.

The actual touch points between the two surfaces could be surprisingly few. However, if extra weight was applied to the top surface, the number of touch points increased.

This was the best explanation of friction until the 1980s, when it became possible to analyze it at the atomic level. Researchers now had electron microscopes that could "see" that the atoms right on the surfaces of the objects (at those touch points) were forming weak electrical bonds. Some of the atoms in the box were forming bonds with atoms in the table! (The dots represent atoms. The squiggly red lines suggest electrical attraction between atoms in the box and atoms in the table surface. Atoms don't look like colored dots and electrical attraction isn't red, but this diagram gets the point across.)

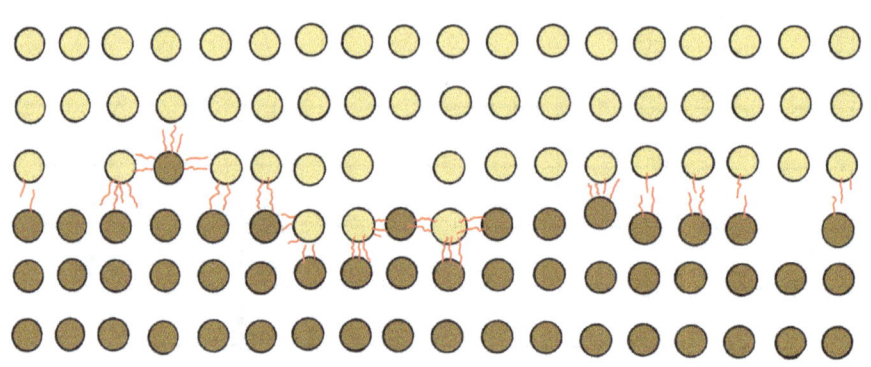

Although friction is all about slowing, stopping, and not moving, it is still considered a "force." It is one of the basic forces that engineers need to consider as they analyze motion. One of the techniques they use to analyze the forces acting on a moving object, is a **free body diagram** (also known as a **force diagram**).

Let's draw a free body diagram for a box sitting on a table. First, we draw a rectangle representing the box. Then, we draw a dot in the middle. (1) The dot represents the center of mass. In our first adventure, we found that when we balanced something on the point of its center of mass, all of the weight of that object seemed to be concentrated at that one point. If an object weighs 100 grams, the point on which it is balancing will feel all of those 100 grams. So in our free body diagrams, we can use a "center of mass dot" to represent the entire mass of the object. Now we can draw a line for the table. (2)

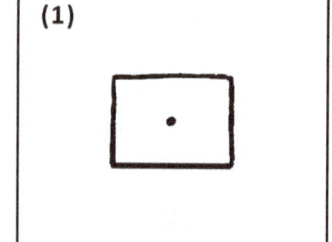 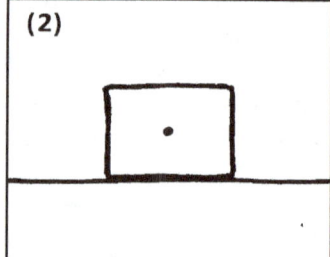

You can probably guess that we need to draw a force line pointing down, to represent gravity. We'll use a capital letter "F" to represent the word "Force," but we're going to have a lot of forces, so we need a way to identify the different forces. We'll use small subscript letters next to the F. For the force of gravity we will use a tiny "g" to represent the word gravity: F_g. The arrow will point straight down because gravity always pulls exactly straight down. (3)

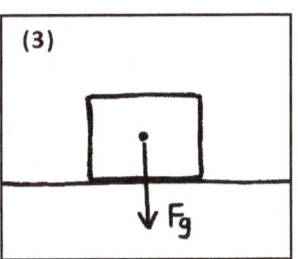

The diagram might look finished at this point, but we are missing a force. The diagram tells us that there is a downward force acting, and nothing opposing it. If we stop at (3), an engineer will interpret this diagram as a box moving (or falling) downward, maybe taking the table with it. But that's not what we intend. We want a box sitting still on a table. Therefore, there must be a force counteracting gravity. Odd as it might seem, the table is exerting an upward force on the bottom of the box. The upward push of the table is of exactly the same strength as the downward pull of gravity. This upward push is called the **normal force**, represented by F_N. In this context, normal does not have its usual (normal!) meaning. Here, normal means "perpendicular."

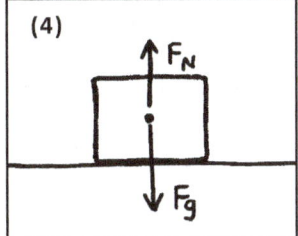

So we draw an arrow pointing up, and label it F_N. Now our diagram is finished. (In Latin, "normalis" means "made using a carpenter's square." A carpenter's square is an L-shaped tool used to make perfectly square corners.)

The lengths of the arrow lines are very important, as they indicate the strength of each force. Here, the forces are equal so we make the lines the same length. The lines don't have to be a particular length (you have some artistic license here) as long as they are the same length. The strength of a force is called its **magnitude**. Forces only have two qualities: **direction and magnitude**.

Now let's draw a diagram for a box sliding across the floor. To draw the free body diagram, we don't need to know what is pushing or pulling the box.

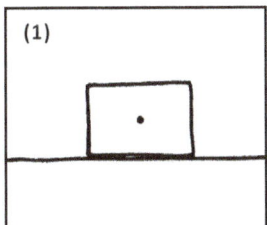 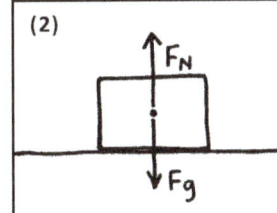 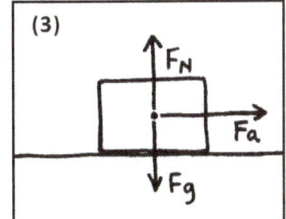 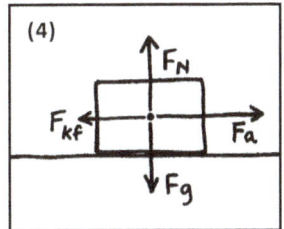

STEP ONE: Draw the floor, the box, and a dot in the center of the box.

STEP TWO: Draw the arrows for the force of gravity (F_g) and the normal force (F_N).

STEP THREE: Draw an arrow for the "applied force" (F_a) which is the push or pull that is moving the box.

STEP FOUR: Draw an arrow for the force of kinetic friction, (F_{kf}). Because the box is moving, we know that the friction force must be less than the applied force, make the friction arrow shorter.

Finally, let's draw a heavy box sitting on the floor.
Someone is pushing hard on the box but it is not moving.

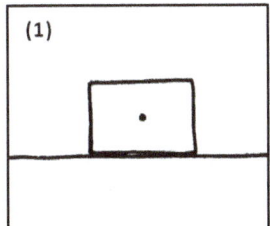 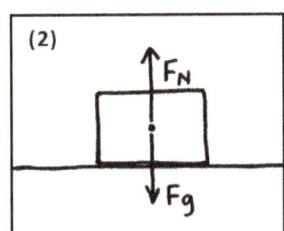 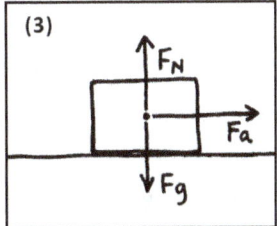 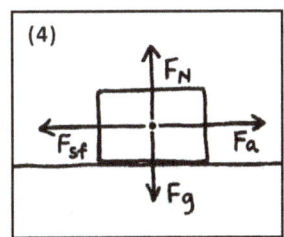

STEP ONE: Draw the floor, the box, and a dot in the center of the box.

STEP TWO: Draw the arrows for the force of gravity (F_g) and the normal force (F_N).

STEP THREE: Draw an arrow for the "applied force" (F_a) of someone pushing or pulling the box to the right.

STEP FOUR: Draw an arrow for the force of static friction, (F_{sf}). Since the box is not moving, we have static friction, not kinetic. We make the arrows the same length to show that nothing is moving.

If opposing arrows are the same length, no one is winning the "tug-of-war."
If one of the opposing arrows is longer, motion is occurring in that direction.

In free body diagrams, whenever there are two arrows of equal length pointing in opposite directions, we call this *equilibrium*. (You can see the word "equal" inside the word "equilibrium," though it is spelled "equal.") We might think of arm wrestlers who are pushing in opposite directions. If they are each pushing with equal force, their hands will stay in the same position no matter how much force is applied. However, if these forces become unequal by even a small amount, we will begin to see motion in one direction as one arm starts to go down. **Motion is the result of unbalanced forces.** Applied forces can be very strong, but as long as they are equal, we have equilibrium resulting in no motion.

ACTIVITY 5.6: Interpret some free body diagrams (answers at the bottom of page 78)

Match the diagrams to their descriptions.

1) The box is moving on a frictionless surface. _____

2) The box is going up. _____

3) Someone is pushing on the box but it is not moving. _____

4) Someone is pushing down on the box, but the normal force is pushing back the same amount. _____

5) Two people are pushing the box with equal strength in opposite directions. _____

6) The box and the table are falling. _____

[Diagrams A through F showing free body diagrams of boxes with various force arrows labeled F_N, F_g, F_a, F_{sf}]

ACTIVITY 5.7: Match vocab words to their definitions (answers at the bottom of page 78)

We learned some important physics vocabulary words in the past two adventures. Can you match them to their definitions?

1) static _____
2) mass _____
3) equilibrium _____
4) inertia _____
5) kinetic _____
6) normal _____
7) weight _____
8) magnitude _____
9) force _____
10) friction _____

A) measuring mass while in a particular gravitational field
B) how large or small something is
C) moving
D) the anti-motion force in the universe, caused by the rubbing of two surfaces
E) perpendicular
F) a push or pull in a certain direction
G) when two opposing forces are equal
H) the measure of how resistant something is to a change in velocity
I) not moving
J) the tendency of something to remain as it is

Adventure 6: Gravity with Galileo
(and Newton, and Einstein)

We now return to the city of Pisa, where Galileo is about to do a demonstration about gravity. For over a thousand years, students all over Europe had been taught Aristotle's ideas about gravity. He said that heavy objects fall faster than lighter ones. Galileo had the audacity to "fact check" Aristotle and run some experiments to see if this idea was correct. And now, Galileo is staging a large-scale demonstration of his experiments. Many of the faculty members at the University of Pisa are gathered today to see Galileo drop things from the top of the Leaning Tower.

So what went wrong with Galileo's experiment? (Of course, people did eventually realize that Galileo was right; other scientists would have started doing this same experiment themselves.) Why did the balls not hit the ground at exactly the same time? There could have been a split second difference in the release of the balls. Or perhaps there was a difference in air resistance between the two balls. Maybe one of them was more affected by air molecules and for that reason fell more slowly. Air resistance is a considerable force and must be taken into account. An extreme example of this would be to drop a solid, heavy object, like a hammer, and something very light, like a feather. You can imagine the feather floating down slowly and gently while the heavy hammer drops quickly and lands with a thud. The feather is very much affected by air resistance while the hammer is not. Therefore, to really test whether all objects fall at the same rate, you'd need to get rid of all the air and do the experiment in a vacuum. Can you think of a place without air? How about the moon?

During the Apollo 15 mission to the moon in the summer of 1971, astronaut David Scott did this experiment. The Apollo mission had planned this in advance and made sure that a hammer and a feather were taken on the mission. They had a video camera ready to film this historic re-do of Galileo's famous experiment. Did it work? Did the feather drop at the same rate as the hammer? Watch and find out!

ACTIVITY 6.1: Watch an astronaut drop a feather and a hammer on the moon

The playlist has a the original video taken by the astronauts of Apollo 15.

ACTIVITY 6.2: A way to test gravity without air resistance

You will need: a hardcover book, a small piece of paper (just a few centimeters square), a solid, relatively heavy object that won't roll (such as a large coin or a small wooden block)

1) Hold the scrap of paper and the other object at shoulder level and drop them at the same time. If your paper was small and light enough, it should have fluttered down gently, similar to what a feather would have done. (You can use a feather if you have one!)
2) Now put the paper and the other object on top of the book. Hold at shoulder height, then drop the book. If you did this correctly, both the paper and the other object should have stayed right on top of the book the whole way down. Thus, the paper and the other object fell at the same rate as the heavy book.

The paper and the other object stayed on top of the book because the book acted as a plow, pushing through the air molecules. The smaller objects "hid" behind the book and thus didn't experience friction due to air. If Aristotle had been right, the book would have fallen faster than the smaller objects, leaving them behind and getting to the ground first.

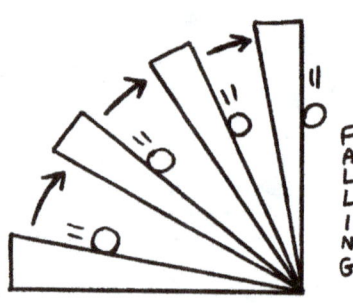

Galileo didn't stop thinking about falling objects after he figured out that they fall at the same rate. He wanted to know more about the way they fell, but it was hard to study them because they fell so fast—he wished he could slow them down a bit. One day, he figured out a way to do this. He was watching a ball roll down a ramp. He imagined making the ramp steeper. The ball would roll down faster. What if he made the ramp steeper and steeper? The ball would go faster and faster. What if he finally made the ramp so steep that it was pointing straight down? We would then say that the ball was "falling" not "rolling." Falling must be the very fastest version of rolling. A ramp really is a way to slow down the effects of gravity!

Galileo constructed a very long ramp with a central track for rolling balls. Timing the rolling balls would be difficult, however. In Galileo's day, there weren't any accurate clocks or watches. Remember, the first pendulum clock was not built until after Galileo's death. Here is how Galileo described his solution to this problem:

"We employed a large vessel of water placed in an elevated position; to the bottom of this vessel was soldered a pipe of small diameter giving a thin jet of water, which we collected in a small glass during the time of each descent [of the rolling ball]. The water thus collected was weighed, after each descent, on a very accurate balance; the difference and ratios of these weights gave us the differences and ratios of the times."

Water clocks had been in use since ancient times, so Galileo wasn't the first person to think of using water as a timer. No doubt, Galileo designed his water timer to suit the needs of this particular experiment. He would start the water timer as soon as he let go of the ball and the water would drip the whole time the ball was rolling. The water ran into a little collection dish. When the ball hit the bottom of the ramp, the water timer was shut off. He then weighed the water to find out how much had been collected. After recording this number, he then did the experiment again. He started the ball at various points on the track and collected all these numbers. When he looked at his results, a definite pattern appeared.

The ball would always start out slowly and pick up speed as it went down. By the time it reached the bottom it was going very fast. When he compared the weights of the water collections, he found that the numbers formed a pattern. He always got this pattern regardless of how steep the incline was or how heavy the ball was.

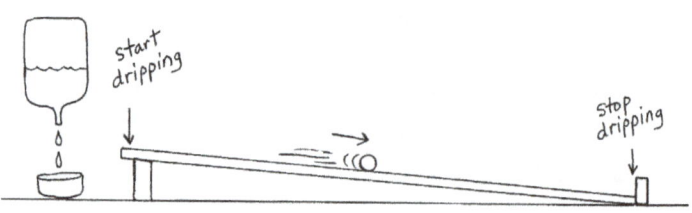

Why don't you try this experiment and find this pattern for yourself?

Answers to 5.6: 1) E 2) B 3) F 4) C 5) D 6) A

Answers to 5.7: 1) I 2) H 3) G 4) J 5) C 6) E 7) A 8) B 9) F 10) D

78

ACTIVITY 6.3: Galileo's famous ramp experiment

The instructions for setting up a ramp are in the teacher's section. If you are unable to set up a ramp yourself, no problem—you can watch a video of it on the playlist ("Discovering Motion 6.3") and use the video data on your own graphs. The graphs you need are on the next page. The teacher's guide also has a sample graph.

Before you do the lab, please watch this video on the playlist: "Galileo's Measure Of Gravity Explained By Jim Al-Khalili" *If the video happens to be missing from the playlist, just search for it. It's a very popular video and there should be several uploads. This video will help us to analyze our data.*

Hopefully, you discovered what Galileo did. Did your graph look like a parabola? Falling objects pick up speed as they fall. The rate at which they pick up speed is "exponential," meaning that squared numbers are involved. Galileo did another experiment to find out more about this exponential rate of falling bodies.

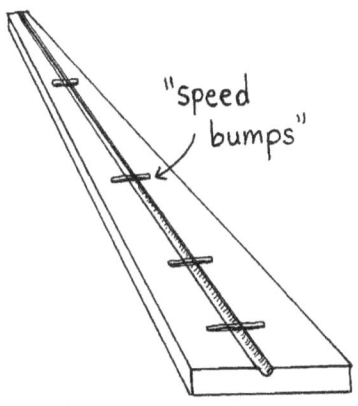

You'll remember that Galileo was a very talented musician. He decided to use his excellent sense of rhythm to help him in his study of falling objects. He made small "speed bumps" for his track, so that as the ball rolled over a bump it would made a clicking sound. He could move these bumps up and down the track until the sound of the clicks made a steady beat. After the clicks were even, he then measured the distances between the bumps. The distance between the first and second bumps was his basic unit of measure. He found that the distance between the second and third bump was 3 of those first units. So the total distance covered in two clicks was 4. The distance between the third and fourth bumps was 5 original units. So the total distance covered in 3 clicks was 9. The distance between the fourth and fifth bumps was 7 times the original distance, so the total distance covered in 4 clicks was 16 units.

Galileo called this pattern "The Rule of Odd Numbers," since each additional distance was always the next odd number (3, 5, 7, 9...). However, more importantly, we should notice that 4 is the square of 2, 9 is the square of 3, and 16 is the square of 4. This is similar to the pattern that Galileo discovered in pendulums—the "Inverse Square Law." As pendulums get longer, the rate at which they swing gets slower by 1 over the square root of the amount by which the length was increased. Falling objects and pendulums both involve squared numbers or square roots. The reason for this connection is that they both rely on gravity for their motion.

ACTIVITY 6.3

Galileo's famous ramp experiment

Galileo wanted to study gravity, but found that objects fell too fast for him to be able to study them. Then he figured out that he could slow down the action of gravity by using a ramp. His ramp was about 7 meters long (23 feet) and had a narrow groove cut along the center. Galileo sanded and polished this groove until it was as friction-free as possible. He used a very smooth bronze ball. For a stopwatch, Galileo used a water timer that dripped droplets at a constant rate. He could start and stop the water drops each time he rolled the ball a certain distance. Then he could measure how much water he had collected on each trial and try to figure out if the ratios followed any kind of mathematical pattern. The pattern he found was very similar to the pattern he found for pendulums.

Record the data from your ramp experiments below. Make sure to label each axis and write what the units represent. Two graphs are provided in case you would like to do the experiment twice, changing a variable such as the steepness of the ramp or the mass of the ball.

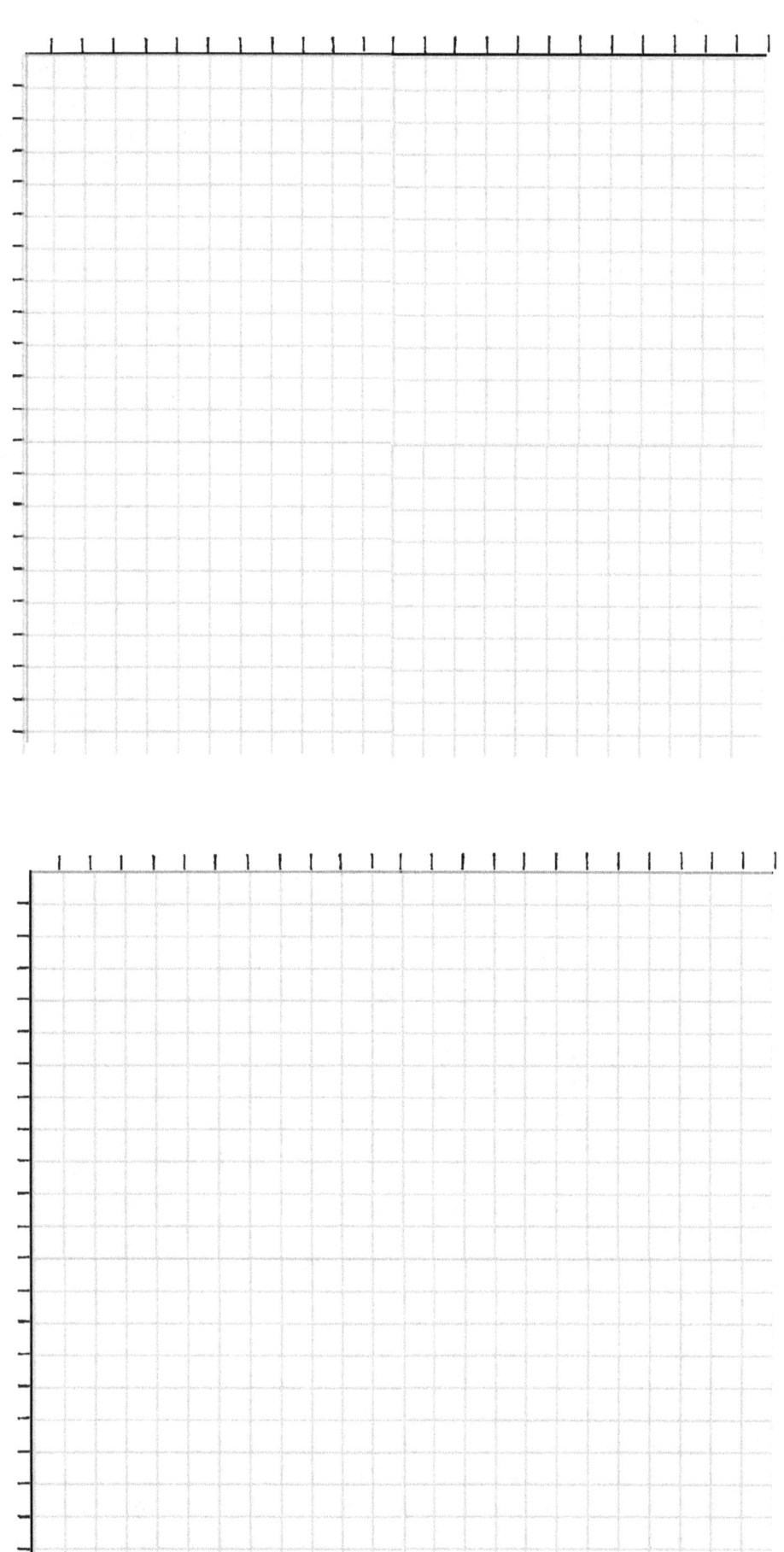

ACTIVITY 6.4: Let's analyze our data

Whether you collected data yourself or used the video data, go to the playlist and watch the video titled: "Discovering Motion 6.3 Analysis" (If you are unable to watch the video, there is a written summary in the teacher's section.)

ACTIVITY 6.5: Do what Galileo did and adjust speed bumps until the sounds are even

You will need: *a long board, thin rubber bands, a marble (or better yet, a steel bearing which is heavier than a marble)*

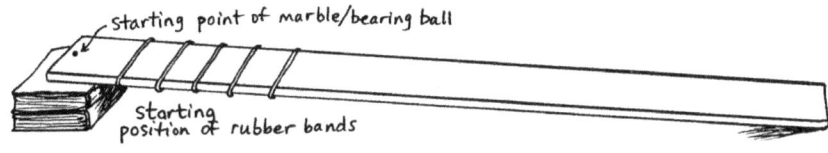

1) Put about 5 rubber bands around one end of the board, then put that end of the board up onto a short stack of books to make it into a ramp. Make a dot at the top of the board, marking where you will start the marble each time.
2) Start the marble at the dot and let it go. Listen to the bumps as the marble rolls over each rubber band. Are the bumping sounds evenly spaced, or do the sounds speed up as the marble rolls down?
3) Adjust the rubber bands until the bumping sounds come at regular intervals. It doesn't matter how long the interval is, only that the sounds are making an even rhythm.
4) Once you have the bands adjusted so that the rhythm is even, try lifting the board to make the ramp steeper. Is the rhythm faster? Do the bumping sounds stay evenly spaced? Try lifting the board even more, then try making it very low.
5) Now measure the space between the starting point of the marble and the first band. You might cut a piece of paper that will fill that gap perfectly. Now use this paper as a measuring unit for the other gaps. How many units will fit between the first and second bands? _____ Second and third bands? _____ Third and fourth? _____ Fourth and fifth? _____

The predicted result is that 3 paper units will fit between the first and second bands, 5 units will fit between the second and third bands, 7 between the third and fourth bands, 9 between the fourth and fifth. This is Galileo's "Rule of Odd Numbers." If you did not get exactly this result, it's okay. (Can you think of any reasons why your results differed from Galileo's?)

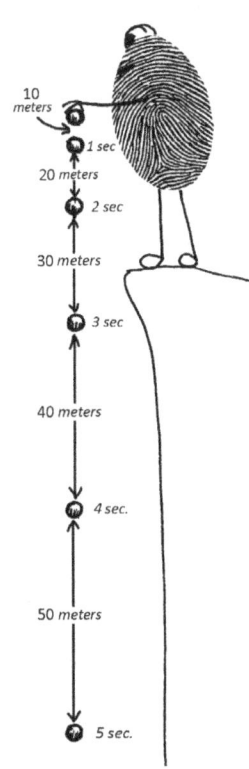

What Galileo discovered is that gravity causes **acceleration**. You know that cars have foot pedals called accelerators. What does the accelerator do? It makes the car go faster, of course. Once you reach the speed at which you want to travel, you no longer have to apply the accelerator (except to counteract friction).

To accelerate means to increase your speed, going faster and faster with each passing second. The word **speed** can be defined with a number and a measuring unit, such as "35 kilometers per hour" or "6 meters per second." It's a steady rate. But you can't measure acceleration this way. Think about a car pulling out onto a road. In the first few seconds, the car isn't going very fast, but by the time 10 or 15 seconds has passed, it can be zipping along. Its speed wasn't constant. Once it reached the highway speed limit, however, then its speed became constant and the driver no longer needed to use the accelerator; (well, just a little, as an opposing force to friction). If someone had asked the driver what his speed was while he was pulling out, he might have said it was 10 km/hr. But then, a few seconds later, the speedometer would not have read 10 km/hr anymore; perhaps it would have said 25 km/hr. To get an answer about how fast he was going, you would need to specify a particular point in time. After 1 second? After 5 seconds?

Acceleration measures the rate at which change is happening. Galileo found that no matter what his actual numbers were, the rate of increase stayed the same. He concluded that gravity applied the same rate of acceleration to all objects. Scientists have calculated this rate of acceleration due to gravity and found it to be about 10 meters per second for each second, or "per second, per second." You write it like this: 10 m/sec^2 "Sec^2" can be read as "per second, per second" or "per second squared."

As our volunteer is showing, this means that an object falls about 10 meters (about 30 ft) in the first second. During the next second, it falls a distance of 20 meters. In the third second, it falls 30 meters. In the fourth second, it falls 40 meters. Then 50. (Our volunteer must be a giant if these distances are meters!)

Acceleration due to gravity on other planets:

Mercury
-3.7 m/s²

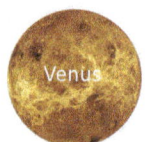
Venus
-8.8 m/s²

Mars
-3.7 m/s²

Jupiter
-24.5 m/s²

Now we need to make some adjustments to our estimate of "10." First, acceleration is a ***vector*** force. By definition, **a vector has both speed and direction**. Speed is not a vector because it only tells you how fast something is going, not which direction it is going. **Acceleration is defined as a vector, so it will tell you not only speed, but also direction.** Our estimate of "10 meters per second per second" gives you the information about speed. The direction will be revealed by a positive or negative sign. Physicists have conveniently defined positive as up (away from the center of the earth), and negative as down (towards the center of the earth). Since we see gravity pulling things down, then we need to add a negative sign. So more correctly, it would be -10 m/sec² (you can say either "minus" or "negative").

And now we need to make *another* small correction. As nice as it is to use the approximate value of -10 m/sec² (because doing calculations with the number 10 is so easy), the truth is that all physics texts use -9.8 (which is shortened from -9.80665). So we'll be multiplying and dividing by 9.8, not 10. Fortunately, we have calculators. But, oddly enough, -9.8 isn't exactly right, either. Scientists have been able to measure the strength of gravity in various places around the globe—on mountains, in valleys, at sea level, over ocean trenches, etc.—and they have found that they don't always get -9.80665. Their experimental numbers are lower on the tops of mountains, and higher in valleys and at sea level. When you are on top of a mountain, you are further away from the center of the earth, so you are slightly less susceptible to its pull. When you are at sea level, gravity will be at its strongest, and sea level gravity at the poles is the strongest of all because of the slightly squished shape of the earth (see image at right). Over ocean trenches, gravity is slightly less because the water over which you are sailing is much less dense than rock; so density also plays a role in gravity. As you go down into the earth (perhaps in a mine), gravity would increase just a bit, though not enough for you to notice. Scientists have averaged all this data, to come up with a standard number that all physicists and engineers can plug into their equations. Thus, you can go ahead and memorize "-9.8 m/sec²" as the acceleration due to gravity (on earth). In equations, this number is represented by the letter "**g**." (It is always lower case.)

The earth slightly wider around the equator, though here the shape is much exaggerated.

ACTIVITY 6.6: Observe the difference between speed and acceleration

PART 1: Calculating an object's speed

You will need: *a "track car" that runs at a steady pace (NOT a remote controlled car where you control the speed and direction-- a track car is one that you turn on then simply set on a track), tape or string you can stick to the floor to make a start and stop line, a measuring tape, a timer (optional: calculator)*
NOTE: *If you can't get a track car, check the playlist for a video of a track car.*

1) Measure a distance of 1 meter on the floor and mark the beginning and ending point with tape or string.
2) Turn on your car and get your timing device ready.
3) Set the car down a little before the start line so that when it crosses the start line it will be going at its top speed. As the car crosses the start line, start your timer. Stop the timer when it crosses the finish line.
4) Now calculate the car's speed. You might already know the formula "D=RT" (distance equals rate multiplied by time). We can rearrange this formula to make "R=D/T" (rate equals distance divided by time). We can use this version of the formula to easily find the rate (the speed) at which our car traveled. The distance is 1 meter. The time will be how many seconds your car took from start to finish. Divide 1 by the number of seconds. You can use a calculator to make this a decimal number instead of a fraction. Your answer: _____ meters per second.
5) Now mark off a distance of 2 meters. Time the car going this distance.
6) Use the same technique to calculate the speed (distance divided by time). Your answer: _____ meters per second.
7) Did you get close to the same answer both times?
8) Extension: Predict how long it will take for the car to go 1.5 meters. For this, you need to rearrange the formula again. D=RT can also be written as T=D/R" (time equals distance divided by rate). So to predict time, divide the distance (1.5 meters) by the number you found in step 6 (or step 4). Your prediction: _____ seconds
9) Now time the car across this distance. Your result: _____ seconds.

PART 2: Observing acceleration

You will need: a copy of the following page printed onto card stock (if need digital patterns to print, you can download patterns at www.ellenjmchenry.com/discovering-motion-printables), scissors, glue stick, a piece of black thread, a paperclip

NOTE: Even if you don't have a track car for the second part of this experiment, you can still do the first part, where you simply hold the accelerometer and watch it as you walk or spin around.

1) Cut out the two pattern pieces. Fold lines are indicted with dotted lines.
2) It is recommended that you score the fold lines before you fold them. Scoring is a light scratching or pressing with something fairly sharp, but not sharp enough to actually cut through.
The point of a compass (the thing you use to draw circles) works very well for this, but you can also use a nail or large needle, or point of a scissor if you can get it to score right along the dotted line.
3) Fold on dotted lines as shown in picture 1.
4) Glue the center fold as shown in picture 2.
5) Fold as shown picture 3 and glue in place.
6) Then glue the half circle onto it, as shown in picture 4.
7) Use scissors to make a tiny snip along the line at the top of the semi-circle (picture 5). The knotted thread will go here.
8) Cut a piece of thread about 15 cm (6 inches). It will be trimmed to the right length later.
9) Tie a large knot in one end of the thread. The best way to do this is to do a standard loop tie, then do it again and again, at least 4-5 times. The knot needs to be large enough that it will not slip through the slot you just cut with the scissors.
10) Slide the thread into the slot and pull until the knot sticks at the back of the slot. Tie a paperclip onto the other end of the thread so that it hangs almost to the bottom of the semi-circle. (Shown in the picture here to the right.)
NOTE: We won't need the numbers on the accelerometer for this activity—they are for activity 8.10. Since you'll be using your accelerometer again, make sure you keep it somewhere safe until then.

<u>What to do with the accelerometer:</u>

1) Before you mount the accelerometer onto the car, try a few experiments. Hold it with your hands and keep your eyes on it as you start walking. Which way does it swing? If you keep walking at a steady pace, does the paperclip go back to hanging vertically? (Make sure that you have minimum friction. The paperclip shouldn't get stuck against the paper.) What happens if you suddenly stop? Does the clip swing the other way?
2) Now stay in one place and just spin around as you hold the accelerometer. Watch it carefully! Does it behave differently than it did when you were walking in a straight line? Does the clip stay up, indicating acceleration? Why is the clip showing acceleration when you are spinning at a steady pace? We'll answer this question in chapter 9.
3) Now mount the accelerometer onto your track car. If you don't have a track car, you can watch the video on the playlist.
4) Turn on the track car but hold it off the ground. Wait until the paperclip is not swinging, then set the car down and watch carefully what happens to the clip in the first few seconds. Which way does the clip move?
5) Now keep watching as the car continues to travel. Is the paperclip still moving after the first few seconds? Once the car reaches stop speed, it will no longer be accelerating, thus the paperclip should stop swinging. The paperclip won't swing again until the car is stopped.

Acceleration is not about how fast you are going. it's about how fast your speed is increasing or decreasing. If you took your accelerometer on an airplane, once you reached cruising speed, the paperclip would hang straight down even though you and the plane are traveling at hundreds of kilometers per hour. You only experience acceleration during take-off and when you land. (Deceleration is negative acceleration.)

<u>Another way to make an accelerometer:</u>

Another way to make a hand-held accelerometer is to tape the thread with the paperclip (or a nut or washer, as shown here) to the underside of the lid of a clear jar. Then fill the jar with water and screw the lid on very tightly. Hold the jar and watch the paperclip as you walk or spin around. The water will dampen the swing of the "bob" to that it doesn't swing back and forth quite as much, making it a bit easier to watch.

HOW TO ASSEMBLE THE ACCELEROMETER

It is recommended that you "score" the fold lines (the dotted lines) before you fold them. Scoring is just a light scratching or pressing of the paper, not cutting. Score will make the folding very easy.

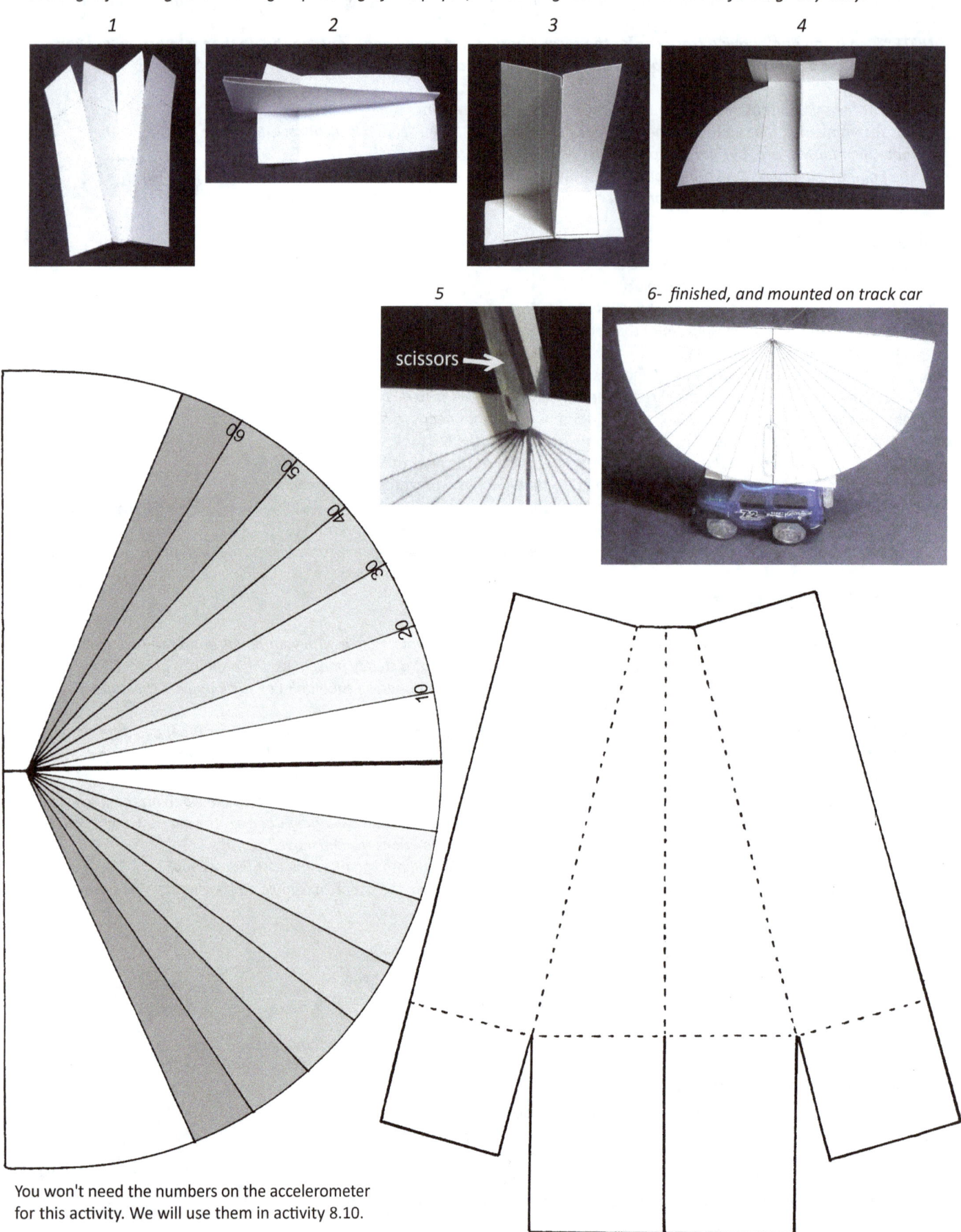

You won't need the numbers on the accelerometer for this activity. We will use them in activity 8.10.

And now back to our discussion about gravity and acceleration...

Gravity causes falling objects to accelerate. Every second that passes, an object adds about 10 meters per second to its speed. However, there is a limit to how fast something can fall if it is inside the earth's atmosphere. This limit is called the ***terminal velocity***. Eventually the force of air resistance becomes so great that the object can't fall any faster. You can feel how strong air resistance is if you put your hand out the window (in a safe manner) when you are in a car that is traveling on a highway. Air isn't nothing—it can apply quite a force!

Skydivers reach terminal velocity as they are falling. Even before they put up their parachute, they can feel the air pressure slowing their fall. When the force of the air pushing them up is equal to the force of gravity pulling them down, they stop accelerating. They are still falling, but their fall is at a steady rate, not an increasing rate. At terminal velocity, they will be falling at about 120 miles per hour (190 km/hr). Then, when their parachute opens, it interacts with a larger volume of air, and their terminal velocity will drop, slowing their fall even more and allowing them to land uninjured.

The shape of an object plays a large role in terminal velocity. The skydiver was wearing his parachute before it opened, so the combined mass of the skydiver and the parachute was the same before and after the parachute opened. The difference was the shape of the parachute; it went from a tight bundle to a large sheet.

ACTIVITY 6.7: Watch a few short videos about terminal velocity (use YouTube playlist)

We see some great examples of air resistance in the natural world. Some plants have seeds that are designed to be able to float for a long time, allowing them to travel quite a distance from the parent plant. In the case of thistle, dandelion, and milkweed seeds, the plant makes a spherical, or almost spherical, ball of incredibly thin filaments. These filaments create a large amount of surface area that can interact with air molecules. Combine this design element with a seed that weighs a tiny fraction of a gram, and you have an object with a terminal velocity that is extremely low. They can stay airborne for many minutes. Maple seeds are comparatively large and heavy, but they use motion to decrease the rate at which they fall, spinning like the blades of a helicopter. Their terminal velocity is much higher than the "puff-ball" seeds, usually reaching the ground in less than a minute.

thistle seeds   *milkweed seeds*   *dandelion seed*   *maple seed*

ACTIVITY 6.8: Engineering challenge: Design for maximum air resistance

In this activity you will be changing the shape of a piece of paper to maximize air resistance. (If it happens to be the right time of year for this type of seed, having some examples on hand to examine would be great.) You could pluck off some of the filaments, or trim them so the ball is smaller, and see how much it affects their terminal velocity.

You will need: *sheets of paper, tissue paper, scissors, tape or glue*

1) You can use the ideas shown below, (perhaps modify them?), or invent your own. Make several mechanisms.

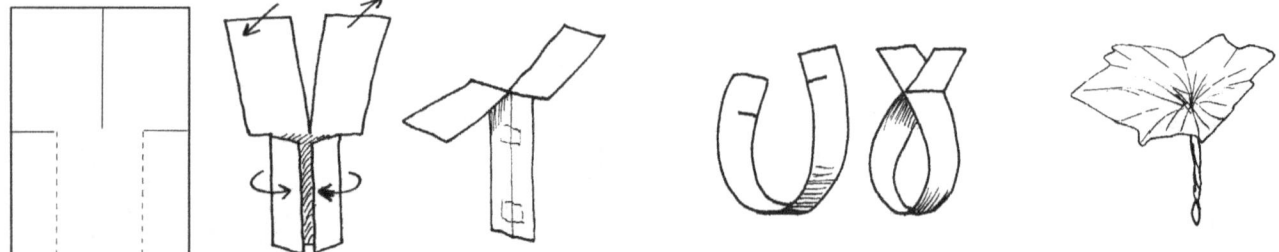

2) Stage a contest between the mechanisms you designed. Drop them from as high as possible, all at the same time. The last one to touch the ground wins!

Isaac Newton knew about Galileo's ramp experiments. By Newton's time, most scientists had accepted the fact that all objects fall at the same rate, and that gravity causes acceleration. Newton's genius was to "connect all the dots," as they say, and realize that the force of gravity not only causes objects to fall to the ground, but also keeps all the planets in their orbits around the sun. In Galileo's day, it was heresy to say that the earth went around the sun (heliocentrism) and not that the sun went around the earth (geocentrism). But by Newton's day, heliocentrism had been generally accepted, so Newton was free to explore ideas about the mechanism that kept the planets in their orbits around the sun.

Let's suppose that the story about Newton watching the falling apple was true, or at least somewhat true. It's possible that he used extrapolation to deduce that gravity was holding the moon in place. Perhaps he imagined the apple tree to be twice as tall as it was. Would the apple still fall the same way? Of course. What if the apple tree were ten times as tall? Would the apple still fall? Yes, of course. What if the apple tree were 100 times as tall?

That would be the height of a small mountain. Newton knew that gravity still worked on mountains. What if the apple tree were a thousand times as tall? There wasn't any reason to think that the apple would behave any differently even at that height. What if the apple tree were so tall... that it reached to the moon? Or better yet, what if the moon were like a giant apple? Could the moon be falling towards earth? Newton realized that the answer was yes, as long as the moon were experiencing another force that was pushing it forward in its orbit. As his colleague Robert Hooke had suggested, curved motions, like the moon going around the earth, might actually be the result of two straight-line forces acting together. It was obvious that the moon was in motion, so perhaps gravity did extend all the way to the moon. But how, exactly, did Newton come up with an equation to describe gravity? (Don't worry if you can't follow all the math listed below. Just appreciate how logical Newton was.)

First, Newton knew the approximate distance between the earth and moon. This had been accomplished by very clever astronomers who measured the size of the earth's shadow on the moon during an eclipse. They concluded that the moon's distance was 60 times the radius of the earth. Additionally, it was common knowledge by this time that the orbital period of the moon (the time it took to go around the earth once) was 27.3 days.

Next, because Newton knew the distance from the earth to the moon, he was able to use the formula C=2πr (circumference of a circle is equal to 2 times "pi" times the radius) to calculate how far the moon traveled in one orbit around the earth. Using a formula that you know, (D=RT), Newton was able to figure out how fast the moon was traveling. The moon speeds along at 1022 meters per second.

The radius of the earth was also known at that time, so Newton had this number in his math toolbox. He used it in a very simple formula, $A_c = v^2/r$ (acceleration equals velocity squared divided by radius). Notice that we need to use the word "velocity" instead of "rate" because we need the letter "r" to represent the word "radius." The velocity is 1022, and the radius of the earth is 384,000,000 meters. Do the math and you get .00272 m/sec^2 (add a minus sign to make it a vector). This is the rate at which the moon falls toward the earth (not -9.8).

That's what Newton was about to find out. When he divided the accepted value for gravity on earth (9.8) by his calculated value for the acceleration the moon was experiencing (.00272) he got approximately 3600. He noticed that 3600= 60 x 60. Now where had he seen the number 60 recently? He remembered that the distance between the earth and the moon is 60 times the radius of the earth. Another inverse square law? Apparently!

Thus, Newton realized that mass and distance both play a role in gravitational force. The moon is much more massive than an apple, but it is also much farther away. He combined his discovery of this new inverse square law with his other ideas about force and motion, and proposed a formula for gravitational force: $F_g = \frac{(m_1)(m_2)}{r^2}$ where "F_g" is the force of gravity, "m_1" is the mass of one object, "m_2" is the mass of the other object, and "r" is the distance between the centers of the objects.

After further thinking and more calculations, Newton decided that he needed to add a very small number to this formula to make it more accurate. Oddly enough, he was not sure exactly what this small number was, so until he figured it out he used the letter "**G**" to represent it. We call it "big G" or "the universal gravitational constant." A constant (as the name implies) is a number that doesn't change. It was not until 100 years later that someone calculated the value of G. More on that in a minute.

The final version of Newton's formula looked like this:

$$F_g = \frac{G(m_1)(m_2)}{r^2}$$

Remember, when two things are right next to each other, like (m₁)(m₂), that means they are multiplied.

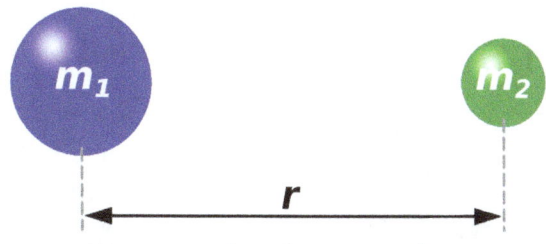

We measure from the centers of mass.

ACTIVITY 6.9: Watch a video about what we just read

Let's take a break from reading and watch a video about what you just read. Our suggestion is "Newtonian Gravity" Crash Course Physics #8. We'll put it on the playlist, but you can also search for it. You will learn some extra tidbits of information about how Newton's Law of Gravitation confirmed an early set of laws about planetary motion formulated by Johannes Kepler, who lived at about the same time as Galileo. (You might want to watch this video more than once, since the narration is so fast.)

Now, what about big G? What is G equal to? Newton knew that this number could be found by doing a very precise and delicate experiment where the force of gravity between two objects could be accurately observed and measured. He understood that *every* object has gravity, not just planets and moons. There is gravity between all the objects in a room, but they don't move closer to each other because the gravitational forces are so small compared to the forces of friction that are keeping the objects in place. But what if you could suspend some heavy balls so that they would be free to move in an almost friction-free environment? Would we observe gravity pulling the balls toward each other?

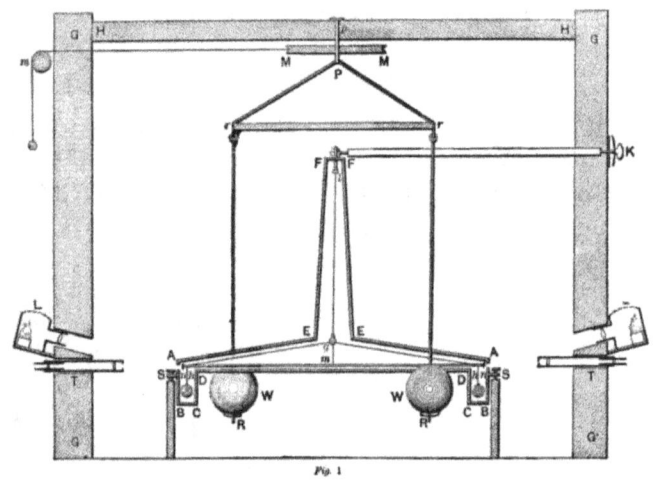

Newton could never figure out a way to do this experiment, but several decades later, a man named John Michell figured out a way to do it. He built a wooden box to house the experiment because he knew that the air around the experiment had to be perfectly still. One slight breeze and the experiment would be ruined. He also had to make sure that it was far enough away from any actions that would shake the ground, such as cutting wood, hammering, plowing, etc. This diagram is a cross section through the box showing the hanging balls inside. There were two "peepholes" in the sides of the box where someone could use a telescope to get an up-close view of the balls and the wires holding them up. Peepholes were necessary because if someone went into the box to take measurements, they would create moving air currents as they entered the box, which would ruin the experiment. So all observation had to be done without going into the box. Sadly, Michell died before he could actually do this experiment. The equipment was passed along to another scientist who turned out to be just the right person to do such a tedious experiment: Henry Cavendish.

Cavendish was a very odd person and was neurotic about details; he could fuss and fuss until something was just right. He had other neurotic traits that made life very difficult for him, (like being unable to talk to women, including his housekeeper), but his obsession with detail made him the perfect person to carry out this experiment. He could devote himself to staring at these balls (through the telescopes in the peepholes) for hours at a time, day after day, week after week, month after month.

Cavendish was not actually trying to calculate big G. In his mind, he was using this experiment to "weigh the world." By using the data he had collected, he could compare the forces of gravity that the balls exerted on each other to the force of gravity exerted by the earth (simply the weight of the balls), and then use the known density of the metal balls to figure out the unknown density of the earth. (If this sounds complicated, don't worry about it. There won't be a quiz.) He then used a little bit of logic and a few calculations to figure out that the earth was (on average) 4.5 times as dense as water. Just multiply this number by the known volume of the earth, and you know the approximate mass of the earth. In later years, other scientists realized that the data that Cavendish had collected could be used to find the value of big G. Ready for it? It is a very, very small number with a lot of units after it.

$$G = \frac{1}{667{,}000{,}000{,}000} \quad \frac{m^3}{kg/sec^2} \qquad \text{meters} \atop \text{kilograms/seconds per second}$$

Scientists like to avoid using fractions when they can, so they have another way to write G:
6.67 x 10⁻¹¹ m³ kg⁻¹ s⁻². Does that look better than a fraction with a huge denominator? The number 11 tells you how many places the decimal point was moved, and the minus sign in front of the 11 tells you that the number belongs in the denominator.

ACTIVITY 6.9: Watch a video about the Cavendish Experiment

Find "Does Gravity Have a Value? / Gravity and Me / Spark" on the playlist, or search for it. The first half of this video does a great job of showing Cavendish's experiment. You can stop the video at this point, or you can watch on to see a demonstration of how gravity affects our bodies on a daily basis. You can do this experiment for yourself if you have a very accurate way to measure your height. The last few minutes of the video are about gravity's affects on our posture and what exercises we can do to counteract it. (There are a few other videos about the Cavendish experiment on the playlist but they are more technical and are not required.)

Newton figured out the effect that both mass and distance have on the strength of gravity. If we know the mass of two objects and we know how far apart they are, we can calculate the force of gravity between them using the Newton's Universal Law of Gravitation: $F_g = G(m_1)(m_2)/r^2$

For small objects, the contribution of big G is very small. This means that we can set this difficult number aside and try some practice problems with just mass and distance as long as we keep these values small. But first, you might find this diagram helpful to see how (gravitational) Force changes when mass and/or distance changes.

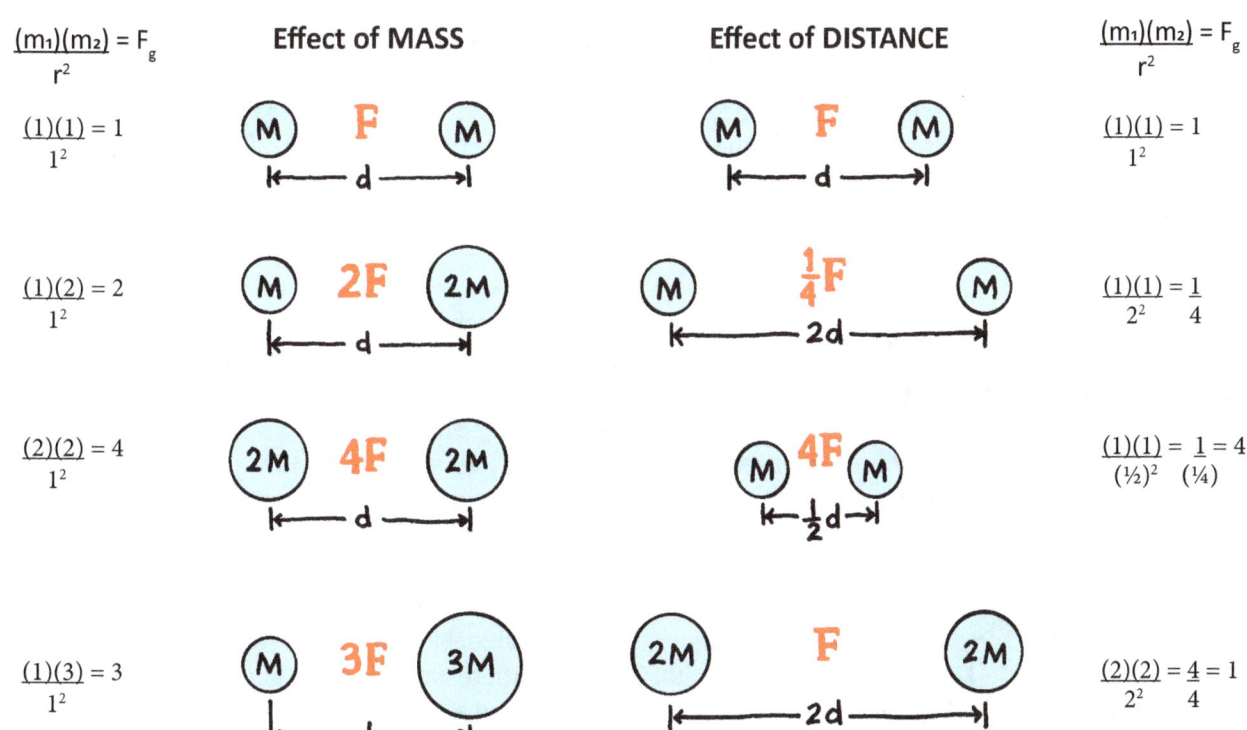

ACTIVITY 6.10: Some thinking questions

We are going to set aside big G for these problems, as we did in the previous diagram. We will focus only on the relationships of mass and distance. Use $F = (m_1)(m_2)/r^2$ By not using G, we won't be able to come up with actual numbers for the strength of gravity, but we will be able to do comparisons. We'll ignore the force of gravity exerted by the earth, pulling the objects down and only consider the gravity between them. (Answers at the bottom of page 92.)

1) Two ping-pong balls are sitting on a table, 1 meter apart. They each have a mass of 5 grams. If one of the balls is replaced by a steel ball (same size) that has a mass of 60 grams, how much will the gravity between them increase?
a) 2F   b) 6F   c) 300F   d) 12F   e) 5F

2) Two people are standing 3 meters apart. They both weigh 70 kilograms. If they walk in opposite directions and increase their distance to 9 meters, what happens to the gravity between them?   a) 1/9 F   b) 1/3 F   c) 3F   d) 6F   e) 9F

3) Two asteroids are floating in space. They start out 100 km apart and then begin drifting away from each other. When they are 200 km apart, what will the force of gravity be, compared to its initial value?   a) F   b) 2F   c) 1/2 F   d) 1/4 F   e) 4 F

4) Those same asteroids continue to drift apart. A year later they are 500 km apart. How much weaker is the gravitational force between them now, compared to the initial value?   a) 1/5   b) 1/10   c) 1/25   d) 1/500   e) 1/50,000

5) A 10 kg bowling ball is sitting on a table and is 10 meters away from a 200 kg steel ball. On another table, two large books are sitting 1 meter apart. One book weighs 5 kg and the other weighs 4 kg. Which pair of objects (the books, or the bowling ball and paperclip) is experiencing a stronger gravitational force (between them), or are they the same?

6) A team of scientists are trying to replicate Henry Cavendish's experiment. First, they try a large ball with a mass of 40 kg paired with a smaller ball of mass 20 kg and place them at a distance of 10 centimeters away from each other. The wires snap when they try to hang the larger ball, so they decide to make the 20 kg ball the larger ball and pair it with a 10 kg ball. At what distance should they place the balls in order to keep F_g the same as it was when using the larger ball?
a) 1 cm   b) 5 cm   c) 10 cm   d) 1/5 cm   e) 25 cm

In this last part of our gravity adventure, we're going to introduce some ideas about gravity that are very challenging to understand. Your brain might feel like it is getting quite a workout. These ideas were a challenge even to physicists when they were first proposed, so don't worry if you have a hard time understanding them. The ideas we are going to discuss are really college-level science, but in today's world of Internet education, many students are at least aware of these ideas even before high school. It just didn't seem right to end our gravity adventure before discussing how most physicists today think about gravity. So buckle your mental seat belts, and here we go....

Newton's ideas went unchallenged for over 200 years. His equations worked quite well for figuring out just about anything in astronomy. There was a tiny problem with explaining the orbit of Mercury (because of how close it is to the sun), but otherwise they worked perfectly. However, no one knew exactly what gravity is. How is it able to reach so far into space? The sun holds on to planets that are hundreds of millions of kilometers away. Newton openly admitted that the did not even have a guess as to what gravity is, and said that this "action at a distance" bothered him. Around the turn of the 20th century, other forces were becoming more understandable. Electromagnetism had been mostly explained by this time, and there were theories about the forces inside atoms. But gravity? No one had a clue.

Now you might think we are going to jump right to Albert Einstein, but we're not. We're going to back up a bit and mention an important proposal made by mathematician Pierre-Simon LaPlace around the year 1800. He proposed that gravity was not a force at all, but rather a giant "field" that was bent and  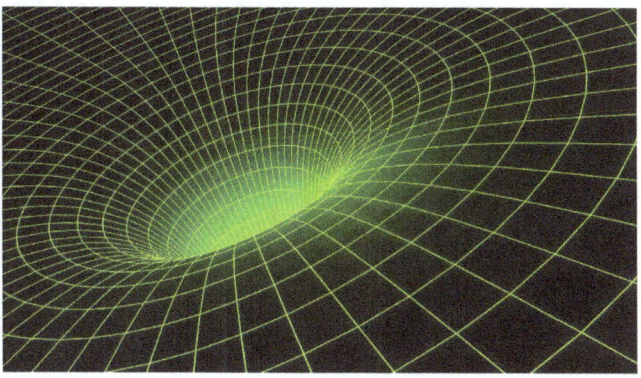 warped by massive objects. Einstein will pick up this idea and run with it, but Einstein wasn't the first person to propose that gravity is not a force. LaPlace suggested that there was some kind of "field" in space that was being bent, or warped, by the sun, a bit like a heavy ball can warp a piece of stretchy fabric. The planets simply followed the curves of this field. LaPlace developed equations that predicted the behavior of the planets according to this theory, and came to essentially the same conclusions as Newton. LaPlace's. However, LaPlace's math wasn't quite as easy to follow as Newton's, so no one saw much benefit to switching over to LaPlace's theory of gravity. Another 100 years would go by until this idea—that gravity was not a force— would again be explored.

The year that Einstein announced his new theory was 1905. This is sometimes called his "miracle year" and is likened to the year that Newton spent at home while his university was closed, when he discovered the nature of light and invented calculus. Einstein wrote four papers in 1905, but one of them ended up being in the top ten list of most famous papers ever published in any area of study. In this famous paper, Einstein proposed his Special Theory of Relativity. He used LaPlace's idea of a gravitational field, but took it a step further and included time into this field, making a fabric of "space-time."

In a nutshell, relativity is based on the idea that each observer in the universe has a unique point of view, and that point of view will determine how an observer interprets both motion and time. For example, an observer on the side of a road sees cars streaming past him. However, the passengers inside the cars see the opposite; they see the landscape moving as they sit still inside their car. Whose viewpoint is "correct"? Einstein said both are correct.

The exact moment when this thought occurred to him (that viewpoint is relative) was when he daydreaming one afternoon, watching a window cleaner standing on a scaffold near the top of a tall building. He imagined what would happen if the man fell off the scaffold. Years later, he called this "the happiest thought of his life." Why? Because it launched him into a thought experiment. He imagined that the man had been holding tools when he fell. Then he extended the imagined fall to a much greater height than a rooftop. What if the man was able to fall for an extended period of time, and during his fall he let go of the hammer he was holding. Einstein could clearly see in his mind's eye that if the man let go of the hammer, it would remain in place, right next to the man's hand, so that he would easily be able to grab it again (because all objects fall at the same rate). He also realized that from the man's point of view, he and his hammer were stationary while the building moved past them.

Einstein realized that if all visual clues were removed (the building, the landscape, the sky), and air resistance were also removed (so that his hair and clothes were not flapping around), the falling man would be in exactly the same state as an astronaut floating weightless in space. Neither of them feels the effects of gravity. Free-fall is essentially the same thing as being in zero gravity. Einstein called this being in an "inertial frame of reference."

But is a falling man really in zero gravity? Can't we time his fall and see gravity accelerating him at 9.8 meters per second every second? Well, we can only time him if we use the building and landscape around him as reference points, and time his motion <u>relative</u> to these. If we can completely isolate the falling man from any surroundings, it's sort of like we've gotten rid of gravity. This is the conclusion that Einstein came to. He proposed that gravity isn't a force, and there isn't any "action at a distance." When you let go of an apple, it immediately goes into an inertial frame of reference, free from gravity. It is the ground that rushes up to meet the apple, not the apple that rushes down to the ground. As soon as the apple hits the ground, its inertial frame of reference is gone and it is subject to Newton's second law, which we will cover in the next adventure.

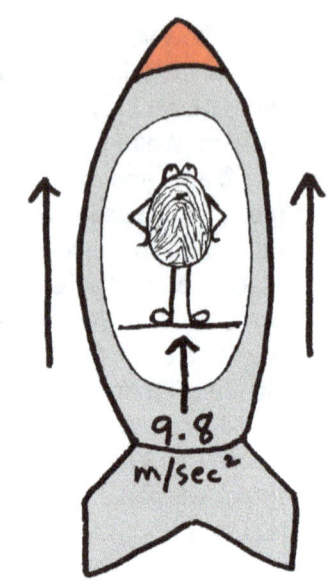

Einstein also realized that if an astronaut were inside a rocket that suddenly began to accelerate at 9.8 m/sec per second, he would feel the floor pushing up on his feet at exactly the same strength as earth's gravity. The astronaut would feel as if he were standing on a floor on earth. To him, there would be no difference between the rocket's acceleration and the feeling of earth's gravity.

Einstein's explanation of gravity hinged on his theory about "space-time." He proposed that empty space isn't nothing. Einstein said that space itself is made of some kind of framework, and time is part of this framework. (The theories about this framework are extremely complicated and far beyond the scope of this book.) Therefore, mass curves this framework just by its mere presence. Very massive objects cause a great deal of curvature which is why the sun has "more gravity" than the moon. Once space-time is curved, objects simply travel along these curves. "Mass tells space-time how to curve, and space-time tells matter how to move." (quote by John Wheeler)

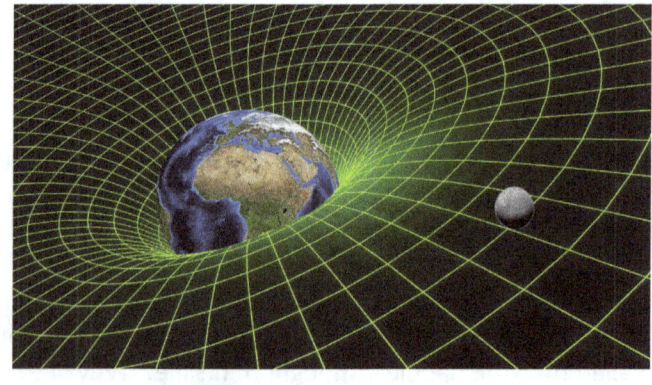

Einstein's equations seem to indicate that time is also affected by gravity, so clocks would run more slowly at sea level than on mountain tops. Motion also affects time, with time slowing down the faster you go. This notion produced the famous "twins paradox" thought experiment where one twin gets on a rocket and travels to a distant star and back, while the other twin stays on earth. Each twin would feel time passing as it normally does, but the twin who traveled to the distant star would appear much younger than the twin who stayed home because time would have slowed down due to the fact that he was traveling so fast. This experiment can never actually be done, of course, so we'll never be able prove this is true.

If your brain is reeling and you are saying, "This can't be right," don't worry, Einstein's ideas came as a shock to most physicists of his time. Experiments had to be done to check some of the predictions made by the theory of relativity. The first major victory came in 1919 during a solar eclipse. Einstein predicted that the sun's gravity would bend the light coming from stars that were behind the sun, making them appear to be slightly shifted from where astronomers knew they should be. As the eclipse progressed, the stars did, indeed, appear exactly where Einstein said they would. Another confirmation was that the theory correctly predicted the orbit of Mercury.

Also in that miraculous year of 1905, Einstein penned the equation he is so famous for, $E=mc^2$. The "E" stands for "energy," "m" is "mass," and "c" is the speed of light. The equation says that mass can turn into energy and energy can turn into mass. The speed of light is about 300,000 kilometers per second. When we square it, we get a very large number, about 90,000,000,000. So when you multiply a mass by this number you get a lot of energy! Famously, this equation was used to predict how much energy would be released by an atomic bomb. Einstein was very upset that his theories were being used to build bombs.

It's okay, no need to panic. A science historian did some research and found out that Einstein wasn't the first person to come up with $E=mc^2$. This equation appeared in a science magazine called <u>Atti</u> in 1903, in an article by a physicist named Olinto de Pretto. Einstein was probably aware of this article. Or, perhaps it was a coincidence that two people came up with the same idea at about the same time. Either way, there is historical evidence that Einstein was not the only person to come up with this equation, so we haven't altered history. Or at least not too much...

ACTIVITY 6.11: Watch a helpful video that reviews what we just learned

Watch "Why Gravity is NOT a Force," by Veritasium (on the playlist, or search YouTube for it).
While you are at the playlist, try this video, too: "General Relativity Explained simply & visually" by Arvin Ash

---

Although the relativity "model" of gravity is the one most generally accepted today, it isn't the only one. In science, a "model" is a way of thinking about something. Models always include some assumptions (starting points that are seldom debated). In the relativity model of gravity, it is assumed that time can, indeed, run faster or slower, or even come to a complete stop if you travel as fast as light. This is believed to be true because the mathematical equations turn out a certain way. Has anyone been able to test this assumption by traveling close to the speed of light? No. The best we can do is perform simple experiments with tiny particles that can be accelerated to very high velocities for very short periods of time, only fractions of a second. We can't actually do the twin paradox experiment. Another assumption that Einstein had to make to get his math to work was that the speed of light is constant and has never changed. However, this assumption is being challenged by a number of physicists today. There is some evidence that the speed of light might have been greater in the past. If the speed of light is not constant, then relativity must be re-examined. Also, Einstein assumed that point of view must be relative because there is no absolute "reference frame" for everyone. In other words, there is no ultimate ruler against which to measure every movement in the universe. However, in the 1960s, the Cosmic Microwave Background radiation was discovered. This is a field of very long radio waves that exist everywhere in the universe. Scientists can use this as the ultimate ruler—everything moves relative to this field of radiation. Einstein did not know about this field of radiation when he made his relativity assumptions.

The biggest problem with relativity is that it doesn't fit with another branch of science: quantum mechanics. The field of quantum mechanics is very well established and has much experimental evidence to back it up. Quantum mechanics deals with light energy, and with atoms and sub-atomic (elementary) particles like quarks. It has theories about how and why atoms hold together. Oddly enough, it was Einstein who made some of the first quantum physics discoveries. Going into detail about why relativity and quantum physics don't fit together is far outside the scope of this chapter. (If you want to know details, try the Arvin Ash YouTube video on this topic.) The fact that relativity and quantum mechanics don't fit together is seen as the biggest problem in physics today.

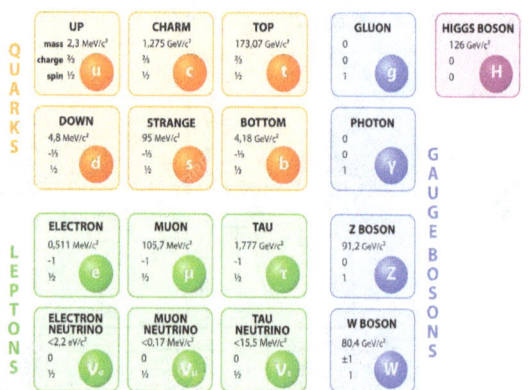

A new branch of physics call Stochastic Electrodynamics (SED), takes another approach to gravity. SED physicists say that time and space are solid and can't be stretched and bent the way Einstein imagines. They don't believe that gravity affects time. They believe that clocks (especially atomic clocks, like the cesium clock shown here on the right) can be affected by gravity, but not time itself. Since atomic clocks are necessary to do the experiments that supposedly "prove" relativity, how can we be sure that time is being affected, and not the clocks? SED physicists have SED explanations for all the "proofs" of relativity, including Mercury's orbit and the bending of light around the sun. And... they have a theory of gravity that fits with quantum mechanics. Gravity, they say, is the result of a very complex interaction between the electrical charges of the elementary particles and the electrical charges of the "virtual particles" that create the "fabric" of space (not space-time). If SED physicists are correct, this will provide one over-arching explanation for all physical phenomena, including gravity.

This atomic clock keeps time using atoms of the element cesium.

How can relativity be in question if its math works so well to explain gravitational observations? However, the same could be said about the Newtonian model. If Newton was so wrong about gravity, thinking it to be a force when it really wasn't, then why does his universal law of gravitation work so well? It seems that the mathematics of a scientific model can work well regardless of whether its assumptions can be proved to be valid.

But now we've really overshot the scope of this book! In our next adventure, we'll come back to cold, hard reality and talk about falling, crashing, and smashing.

ACTIVITY 6.12: Review activity: Online simulation of Universal Law of Gravitation

Here is a link to an online simulation where you can adjust the masses of two objects and the distance between them and see how this affects the force of gravity between them.

ACTIVITY 6.13: Review questions
(Answers are in the teacher's guide section.)

1) Galileo used a _____ _____ to time the rolling balls in his famous ramp experiment.
2) The speed of the balls as they rolled down Galileo's ramp was: a) linear   b) exponential   c) even
3) What did Galileo call the pattern he discovered? The Rule of _____ _____
4) TRUE or FALSE? Acceleration is rate at which velocity is changing.
5) Which has a direction, speed or velocity? _____
6) Which planet has the highest acceleration due to gravity? _____
7) The acceleration due to gravity on earth is: a) 9.8 m/sec^2 b) -9.8 m/sec^2
8) TRUE or FALSE? The acceleration due to gravity on earth is exactly the same everywhere on the planet.
9) TRUE or FALSE? Accelerometers show you how fast something is going.
10) TRUE or FALSE? When skydivers reach terminal velocity, they stop accelerating.
11) TRUE or FALSE? The shape of an object plays a large role in terminal velocity.
12) Which one of these did Newton NOT have to know, in order to create the Universal Law of Gravitation?
   a) radius of the earth   b) time of moon's orbit around earth   c) formula for circumference of a circle
   d) time of earth's rotation on its axis   e) distance between the earth and the sun
13) The distance between the earth and the moon is _____ times greater than the earth's radius.
   a) 3600   b) 600   c) 60
14) What was Cavendish trying to find out when he did his famous experiment in the shed?
   a) the value of big G   b) the force of gravity on earth
   c) the density of the earth   d) the weight of the earth
15) TRUE or FALSE? Scientists definitely know what gravity is.
16) Who was the first person to propose that gravity might actually be a giant "field"? _____
17) Who was the first person to officially propose that gravity is not a force? _____
18) The Theory of Relativity says that gravity is a result of:
   a) the slowing down of time   b) the curvature of space-time   c) the curvature of space
19) TRUE or FALSE? All scientists believe that gravity affects time.
20) TRUE or FALSE Relativity helps to explain all branches of science.

ACTIVITY 6.14: Put these historical figures in chronological order

This list is in alphabetical order. Can you put them in chronological order? If you need to check the Internet to get some birth dates, that's okay. (Answers at the bottom of page 98.)

Archimedes, Aristotle, Cavendish, Da Vinci, Einstein, Euclid, Galileo, LaPlace, Michell, Newton

1) _____   6) _____
2) _____   7) _____
3) _____   8) _____
4) _____   9) _____
5) _____   10) _____

ACTIVITY 6.15: Math review   (Answers are in the teacher's guide section.)

Find the mechanical advantage of these levers.

1) _____

9 m | 54 m

2) _____

8 cm | 32 cm

3) What is the length of the long arm? _____

12 g ... 3 g
5 m | ?

4) What is the mass of each blue ball? _____

? ... each red ball is 21 g
3 m | 8 m

5) If the handle of a screwdriver is 24 millimeters in diameter, and the shaft is 6 millimeters in diameter, how much mechanical advantage will the screwdriver give you? _____

6) In a set of gears, the follower has 28 teeth and the driver has 7. What is the mechanical advantage? _____
Does this give you a speed advantage or a force advantage? _____

7) Galileo started with a pendulum whose period was 60 cycles per minute. He then made the pendulum four times as long. What was the period of the new longer length? _____
When he made it 9 times longer, what was the period of the pendulum? _____

8) Convert 1200 grams to kilograms: _____
10) Convert .15 kg to grams: _____

9) Convert 60 cm to meters: _____
11) Convert 3 meters to kilometers: _____

12) A sandpaper-covered wood block began sliding down a ramp when the vertical height reached 28 cm. The horizontal length of the board at that point was 70 cm. What is the coefficient of friction? $\mu=$ _____

13) Two heavy balls are sitting on the floor 2 meters apart. One of them is a bowling ball with a mass of 8 kg. The other is a steel ball with a mass of 22 kg. Find the amount of gravity between them using Newton's Universal Law of Gravitation, but keep the letter G the way it is, so your answer will be something times G. _____

14) Two asteroids are drifting in space. They start out 50 km apart and then begin moving in opposite directions. When they are 150 km apart, what will the strength of gravity be compared to its original value? _____

15) The acceleration due to gravity on a exo-planet is three times what we have here on earth. What is the acceleration due to gravity on this planet? _____

16) If a rock falls off a cliff, how far does it go in one second? _____
How far does it travel in the next second? _____   In the third second? _____

BONUS POINT:
Draw a free body diagram of a box
that is being dragged across the floor:

Adventure 7: Calculating and colliding with Newton
(Newton's 2nd and 3rd Laws)

You guessed it! There is more to force than just mass. A cannonball has a lot more mass than a marble, but if you got hit with a cannonball that was only going at a speed of 1 km/hour (slower than the speed of walking), you would easily be able to catch it and not get hurt. On the other hand, if a marble came at you as fast as a bullet, it could cause some serious injury to your body.

Newton also knew this to be true. Since childhood, he had watched things fall and knew that an apple dropped from knee-height would hardly be bruised, but an apple dropped from the top of a roof would smash and splatter as it hit the ground. He knew that the longer something fell, the more acceleration it would experience. Every second an object falls, it picks up additional 9.8 meters per second. He figured out that the relationship between force, mass and acceleration is actually a fairly simple one. He addressed the relationship between force and mass in his famous Second Law. Unfortunately, the way he wrote this second law is a bit harder to understand than his first and third laws. Here is how he stated his **Second Law**:

LEX II: *The alteration of motion is ever proportional to the motive force impressed; and is made in the direction of the right line in which that force is impressed.*

He went on to say, "If any force generates a motion, a double force will generate double the motion, a triple force, triple the motion, whether that force be impressed altogether and at once, or gradually and successively." This statement, along with other things Newton said, has led scientists to a mathematical way to express this second law. There is general agreement that the best way to summarize **Newton's Second Law** is like this:

$$\text{FORCE} = (\text{Mass})(\text{Acceleration})$$
$$F=ma$$

Remember, when two letters or numbers are next to each other (with or without parentheses) this means they are being multiplied.

Force equals mass times acceleration. This is how you see it stated in many textbooks. This accomplishes two important purposes. First, it makes the law much easier to remember, and second, it gives it an easy and obvious application. You can use this simple formula to calculate force.

But first, let's review how we measure force. The unit of force is the **newton, N**. One newton is the force it takes to give 1 kilogram of mass an acceleration of 1 meter per second per second. A pineapple has about 1 kg of mass. The force it takes to move a pineapple (across a frictionless surface) the distance of one meter in one second is 1 newton of force. If that 1 N of force kept going after the first second, the pineapple would move 2 meters in the following second and 3 meters in the third second.

Gravity gives an acceleration of 9.8 m/sec², so 1 N (only 1 m/sec²) is not much force compared to gravity. Even if we just hold the pineapple and don't drop it, the pineapple is still experiencing the acceleration due to gravity. That's one of the trickiest parts of understanding gravity. We still call it "acceleration due to gravity" even if the object is resting on a surface, not falling. The formula F=ma tells us that we can multiply the mass of an object times the acceleration due to gravity and get the downward force of the object in newtons. For the pineapple, 1kg x 9.8m/sec²= 9.8 newtons. When you hold a pineapple, you are feeling a force of 9.8 newtons pushing down on your hands.

Our formula, F=ma, tells us that Force is proportional to mass. If we keep the acceleration at 1 m/sec², this will give us some very easy math because if a=1, then we have just F=m. (Multiplying a number by 1 doesn't change the number, so "m" times 1 is just "m." This means we can think of "a" as not being there.) "F=m" tells us that we need 2 newtons of force to accelerate a 2 kg weight, 3 newtons of force to accelerate a 3 kg weight, and so on. Any increase in mass will be directly proportional to the amount of force needed.

Another thing we can notice about F=ma is that there are many ways to make the same amount of force. If we set the units aside and just look at numbers, and we let force equal 24 N, we can see that as mass increases, acceleration will decrease, and vice versa. F=24   24=(6)(4)   24=(4)(6)   24=(8)(3)   24=(3)(8)   24=(12)(2)   24=(2)(12)
If mass is large, acceleration will be small. If acceleration is large, mass will be small.

F=ma works in zero gravity (free fall), too. Scientists on the International Space Station set up a way to demonstrate this. They used a bungee cord strung across the cabin to make sort of a "sling shot" that could provide force. They kept the force the same by always pulling back the cord the same amount. Then they launched objects of different masses and watched the difference in acceleration. In fact, why don't you watch this right now?

ACTIVITY 7.1:   Watch the ISS demonstration

Use the Youtube playlist, or search for "STEMonstrations: Newton's 2nd Law of Motion"

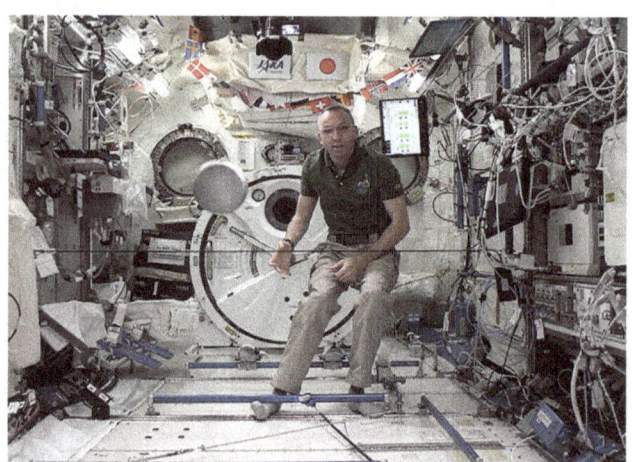

ACTIVITY 7.2:   Do your own demonstration of F=ma

You will need: *a corner cut from a cereal box, a meter stick, a few books, a sheet of cardstock, tape, scissors, colored pencils, at least three balls of different masses (examples: small and large glass marbles, steel marble/bearing, ping pong ball, lightweight wooden ball), and three copies of the 1-cm graph paper pattern page in the teacher's section for this chapter. NOTE: If you constructed a 2-meter ramp for experiment 6.3, you can use half of this ramp instead of making another one.*

How to set up:
1) Turn a meter stick into a ramp. Cut thin strips of card stock paper and tape them along the sides to make "railings" so that the balls won't jump off the track. Make marks every 25 cm along the track, so you have 25, 50, 75 (and the end is 100).
2) Cut a corner from a cardboard box to make the shape shown here on the right. The corner needs to be large enough to fit your largest ball/marble.
3) Make three copies of the graph paper in the teacher's section for chapter 7. Cut and tape them end to end to make one long piece of graph paper. Label every fifth block (5, 10, 15, 20, etc.) so it will be easy to measure how far the cardboard corner has been pushed along the graph paper.
4) Set the meter stick ramp up on a book, or two books, and do some experimenting with your selection of balls. Choose a ramp height that will give at least a little bit of displacement even for the lightest ball rolling from the 25 cm mark (traveling down just the last 1/4 of the ramp), but won't give too much displacement for your heaviest ball going all the way down the ramp. Ideally, all of your data will stay on the graph paper.
Here is a photo showing how your set-up should look. The tip of the cardboard corner is lined up with edge of the graph paper. Tape your graph paper to the floor if it moves around.

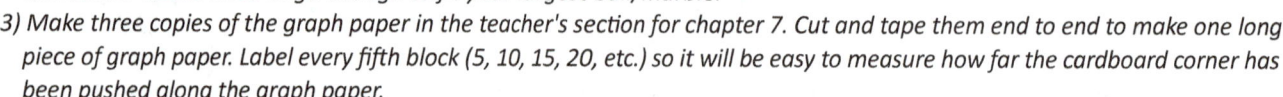

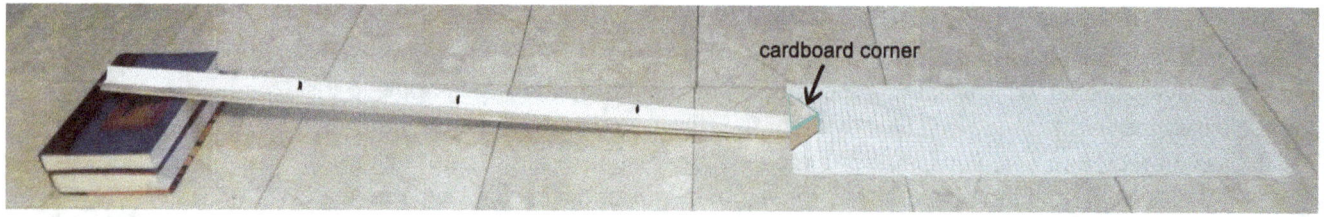

Answers to 6.14: 1) Euclid 2) Archimedes 3) Aristotle 4) Da Vinci 5) Galileo 6) Newton 7) Mitchell 8) Cavendish 9) Laplace 10) Einstein

How to run the experiments:

In every case, the final result will be a measure of Force. Trying to measure force in newtons will be too difficult in this experiment, so we we're going to use our own custom-designed measurement. We're going to determine force by letting the ball go into the cardboard corner and give it a push. Then we'll measure how many centimeters the cardboard corner moves from its starting point. So "Force" will be determined by the number of centimeters the corner is displaced.

1) Assign colors to your chosen balls. There are blocks to record four balls. If you only have three, just don't use the last block. Color in the first block and write which ball it is after the equal sign. Then do the same for the other balls.
2) Put your first ball at the top of the ramp (at 100 cm) and let it go. Then check to see how far the cardboard corner was displaced. Record the number of centimeters using the correct color for that ball.
8) Replace the corner so it is again flush with the starting point. Now place the ball at the 75 cm mark and let it go. Record the result on the graph and return the corner to staring point.
9) Do this for 50 cm and 25 cm. Record results.
10) Now switch to another ball. Place it on the 100 cm mark and let it go. Record the result on the graph using the correct color.
11) Run this ball at 75 cm, 50 cm, and 25 cm. Record results on graph.
12) Do your last ball(s) using their colors and record results at each height.
13) Make "best fit" curves for each color, not simply connecting dots, but making a curve that indicates the trend of the data.

Analyzing your data

1) Do your "best fit curves" look more like lines or more like curves? We don't have very many data points, so it could be either. What did you expect your best fit curves to look like? Did you expect them to look similar to the curves we've seen in previous graphs? This would be a good guess, since the rolling balls are experiencing the acceleration due to gravity, and it's gravity that gives us those exponential curves.
2) Did your results support the idea that both mass and acceleration affect force? Did the force go up when mass went up? Did force go up when acceleration went up?
3) Can you find a point where two different combinations of mass and acceleration produced the same force?
To find these combinations, choose a force from the lower axis (centimeters traveled) and then trace directly upwards until you run into a line. Look to the left and see how many centimeters down the ramp that ball would have gone. Then continue straight up until you meet another line and look to the left to read centimeters on track. Then you can say something like, "The glass marble rolling 80 centimeters produces the same force as the steel marble rolling 25 centimeters." Your numbers will be different than these, but you get the idea. Two different combinations can produce the same force.

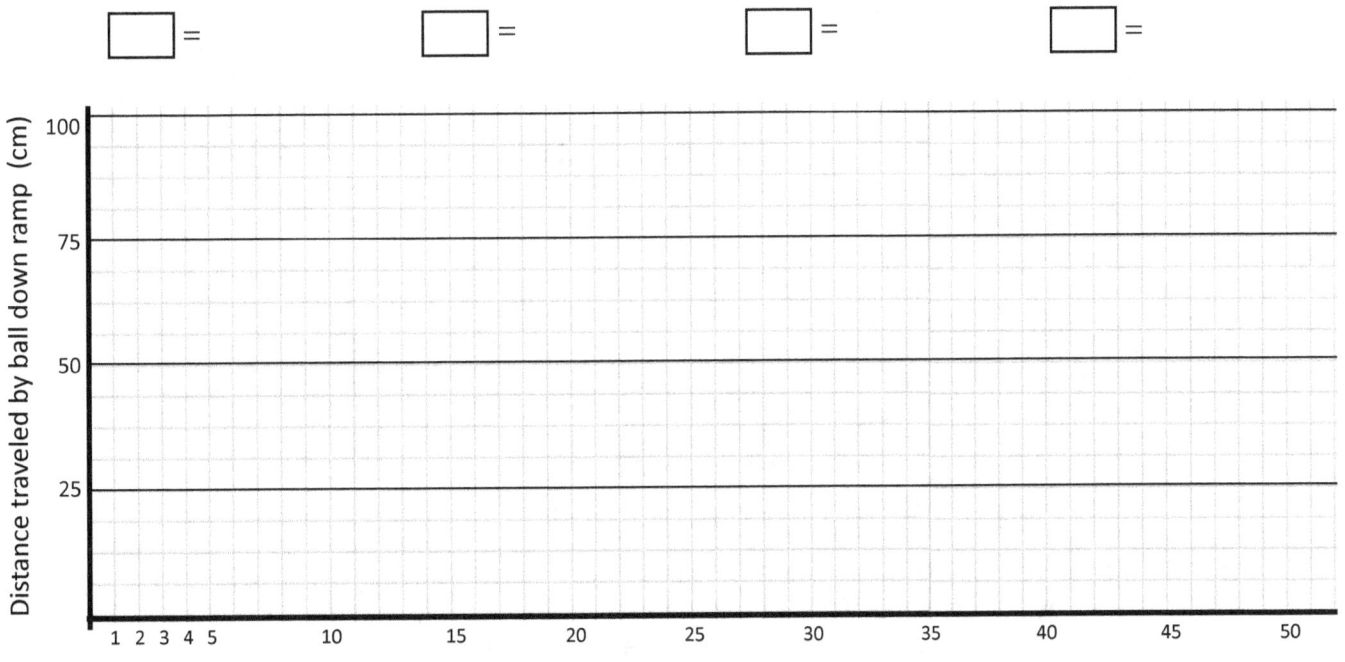

Questions for Activity 7.2: *(discussion and suggested answers in teacher's section)*
1) Did the more massive balls produce more force?
2) Did acceleration produce more force?
3) Which had more effect on force—the height of the ramp or the position of the ball on the ramp?
4) Were the shape s of the lines/curves similar for both graphs? Why this result?
5) Did your experiments confirm Newton's assertion that both mass and acceleration affect force?
6) Were there any instances where a small amount of acceleration and a large mass gave about the same force as a large amount of acceleration and a small mass?

We can use the formula F=ma to solve problems that ask you to find any of the variables: force, mass or acceleration. If you know two of them, you can find the third.

For example, let's consider a mass of 5 kg and an acceleration of 10 m/sec². How much force will be needed to accelerate a 5 kg watermelon at the rate of 10 m/sec²?

F= (5 kg)(10 m/sec²)= 50 newtons.

Using newtons as our measurement of force lets us avoid writing out "kg(m/sec²)" because newtons are, by definition, in kg(m/sec²).

5 kilograms is about 10 pounds, which is pretty big for a watermelon, but some do grow to this size. You may have noticed that we used 10 m/sec², which is approximately the same as the acceleration due to gravity. If someone on the space station launched a 5 kg watermelon with an acceleration of 10 m/sec², it would hit the wall with about the same force that it would if it had been dropped on earth.

We can also use the formula to find either mass or acceleration. For example, if we know the force being applied is 100 newtons, and we know the acceleration is 5 m/sec², what is the mass of the object? We set it up like this: 100=(m)(5). What is "m"? We know that 20 times 5 is 100, so "m" must be 20 kg.

Or, if we know the force and the mass, we can find acceleration. If the force is 24 newtons, and the mass of the object is 8 kg, how much acceleration will this amount of force give to this object? 24=(8)(a). We know that 8 times 3 is 24, so the acceleration must be 3 m/sec².

Another use for F=ma is to calculate weight using newtons instead of kilograms. You'll remember that your spring scale was labeled in both grams and newtons. Calculating your weight in newtons is very easy if you use F=ma. You can measure your mass in kilograms (an inertial balance would be ideal, but a regular scale can be used), and you know that the acceleration due to gravity is 9.8 m/sec². Just multiply your mass by 9.8 and your answer will be in newtons. If someone has a mass of 100 kilograms, their weight would be 980 newtons.

How much is 1 N? A candy bar with a mass of 102 g (the candy shown here has "102 grams" marked on the wrapper) has a weight of almost exactly 1 N (.102 x 9.8).

Yep, that's what we're saying. Ideally, we'd all measure our *mass* with inertial balances, but this is not practical, so we use scales that make use of gravity. But because there isn't really any difference between mass and weight while you are on planet Earth, we can fudge a little bit and mark scales with kilograms or pounds. But to be technically accurate, our bathroom scales should be marked in newtons.

ACTIVITY 7.3: A few simple math problems with F=ma

1) The worldwide average weight of a human is 62 kg. How many newtons does the average human weigh? _____

2) How much force will be needed to accelerate a 20 kg cannonball at 30 m/sec²? _____

3) If someone's weight is 490 N, what is their mass in kilograms? _____

4) A trebuchet launches a large 15 kg pumpkin with a force of 135 newtons. What is the acceleration? _____

5) A lab mouse tips the scale at 8 grams. What is its weight in newtons? _____ (Remember to convert g to kg.)

6) How much force is needed to accelerate a 1,700 kg car at 3 m/sec²? _____

7) A very bad dog jumped up on the kitchen table. The dog is a large breed—his mass is 60 kg. How much force is the table experiencing because of the dog? _____

Even with simple problems like these, you always have to read carefully and pay attention to the units being used. We did not throw anything tricky at you (only converting grams to kilograms) but a teacher might give you a problem in miles per hour, and you will have to convert it to meters per second. Or if a weight is given in pounds or grams, you will have to convert to kilograms. **The answer MUST be in kilograms, meters, and seconds** (which are the units involved in newtons).

Another piece of trickery that teachers might throw at you is making you calculate acceleration instead of giving it to you. For example, they might tell you that the box started out at rest, but after 4 seconds it is now moving at the speed of 2 meters per second. We don't know the velocity at the 1-second mark, or the 2-second mark, or the 3-second mark—only the 4-second mark. So we just divide the final velocity, 2 m/sec, by the number of seconds it took to reach that velocity, 4 seconds. 2/4 = .5 So the acceleration was an average of .5 meters per second for each of those 4 seconds. We don't need to know all the details about what happened at each second, just those final numbers. The mathematical way to write this is: a= v/t $$\text{ACCELERATION} = \frac{\text{Velocity}}{\text{Time}}$$ Acceleration equals velocity divided by time

Here are some more problems that are based on F=ma, but you might have to do some unit conversions or figure out acceleration. Always convert to kg, m, and seconds.

ACTIVITY 7.4: More math problems with F=ma   1 pound = 16 ounces = .45 kg   1 mile= 1.6 km

1) On the ISS, astronauts rigged up a sling shot that gives exactly 1 N of force every time. If they put a 1-pound object into the sling shot, what acceleration can they expect to see? _____

2) A nerf gun can shoot darts with a force of 2 N. If a dart weighs 3 ounces, how much acceleration will the dart have as it leaves the gun? _____

3) An airplane started out at rest on a runway, and at the end of 10 seconds it was going 200 km per hour. If the mass of the airplane is 250,000 kg, how many newtons of force are the engines producing? _____   (a=v/t)

4) An 8-ounce ice hockey puck slides across a friction-free ice surface. A 2.7 N force is being continuously applied to the puck. What will the acceleration of the puck be? _____

5) In the frictionless environment of outer space, a space station launches a small probe with a force of 10 N. After 50 seconds, the probe's speed is measured at .25 km per second. What is the mass of the probe? _____ (Use a=v/t to find a. Don't forget to convert km to m!)

6) An astronaut in the ISS pushes a 200 gram wrench toward the wall. It hits the wall in exactly 2 seconds and the wall is 10 meters away from the astronaut. The wrench hits the wall with about how many newtons of force? _____

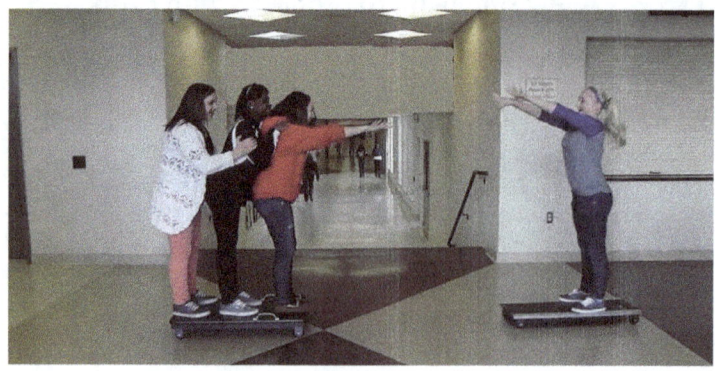

ACTIVITY 7.5: F=ma with skateboards   (This activity also gives you a preview of Newton's third law.)

You will need: 2 skateboards (or 2 of anything with wheels that you can stand or sit on), a few friends, heavy backpack
NOTE: If you can't do this experiment yourself, you can watch it as a video lab on the playlist.

Another way to do this lab is to use two identical small object with wheels, and attach identical strong magnets to each, with repelling sides of the magnets facing each other. Bring the magnets very close, then suddenly let go and allow the repelling force to drive the objects away from each other. Then add mass to one of the object but not to the other, and try it again.

In this experiment, force is assumed to be constant. Force will be applied by a physical push.

1) Start with one person on each skateboard. Position the boards close together with the riders facing each other.
   Ideally, the "riders" should have about the same mass. (You also might want to mark the starting point with tape on the floor.)
2) Riders should hold up both hands, and touch the other persons' palms. Make sure your arms are in a V-shape, elbows down.
   You want to be able to extend your arms and give a push.
3) Riders will both push against each other at the same time.
4) if the riders have the same mass, and they both push equally hard, the skateboards should drift apart the same amount.
   (The distance traveled will give you an approximation of the acceleration each board experiences. If you marked the starting point on the floor, you can actually measure the distance traveled by each board.)
5) Now add one rider to one of the skateboards. You'll have two riders on one board and one rider on the other. Bring the boards together and have the first riders repeat what they did in step 3. Give the same push against each other.
6) The result should be that the board with two riders will travel about half the distance as the board with one rider.
7) If you can fit three riders on one board, try it with three. If not, try giving a rider a heavy backpack. Repeat step 3.
   With the added mass, the board should travel an even shorter distance.

After a lot of observation and thinking, Newton came up with his third law of motion. Here is the English translation of what he wrote in Latin in his *Principia* book:

LEX III: To every action, there is always opposed an equal reaction; *or the mutual actions of two bodies upon each other are always equal, and directed to contrary parts. Whatever draws or presses another is as much drawn or pressed by that other. If you press a stone with your finger, the finger is also pressed by the stone. If a horse draws a stone tied to a rope, the horse will be equally drawn back towards the stone.*

You've probably heard this paraphrase of the third law: **"For every action, there is an equal and opposite reaction."** It is fairly easy to see this law at work when you watch a rocket launch, or when you watch a balloon flying about the room as the air escapes; but the amazing thing is that Newton came up with this idea without watching rockets or balloons. Perhaps he watched someone paddling a boat and observed the paddle going backwards while the boat moved forwards? Or maybe he watched a cannon being fired and saw the cannon jerk backwards as the ball shot out of the barrel? We don't know what got him started thinking about forces, but the more he thought about it, the more he realized that the principles behind forces work everywhere all the time, even if you can see anything moving. The final result was his assertion that as he pressed on a stone, the stone was pressing back. As his feet pressed down on the floor, the floor was also pressing back on his feet. Stones and floors don't move, but according to Newton, they are able to press back. This must be true, said Newton, because basic principles work the same way everywhere, all the time. They aren't true just some of the time, or in particular situations—they are always true all the time. Thus, the basic principles of force must be operating even when we can't actually see them working.

Perhaps Newton observed a cannon jerking backwards as the ball shot out of the barrel.

We can elaborate on this third law and identify four parts:
> **1) Forces always come in pairs.**
> **2) These paired forces are always equal in magnitude (strength).**
> **3) The direction of these paired forces is always exactly opposite.**
> **4) The paired forces always occur in two separate bodies (they are not internal forces inside a body).**

The first two ideas are sometimes hard to see. Think of someone hitting a baseball. As the bat smacks the ball and sends it flying, it seems that the bat is providing all the force. The ball goes flying, apparently the receiver of all that force. Where is the opposing force? However, the batter has a different perspective. The batter can feel the force exerted by the ball. His hands and arms feel the "thud" as the bat and ball collide. Rarely, a bat will even crack in half as it collides with the ball, showing the extreme force experienced by the bat. So the ball really does exert an equal and opposite force. But the ball still goes flying off. Obviously, the bat transfered quite a bit of energy to the ball. Where is the opposing force? Newton's third law is about pairs of objects, not single objects. The flying ball is a single object. While it is in the air, we can't apply the third law to it. However, the ball will eventually come into contact with either the fence at the back of the outfield, or the ground, and then the third law can be applied. The fence or the ground will provide a reaction force.

Another strange aspect of Newton's third law concerns gravity. We assume that massive objects have "more gravity" because of their mass. We know that we weigh more on earth than we do on the moon because of earth's greater acceleration due to gravity. However, Newton's third law tells us that forces are always equal and opposite, so if we allow that gravity is a force (setting aside relativity for the moment), this means that the earth and the moon must be pulling on each other with equal and opposite force. This idea doesn't sit well with our intuition. Surely, the earth must be pulling harder

than the moon does because it is more massive. However, we learned previously how to calculate the gravitational force between two objects. We use the formula $F = G(m_1)(m_2)/r^2$. This will give us the gravitational force acting between the objects. They are both experiencing this force regardless of their size. (The formula takes into account the sizes of the objects.) The confusion comes because we don't consider inertia. Though the force might be the same, the effects can be quite different due to the inertia of the objects. The moon has much less inertia than the earth does because it has less mass. The same force that causes the moon to move around in its orbit is also felt by the earth, but since the earth has much more inertia, it doesn't move very much. However, it does move a little. The common point around which the moon and earth orbit, known as the **barycenter**, is somewhere in the mantle of the earth. So, the moon doesn't orbit the earth—they orbit each other.

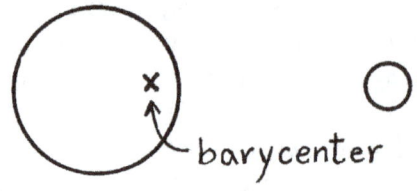

We saw this diagram in the first adventure.

And speaking of misconceptions...`

The ground has nothing to do with the opposite reaction that is propelling the rocket upward. If a rocket fires its engines in outer space, the fuel still goes one way and the rocket still goes the other way. The action/reaction pair isn't the rocket and the ground; it's the rocket and the fuel molecules. However, this erroneous statement about the rocket and the ground has actually appeared in teaching materials for kids.

Another common misconception is that the reaction occurs in response to the action. That is, that the reaction occurs a split second later. However, the action and reaction occur at exactly the same time. Likely, the confusion occurs because of the words "action" and "reaction." Perhaps we should call these forces an "interaction" pair.

Even more common than these misconceptions is a belief about something called the **normal force**. We mentioned the normal force in our adventure with friction (page 75). In physics, the word "normal" means "perpendicular." A table leg is normal to the floor (assuming it is not broken). The wooden peg in this photo is normal to the wall. If something is normal to a surface, that means that all the angles between it and the surface are 90 degrees (square corners).

In the context of the third law, the normal force is often described as the opposing force to gravity, meaning the force of the ground pushing up on your feet, neutralizing the force of gravity pulling you down. However, this is technically not correct. There are actually two pairs of opposite forces. The *gravitational* interaction pair is the force of the earth pulling on you (F_{g_2}), and the force of you pulling on the earth (F_{g_1}). Newton tells us that these forces are equal in magnitude, though this may seem strange since we are so much smaller than the earth. The other pair of forces involve *pressure*, not gravity. There's the pressure that the ground feels as your foot presses down on it (P_1), and the pressure that your foot feels as the ground presses back (P_2). This type of (pressure) force does not need gravity. You can push sideways on a table with your hand and you are applying the same type of pressure to the table that your foot is applying to the floor. (It is this pressure force that bathroom scales actually measure, not the force of gravity.

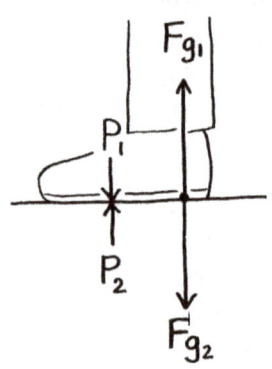

They measure P_2, how much force is needed to counteract your P_1 force.) However, because these two force pairs are related (the pressure being caused by gravity), P_2 is often renamed F_n, the normal force, and is paired with F_{g_2}. We'll learn to label the normal force in some diagrams, but first, some videos and activities.

The P and F forces in the foot diagram represent two very different types of forces. The P forces belong to a category of forces called **contact forces**. Contact forces are caused by (unsurprisingly) physical contact between two objects, and include friction, tension, pressure, and spring force. The other type of force is a **field force**. Gravity is the only field force we will learn about in this book. Electricity and magnetism and also field forces. With field forces, the objects do not have to touch each other. Magnets can attract or repel each other without touching. You can't pair a contact force with a field force. Gravity can't be paired with pressure force. However, engineers find that it works out okay in free body diagrams if you simplify things and just use F_n and F_g (as we did on page 75).

ACTIVITY 7.6: Take a video break!

Let's take a break from reading and watch some videos about Newton's third law. Check the playlist for some short videos that review what we've just learned. (Don't miss the video showing the water-powdered jet pack that lets you "fly" above the water like something in a science fiction movie.)

ACTIVITY 7.7:   A third law demo toy: "Clackers"

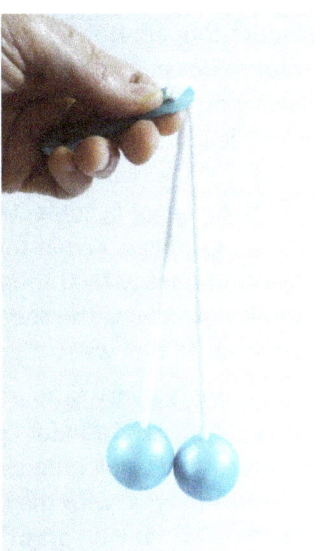

NOTE: You may want to purchase these instead of trying to make your own.

You will need: *a piece of string about 30 cm long (length can vary), two large plastic or wooden beads, marker or pen (optional: metal key ring)*

1) Push each end of the string through the hole in one of the beads, and then tie a knot in the ends of the string so the beads can hang as shown in photo.
2) Measure exactly halfway between the beads and make a dot with a marker.
3) Tie a knot exactly where the mark is, or tie the ring at this spot.
4) Hold the knot/ring and let the beads hang down. If they are not exactly even with each other, adjust your knot.
5) Move your hand up and down until the beads start to clack together. Keep this motion going, increasing the swing of the beads and the strength of their collision.
6) With a little practice, you might be able to increase the swing of the balls so much that they go all the way up and clack together over your hands. You can get a very fast over-under-over-under clacking pattern that is double the speed of the regular clacking.

TIP: If you need help figuring out how to use your clackers, search "how to use clackers" on a video platform like YouTube.

ACTIVITY 7.8:   Use a skateboard to demonstrate the third law

When you jump, you push on the floor. Newton's third law says that the floor must push back. Does it, really? What if the floor did not push back? What would happen? In this experiment you will push off a surface that doesn't push back.

You will need: *a skateboard*

1) First, stand in a place that gives you plenty of free space for jumping. Do a "standing broad jump" (where you jump from standing position, without any lead-up steps). You can feel yourself pushing against the floor as you begin your jump. But does the floor really push back?
2) Now stand on the skateboard and try the same jump. (Be careful as you jump! Don't hurt yourself.) There will be an equal and opposite reaction as you jump, but in this case you will be able to see that opposite reaction as the skateboard moves backward. How far forward did you jump this time? Likely, you did not move forward at all, or if you did it was a very small amount. The floor had two things that the skateboard did not: friction and inertia.

ACTIVITY 7.9:   Use your track car to demonstrate the third law

You will need: *your track car, at least a dozen pencils, a long piece of cardboard*
NOTE: *If you don't have a track car, check out the video of this on the playlist.*
This activity is very similar to 7.8 with the skateboard.

1) First, let your track car move along the floor or table. Newton's third law says that as the car moves forward, the table is receiving an backward force of equal magnitude. However, we can't see this force in the table because the table doesn't move.
2) Now place the long piece of cardboard on the table and put all the pencils underneath it, as shown.
3) Turn on the track car and set it onto the cardboard. What happens?
   The car's wheels will cause the cardboard to move backwards. The car might not move ahead at all.

Let's go back and take a closer look at the batter hitting the baseball. We noted that Newton's second law can be applied as the ball hits the bat. The bat and ball feel equal and opposite forces. However, there is something else going on here, too. The bat was able to reverse the direction of the ball as well as making it sail high up into the air. While the ball was in contact with the bat, it experienced acceleration due to the force of the bat, but after the ball was no longer in contact with the bat, the acceleration stopped. However, the ball still had **momentum**. Momentum is just **movement in a particular direction** and does not require speeding up or slowing down like acceleration does. Was the ball moving at a constant speed while it was flying through the air? This must be the case because it was the bat that was applying the force that caused acceleration. If the ball is no longer in contact with the bat, that accelerating force is no longer being applied. (What the bat did to the ball is called **applied force**.) The ball continues to fly through the air because it has momentum. (Gravity is not slowing the speed of the ball, and friction with air molecules is so small that we can ignore it for all practical purposes.)

There is a very simple formula for determining momentum. The letter "p" is used to represent momentum ("p" is for "pellere," Latin for "push") so the formula looks like this: **p = mv** (momentum= mass times velocity). Mass is recorded in kilograms, kg, and velocity is in meters per second, so momentum is (kg)(m)/sec. (This formula reminds us of F=ma, with the difference being that velocity uses just seconds, whereas acceleration uses seconds per second. So p=mv is sort of like a "step down" from F=ma.)

Why bother even mentioning momentum? Partly because it is going to help us explain the behavior of the ball, and partly because Newton didn't actually write down F=ma. He actually wrote something more like this: Force equals the change in momentum divided by the change in time (i.e. the amount of time the force is applied). Here is how we write this in mathematical terms:

$$F = \frac{\Delta p}{\Delta t}$$

The triangle is the letter "D" in the Greek alphabet, and is pronounced "delta." The delta stands for "change." This formula becomes more helpful in our discussion of the batter hitting the baseball if we can rearrange it a bit. The rules of algebra say that if we multiply each side of the equation by the same thing, we won't change the equation. So if we multiply each side of this equation by Δt, we get:

$$F(\Delta t) = \Delta p$$

This tells us that the change in momentum (Δp) of the baseball will be determined by the amount of force (F) applied by the bat multiplied by the time interval (Δt) that the bat is in contact with the ball. $F(\Delta t)$ is called the **impulse force.** The size of the impulse will determine the momentum of the ball.

The ball needs to change its momentum from motion towards home plate to motion away from home plate. How big that change will be partly depends on how long (Δt) the bat is in contact with the ball. This is why "follow through" is so important when batting a baseball or hitting a golf ball. The longer the bat or club is in contact with the ball, the more momentum the ball will have.

In certain situations, it is to the team's advantage to have the batter try NOT to give the ball a lot of momentum. They want the ball to stop about halfway to the pitcher's mound. It takes enough time for the catcher to run out and pick up the ball that the runners already on bases will have time to advance to the next base. The batter tries to decrease the momentum on the ball by decreasing the interval of time (Δt) that the bat is in contact with the ball. This is called "bunting" (shown in photo). There is no follow through when bunting. In fact, the batter might even let the bat come backwards a bit, almost like a reverse follow through.

The time interval (Δt) is also important in collisions. The change in momentum in a collision goes from very fast down to zero in an extremely short amount of time. The exact amount of time is very important—fractions of a second can mean the difference between something breaking or not breaking. For example, if you hold up an egg and drop it onto a cement floor, the egg will stop very suddenly due to the stiffness of the floor. Bam! The egg hits zero velocity almost instantly. But if you drop the egg into a bucket of water, the egg will likely survive the fall because the egg's velocity will be slowing down from the time it first touches the water until the time it reaches the bottom of the bucket. Both the floor and the water brought the velocity of the egg to zero, but the floor kept Δt very, very small, while the water allowed for a larger Δt. That little bit of extra time made all the difference.

Our visual processing isn't fast enough to be able to appreciate the difference between going to zero velocity in .2 seconds and going to zero in .1 seconds. However, this can be the difference between life and death. Cars are designed to increase Δt as much as possible, slowing the collision time in order to lessen the force of impact. The front of a car (everything from the windshield forward) is designed to crumple. A car that can't crumple is like the egg hitting the floor, going to zero velocity almost instantly. Air bags can add even more Δt. They only add a tiny fraction of a second to Δt, but that split second makes a huge difference in the outcome.

ACTIVITY 7.10:   Try a few easy math problems about momentum

NOTE: Oddly enough, physicists never came up with a name for the units of momentum. For force we have newtons, but for momentum we have to write out "kg(m)/sec." There are so many unit names in physics—newtons, joules, watts, volts, amps—you'd think they would have come up a name for momentum units. But nope, they didn't.

1) A golf ball has a mass of .05 kg. If a golfer hits the ball giving it a velocity of 30 meters per second, what is the momentum of the ball? _____ (Use p=mv. Your answer will be in kg(m)/sec.)

2) A baseball has a mass of 145 grams. A pitcher throws the baseball hard enough to give it a momentum of 3.75 kg(m)/sec. What was the velocity of the ball? _____ (Use p=mv. Don't forget to convert grams to kg.)

3) A tennis player hits a ball and gives it a momentum of 1.5 kg(m)/sec. (The change in momentum is from zero to 1.5, so we can just use the number 1.5.) The racket is in contact with the ball for .2 seconds. (The change in time is from zero to .2 seconds, so we can just use the number .2.) How many newtons of force will be the result? _____ (Use F= Δp/Δt)

4) A baseball pitching machine can pitch balls with a force of 50 N. We know that the momentum of the balls as they come out of the machine is 5.5 kg(m)/sec. How long (Δt) were the balls in contact with the pitching mechanism? _____ (Use F(Δt)= Δp, or you can rearrange it by dividing both sides by F, to make (Δt)= Δp /F. Your answer will be in seconds.)

5) An average hockey puck weighs about 160 grams. If a puck is coming at the goalie with a momentum of 3.2 kg(m)/sec, what was the velocity of the puck as it left the stick of the player who hit it? _____ (Don't forget to convert grams to kg. Use p=mv. Your answer will be in m/sec.)

6) A car's momentum went from 5 kg(m)/sec down to zero in one quarter of a second because it hit a wall. How many newtons of force was this collision? _____ (Use F= Δp/Δt.)

One idea we'll be exploring in the next activity is called ***conservation of momentum***. (The word "conservation" means to preserve or to prevent change.) This is the idea that occurred to Descartes, though he could not prove it. The idea was debated and refined through the centuries and then became an established law of physics somewhere in the middle of the 1800s. Conservation of momentum is a key concept when studying collisions. For a collision occurring between two objects in an isolated system (meaning they are not being affected by any other forces), the total momentum of the two objects before the collision is equal to the total momentum of the two objects after the collision. The momentum lost by object 1 is equal to the momentum gained by object 2. This is what we will be looking for in the following activity, though we might be using more than two objects. Our "system" will be two or more marbles colliding with each other inside a paper track.

But, wait—before you start the next activity, here is something else to look for: ***elastic and inelastic collisions***. In elastic collisions, not only is momentum conserved, but motion energy (***kinetic energy***) is also conserved. Billiard balls are designed to give elastic collisions. (Technically, not perfectly elastic, though. Almost no real-world collisions are perfectly elastic. There is always a small amount of energy lost due to friction, sound and heat being produced in the collision.) If the cue ball hits another ball exactly at its center, almost all of the momentum and kinetic energy will be transferred to the other ball. "Motion in" equals "motion out," even if the motion gets divided up between two or more balls. When you "break" at the beginning of the game, the momentum and kinetic energy of the cue ball gets distributed randomly to all of the balls because they are all in contact with each other. (Again, these are not perfectly elastic collisions, but we go ahead and give them the benefit of the doubt and call them elastic anyways.)

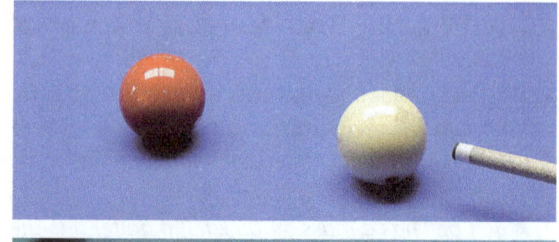

Inelastic collisions must conserve momentum (it's the law!) but they can have an observable loss of kinetic energy. Imagine throwing a ball of clay at a wall. The kinetic energy disappears because the wall didn't move and the clay ball didn't bounce back. How is momentum conserved? If you look at the clay ball after it hits the wall, you'll see that its shape is greatly deformed. The momentum went into the molecules of the clay. Inelastic collisions result in the colliding objects being joined together. The classic example of this is two train cars going in the same direction, with the rear car going much faster. The rear car catches up, then attaches to the forward car.

The clay deforms.

ACTIVITY 7.11: Observing collisions

You will need: about 10 marbles, 2 sheets of card stock, scissors, tape, meter (or yard) sticks, two stacks of books
Optional: other small balls you might want to try

1) Cut a sheet of card stock in half the long way. Then cut these halves in half the long way. Now you will have 4 long strips. Do this to the other piece of card stock, also.
2) Fold the strips in thirds, the long way, so that you have paper troughs perfectly sized for holding marbles.
3) Tape 6 of the troughs end to end, making sure the transitions between troughs are smooth (no bumps to stop marbles from rolling smoothly).
4) Tape the two end troughs to the meter sticks, with two troughs in the middle, as shown in the diagram below. Place a stack of books under each meter stick, making sure the stacks are the same height.

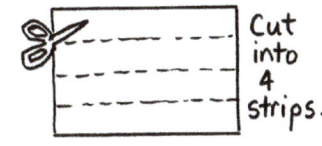

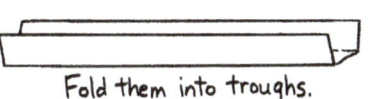

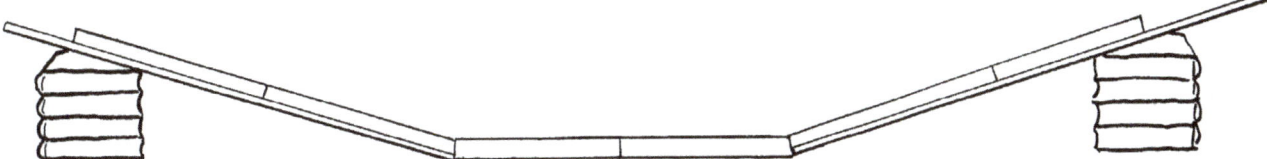

5) Now you are ready to begin your experiments. You will be trying out as many combinations as you can, with different numbers of marbles rolling down the troughs or sitting in the middle trough. You will need to make sure that when you let marbles go on the ramps that they are released at exactly the same time. For many of your experiments, you'll also want to make sure the marbles are released from the same point on the ramp, but you can control acceleration by the height of the staring point.

Here are some suggestions for combinations to try:

left ramp	middle	right ramp
1	0	1
2	0	2
1	2	0
1	3	0
1	4	0
1	5	0
1	2	1
2	2	0
2	3	0
2	4	0
3	4	0

6) Can you engineer an inelastic collision?
   In an inelastic collision, both objects are in motion going the same way, but the rear object is going faster. After the collision they stay in contact with each other and keep going as if they were a single object.

None of your collisions will be perfectly elastic or inelastic, of course. We can't get away from friction and all the other little imperfections of real life.

ACTIVITY 7.12: "Collision Carts" an online simulation

You can experiment with momentum and collisions by going to:
https://www.physicsclassroom.com/Physics-Interactives/Momentum-and-Collisions/Collision-Carts/Collision-Carts-Interactive

Notice that the velocities are shown as (+) and (-) because velocity is a vector, meaning it has direction as well as magnitude. They decided to indicate direction using (+) for direction to the right, and (-) for direction to the left.

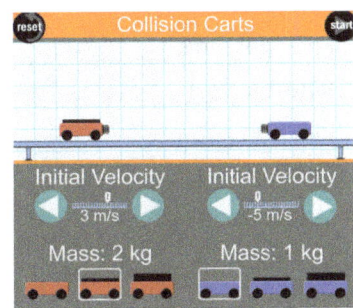

ACTIVITY 7.13: Review questions
(Answers are in the teacher's guide section.)

1) TRUE or FALSE? Newton wrote F=ma in his Principia.
2) TRUE or FALSE? Newton wrote in English.
3) What measurement unit is kg(m)/sec²? _____
4) What is measured using kg(m)/sec? _____
5) What is the average mass of a pineapple? _____ And how many newtons is this? _____
6) Does F=ma work in zero gravity (free fall)? _____
7) Does increasing acceleration increase force? _____
8) How much does a 100 kg person weigh in newtons? _____
9) TRUE or FALSE? The earth and the moon pull on each other with equal strength.
10) TRUE or FALSE? The earth has more inertia than the moon.
11) The center of gravity around which two bodies orbit is called the _____.
12) TRUE or FALSE? The action and reaction occur at exactly the same moment.
13) What word do physicists use for perpendicular? _____
14) A soldier is standing at attention. How many action/reaction pairs are there in his feet? _____
15) There are two kinds of forces. When objects touch, these are _____ forces. When objects do not touch, these are _____ forces.
16) F(Δt) is the: a) interval of time   b) impulse force   c) momentum   d) velocity   e) applied force
17) To prevent something from breaking, do you want Δt to be large or small? _____
18) Who was the first person (as far as we know) to suggest that momentum is conserved? _____
19) TRUE or FALSE? Conservation of momentum occurs even if the objects in a collision stop moving.
19) Can you ever witness a 100% completely elastic collision? _____

ACTIVITY 7.14: Matching definitions
(Answers are in the teacher's guide section.)

1) impulse force _____
2) momentum _____
3) conservation _____
4) field force _____
5) mass _____
6) speed _____
7) velocity _____
8) terminal velocity _____
9) acceleration _____
10) normal force _____
11) inertia _____
12) kinetic _____
13) weight _____
14) magnitude _____
15) elastic _____

A) the measure of how resistant something is to a change in velocity
B) the perpendicular force of an object pressing on a surface
C) the highest velocity an object can achieve while falling
D) the measure of how fast something is going
E) the measure of how fast and in which direction something is going
F) moving
G) how large or small something is
H) no kinetic energy is lost
I) force times the time interval in which the force is applied
J) velocity multiplied by mass
K) the increase in velocity
L) the tendency of something to remain as it is
M) measuring mass while in a particular gravitational field
N) no change, or preventing change
O) force that operates without objects touching each other

Adventure 8: Flies and cannonballs

We'll get to the cannons eventually, but first, we need to stop in and see René Descartes again. (Remember, it's "Day-cart." Don't say the "s.") This time we find him lying in bed, staring at a fly on the ceiling.

Descartes did a lot of deep thinking. He is best known for his philosophical writings (such as "Meditations on First Philosophy, in Which Is Proved the Existence of God and the Immortality of the Soul"), but he also was a math and science genius. He lived alone most of his life and moved constantly, but because he became a famous intellectual, he was invited to meet many other famous people (poets, artists, authors, and mathematicians), including members of royalty in several countries. When he was in his early 50s, Queen Christina of Sweden invited him to come and live in her royal palace and be one of her tutors. She was only 22, so she was still pursuing her education. Queens were not allowed to attend universities, so she had to hire teachers to come to her palace. Christina was an early riser and loved to get up before the sun. Descartes like to stay up late and then sleep in. Christina insisted that Descartes get up at 5 AM to begin her philosophy lessons. Unable to get to sleep early enough, Descartes began to suffer from lack of sleep. This led to him catching a terrible cold that turned into pneumonia, and he died of it ten days later. (Make sure you are getting enough sleep!)

René Descartes

As a philosopher, Descartes was optimistic, practical, religious, and liberal. What a combination, eh? He sometimes wrote in French instead of Latin so that everyone could read it, not just the intellectual elites. He wanted even women to read his books, which was a very "liberal" opinion during the 1600s. He believed in God, but didn't want to be either Catholic or Protestant, the two main choices during that time period. He didn't understand why these two groups could not get along since they worshiped the same God. He was also a great believer in the power of the human mind, and would likely have agreed with the Renaissance belief that "man was the measure of all things." He believed that humans were essentially good, and if they were instructed in good morals, they would follow them, and live moral lives. He believed that people could achieve salvation through being good. The other famous philosopher and mathematician of that day, Blaise Pascal, strongly disagreed with Descartes and said that humans needed God's grace, and could never earn it by being good enough. (Pascal was also a math genius, though his work doesn't directly affect the physics we are studying, so we won't get to S.N.O.O.P. on him.) Philosophy and religion were important topics to mathematicians in those days. Newton, in his later years gave up writing about science and devoted himself to studying the Bible. It is common even today for mathematicians to be very interested in religion and philosophy.

Pascal and Descartes might have had religious disagreements, but one thing they did agree on was that Aristotle had been wrong about many things, and that universities should stop relying so heavily on Aristotle for their curriculum. Galileo and Newton were also sharply critical of Aristotle, and all of these men (Pascal, Descartes, Galileo, and Newton) advocated a new approach to science: conducting experiments and learning from observation, instead of taking Aristotle's word for it.

Descartes came to the conclusion that perhaps the most important contribution he could make to society was to further the progress of technology. He dabbled in anatomy a bit and dissected animals, but he missed the mark a lot of the time when it came to speculating about what certain human body parts do. We'd laugh at what he thought about the function of various body parts, but he was right when he said that bodies were very mechanical and they functioned like living machines. It turned out that, in the end, his best contributions to technology were his mathematical ideas.

Aristotle

Descartes' biggest contribution to math was the way he combined algebra (equations) and geometry. This blend is called **analytical geometry**. Descartes realized that not only could he describe a fly's position on this 2-dimensional graph, but he could also draw lines, circles, and curves (like the parabola and the sine wave). If you have studied any algebra, or even pre-algebra, you have undoubtedly been introduced to the **Cartesian coordinate system**. This is the graph that Descartes imagined on his ceiling. (Named in honor of Descartes, though only the "cartes" part of his name is used.) Here are some examples of what this grid system allows you to do:

Graph lines using equations:

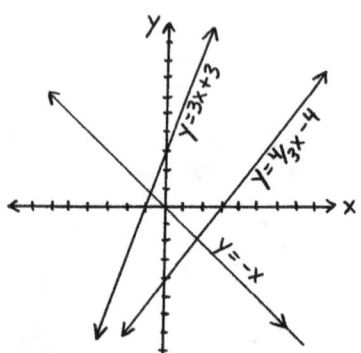

Graph circles:

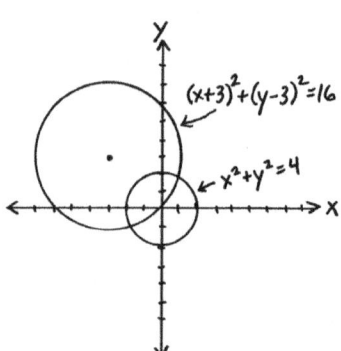

Graph parabolas:

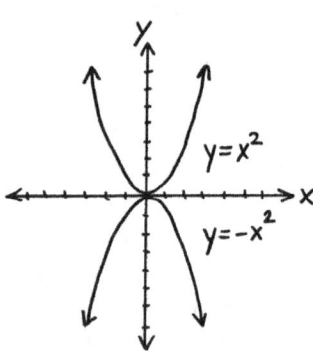

We won't be doing any graphing like this, though we will be looking at parabolas. Right now you just need to understand that measurement from left to right is along the "x-axis," and measurement up and down is along the "y axis." Look back at the previous page and note that the fly is sitting at the point (10, 7). Points are always written as (x,y). The fly is 10 units to the right on the x-axis, and 7 units up on the y-axis. (Eventually Descartes went beyond just his ceiling graph, and made the x and y lines go down into negative numbers.)

ACTIVITY 8.1: Point practice   (Answers on page 114)

Match the points shown on the Cartesian grid to these (x,y) pairs:

1) (-2, -1) _____
2) (1, 5) _____
3) -4, -2) _____
4) (-4, 2) _____
5) (3,2) _____
6) (4, -5) _____
7) (2, 1) _____
8) (3, -1)) _____

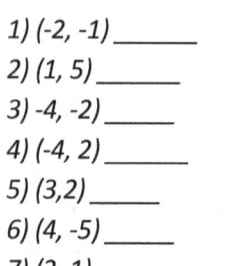

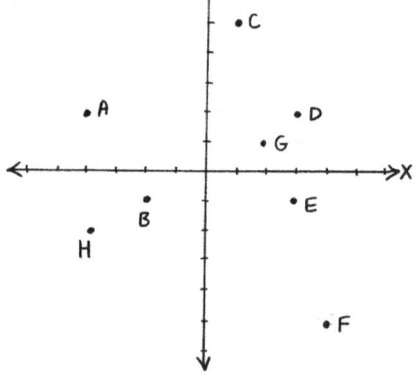

Are these all flies?

Better hurry up, guys. We need to visit Galileo very soon!

So far, our adventures have mostly concerned motion that is going along the x-axis or the y-axis. When we drop something, gravity pulls down in the y direction, so the object drops straight down along an imaginary y axis. When we are studying the motion of a rolling ball or a car, we are looking at motion in a straight line along the x-axis. In a few minutes, we are going to look at a motion that combines movement in both x and y directions. But first, let's look at how the x-axis can be used to represent time.

Let's have the x-axis represent time in seconds and the y-axis represent distance in centimeters. We are going to graph a short journey taken by our fly as it crawls on a wall or ceiling. We're going to specify that our distance measurements are not merely distance, but also specify a direction. This type of measurement is called ***displacement***. We'll determine a certain starting point and keep track of how far away from that starting point the fly ventures out. On the graph, we will represent that starting point at the (0,0) in the bottom left corner. This point is called the ***origin***.

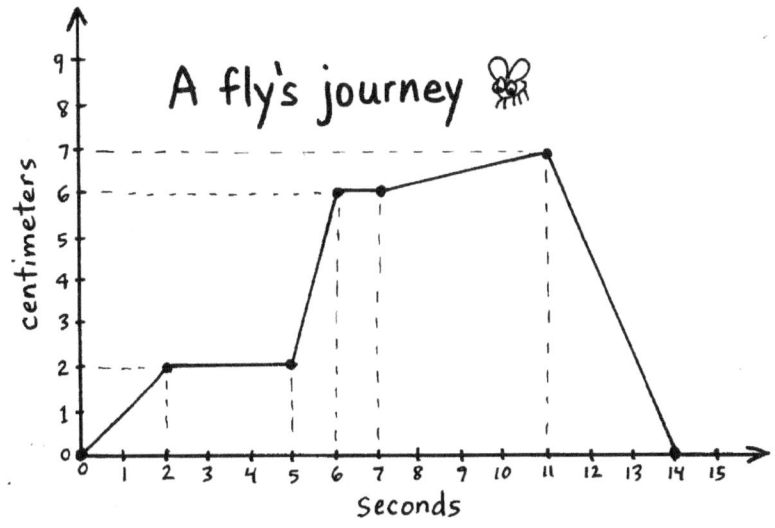

We watch the fly crawl 2 cm away from its starting point in 2 seconds, so we put a dot at the (2,2) mark. Now the fly stays still for 3 seconds, so for the next three seconds (from 2 to 5) the line will be flat because we don't have any additional distance to record. Then the fly starts crawling again and this time it covers a distance of 4 cm in just 1 second. (That's sprinting for a fly!) Can you see this on the graph? It's the steep line between 5 and 6 seconds. Then the fly sits still for a second, which is another flat line. Then the fly crawls slowly, going only 1 cm in 4 seconds. Finally, it turns around and comes back to its starting point, which is why the line goes down instead of up. "Up" signifies away from the starting point and "down" signifies towards the starting point. It comes back to where it started in only 3 seconds.

This type of graph is called a "displacement versus time" graph. We can use the graph to figure out how fast the fly was traveling in cm/sec. In the first leg of its trip it went 2 cm in 2 sec, so that's 2/2. Since 2 divided by 2 is 1, the fly traveled at a speed of 1 cm per second. Obviously, when it sat still it was going 0 cm/sec. Between seconds 5 and 6 it went 4 cm/sec. In the third leg of its trip, it went 1 cm in 4 seconds. 1/4= .25, so it was averaging .25 cm per second. Lastly, when it turned around and came home, it covered 7 cm in 3 seconds. 7/3= 2.3. So it averaged a speed of 2.3 cm/sec. All of these calculations are based on the idea that distance equals rate times time (**d=rt**).

Notice that the steepness of the line corresponds to the speed:

steeper=faster   less-steep=slower   flat=no speed

ACTIVITY 8.2: Displacement-time graphing practice   (Answers in teacher's section)

Draw graphs for each of these scenarios. Make sure you label the x and y axes with appropriate units.

1) You let your dog out the front door. After 5 seconds, she is 5 meters from the door. Then she stops and sniffs for 4 seconds. Then she begins running and covers another 5 meters in 3 sec. Again, she stops and sniffs. She spends 3 seconds sniffing. Then you call her and she comes running back to the front door in 5 seconds.

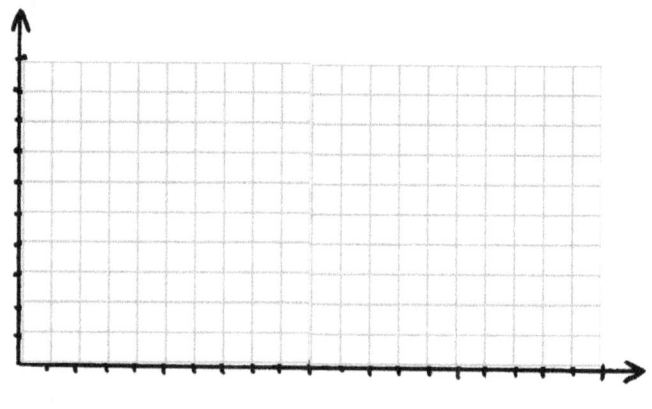

2) A helicopter leaves an airport and flies 3 km in 1 hour. At that point, it reaches a location where it has to hover for 15 minutes (so it doesn't actually go anywhere even though it remains in the air). During the next hour, the helicopter flies another 3 km. Then it again hovers for 15 minutes. Then it turns around and flies 2 km back towards the airport in 30 min., then stops and hovers for another 15 min. Then it turns around and begins flying in its initial direction, away from the airport and covers 6 km in 45 minutes. Finally, it turns around and flies all the way back to the airport in 1 hour.

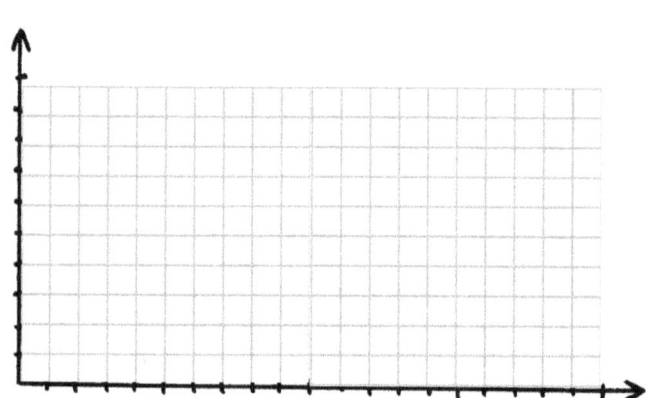

Now we're going to change what the y-axis represents. This time, we'll take it up a notch and have the y-axis represent velocity, not distance. Remember, the difference between speed and velocity is that speed is a **scalar** quantity, meaning it does not indicate direction, and velocity is a **vector** quantity, meaning it has a certain direction.

SPEED has only magnitude   VELOCITY has both magnitude and direction

For example, if we say that someone is walking at a speed of 5 km/hour, we don't know anything about which direction they are walking. To convert this into a velocity, we'd have to specify that they are walking east or north (or any other direction), or towards or away from a certain point (like our fly). Although, having said that, you'll find that many physics problems use the word "velocity" but don't specify a direction, or just assume that you know that the direction is down toward the earth (for falling objects). So although textbooks make a big deal about the distinction, very often the direction of the velocity is not stated and isn't really necessary to solve the problem.

In this graph, the x-axis is still time, but the y-axis is velocity (meters per second). This is the flight of a small drone that you are controlling remotely. in the first 30 seconds, its velocity goes from 0 to 2 m/sec. Since there was an increase in velocity, the drone was accelerating. Then, from seconds 30 to 50, its velocity is a steady 2 m/sec. From 50 to 60 seconds, it goes from 2 m/sec to 5 m/sec, so we again have acceleration. It stays at 5 m/sec for 10 seconds, then slows down for the last 20 seconds, decelerating as it comes back to you and lands at your feet. When it lands, both its velocity and its position are back to zero.

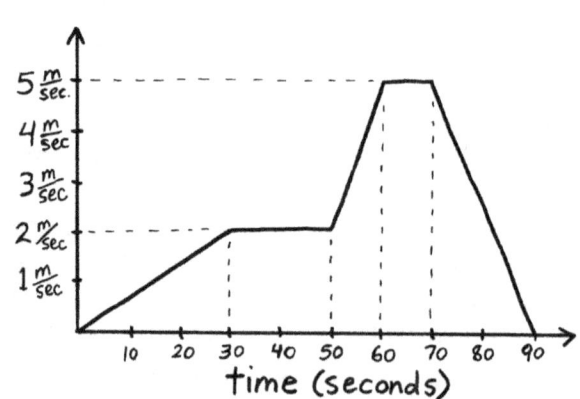

Answers to 8.1: 1) B 2) C 3) H 4) A 5) D 6) F 7) G 8) E

Here is another velocity-time graph. We can imagine the flight of a drone again, but this flight will be much shorter, only 12 seconds. The drone goes from zero to 5 meters per second in 4 seconds, then it maintains steady velocity and flies at 5 m/sec for 4 seconds. Finally, it decelerates for the last 4 seconds, going from 5 m/sec down to zero. The flight has been broken into 3 sections so we can analyze them separately.

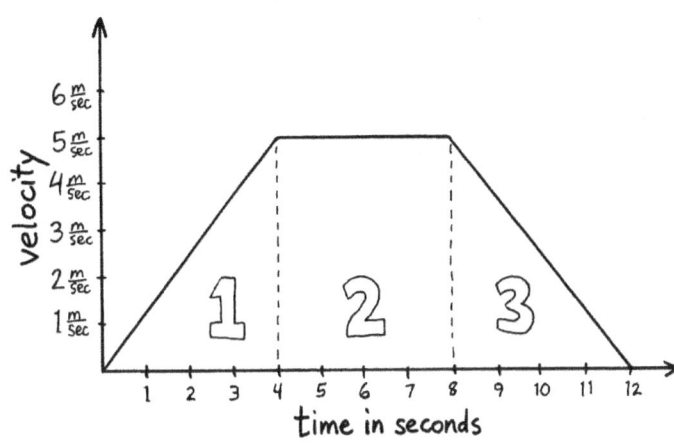

An interesting feature of velocity-time graphs is that the area under a section of line is equal to the distance traveled. You don't need to figure out area in order to figure out distance, but seeing how area can be used for distance will help us to understand how an important formula is derived.

Let's look at section 2 first. Notice that it is a rectangle. You can calculate the area of a rectangle by multiplying the width by the height. For section 2, this would be 4 seconds times 5 m/sec. 4 times 5 is 20, so the distance traveled by the drone during this time is 20 meters. Notice how this looks if we write the calculation properly. Keys things to remember are that anything over 1 is just itself, so you can put something over 1 any time you want to make it into a fraction. Also, things that are the same top and bottom will cancel out because anything over itself is just 1. This works for numbers (4/4=1) but it also works for units. If you have sec/sec, the seconds cancel out. You can show canceling by drawing a line through them.

$$\frac{4 \text{ sec.}}{1} \times \frac{5 \text{ m.}}{1 \text{ sec.}} = \frac{4 \times 5 \text{ m.}}{1 \times 1} = \frac{20 \text{ m}}{1} = 20 \text{ m.}$$

So the area of rectangle 2 is 20. This is also the distance traveled in 4 seconds at 5 meters per second. Now look at section 1. It is a triangle, not a rectangle. You can find the area of a triangle by calculating 1/2(b)(h), one-half times the base times the height. The base line of section 1 is 4 seconds long. The height of the triangle is 5 meters per second. Using 1/2(b)(h), we get:

$$\frac{1}{2} \times \frac{4 \text{ sec.}}{1} \times \frac{5 \text{ m.}}{1 \text{ sec.}} = \frac{1}{2} \times \frac{4}{1} \times \frac{5 \text{ m.}}{1} = \frac{20 \text{ m.}}{2} = 10 \text{ m.}$$

Again, we can cancel anything that appears both top and bottom, so the seconds cancel out leaving us with just meters. The area of this triangle is 10, and the distance covered in 4 seconds by an object going from zero to 5 m/sec is 10 meters.

ACTIVITY 8.3: Velocity-time graphing practice   (Answers in teacher's section)

Draw the graph for the following scenario. Label the x and y axes with appropriate units.

Someone takes their horse out for exercise. After leaving the barn, the horse begins walking and gradually speeds up. At the 1-minute mark the horse is going 1 m/sec, and keeps this steady pace for 4 min. Then the horse increases its velocity until it is going 3 m/sec at the 8-minute mark. In the following minute, it goes from 3 m/sec to 10 m/sec. It maintains this velocity for 2 minutes. Then the horse turns around and takes 3 minutes to slow down to 7 m/sec, and then takes another 3 minutes to slow down to 1 m/sec. The horse maintains this pace for 3 minutes, then stops.

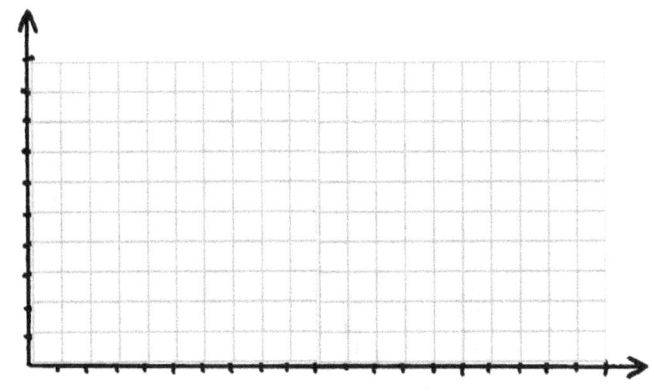

Now for the big leap...

Here we have another velocity versus time graph, but this time it is only labeled with the words velocity and time, not with actual units. Even though units are not given, we can still apply the same ideas we used on the last page. The area of the triangle will give us the displacement (distance) that an object travels while going a certain velocity for a certain time. So we can say that 1/2(b)(h) for this graph is 1/2(t)(v). (The measurement of the base of this triangle is just "t" and the height is "v." It doesn't matter that we don't have numbers.)

Now that we have d=1/2(t)(v), we can use another formula that we've seen previously: a=v/t (acceleration equals velocity divided by time). We used this formula on page 101. Look at how we rearranged this formula here on the right. We started with a-v/t. Then we wrote them as fractions and multiplied each side by "t." We wrote "t" as "t/1" since anything over 1 is just itself. If we multiply both sides of an equation by the same thing, we are not actually changing the equation; we are only changing the way it looks. So then we can cancel the t's on the right side, since they appear both top and bottom. On the left side, we just "a" times "t." Then we can take them back out of fraction form and we end up with the equation at=v.

Let's look at what is happening to the left of the line. We start with our formula 1/2(b)(h). Then we put in the letters "t" and "v," as we discussed at the top of this page. Then we notice that "at" is the same thing as "v," so we can remove the "at" and write in "v," instead (shown by arrow). Then we can rearrange the letters (and this will not change the equation) and we end up with d- 1/2at².

We know that earth's gravity gives an acceleration of -9.8 m/sec², and we represent this as "little g." So when the acceleration in a physics problem is the acceleration due to gravity, we can use "g" as the acceleration, and the formula becomes:

$$d = \tfrac{1}{2} g t^2$$

We're going to learn how to use this formula, but first, we need our assistants to take us back to see Galileo again at the Leaning Tower. How is it going guys?

116

We need to go back to the topic of gravity and falling bodies. We'll show you how to use that formula to calculate the height of the Leaning Tower.

Galileo must have been standing on the lower side of the Leaning Tower (assuming this is a true story, which it might not be) so that the balls would be able to fall to the ground without hitting the side of the tower. If we can accurately time how long it takes for the balls to hit the ground, we can use $d=1/2gt^2$ to calculate the displacement (the distance the balls fell) which is also equal to the height of the tower.

Our assistants timed the balls and found that they took 3.37 seconds to hit the ground. We know the acceleration due to gravity: -9.8 m/sec². If we know these two things, we can use the formula to find "d." (Remember, the negative sign on the 9.8 means that the direction of the acceleration is downward. In this case, where we are just trying to find the height of the tower, we know that the answer can't be negative because height can't be a negative number. So for now, we can ignore the negative sign and just use 9.8.)

You'll need to use a calculator for this. $(3.37)^2$ is 3.37 times 3.37. You should get something close to 11.357. Then multiply by 9.8, then divide by 2. (Multiplying by 1/2 is the same thing as dividing by 2.) Your final answer should be 55.65 m. Most websites cite 55.8 meters as the official height of the lower side of the tower, so we're pretty close. Our limiting factor here is accuracy in timing. It is very difficult to be accurate down to hundredths of a second. The exact time might be more like 3.372 or 3.375, but this would be very hard to determine experimentally.

$$d = \tfrac{1}{2}gt^2$$

$$d = \tfrac{1}{2}(9.8)(3.37)^2$$

$$\phantom{d=\tfrac{1}{2}} 4.9 \quad\ 11.357$$

$$d = 55.65 \text{ meters}$$

But what if we know the height of the tower already—can we use this same formula to calculate the number of seconds of the fall? This is how we can rearrange the formula:

First, multiply both sides of the equation by 2. When we do the same thing to both sides, we don't actually change the formula itself.

$$d = \tfrac{1}{2}gt^2$$

$$(2)d = \left(\tfrac{2}{1}\right)\left(\tfrac{1}{2}\right)gt^2 \quad \text{multiply each side by 2}$$

Then, we can divide both sides of the equation by g.

$$\frac{2d}{g} = \frac{gt^2}{g} \quad \text{divide each side by } g$$

So now the equation looks like this.

$$\frac{2d}{g} = t^2 \quad \text{Now,}$$

Then we take the square root of each side. The square root of t^2 is just t. (For example, what is the square root of 5^2? 5 squared is 25, and the square root of 25 is 5.)

$$\sqrt{\frac{2d}{g}} = \sqrt{t^2} \quad \text{take the square root of each side.}$$

This gives us an equation for t. We already know that g=9.8, so if we know what d is, we can use a calculator to find t.

$$\sqrt{\frac{2d}{g}} = t \quad \text{And now we have a way to find } t!$$

For example, the tallest building in the world as of the writing of this book is the Burj Khalifa (in Dubai, UAE) at 828 meters. If someone drops a baseball from the top of this building, how long will it take to hit the ground? (We're assuming no air resistance. Many objects would reach terminal velocity during the fall, but we are going to ignore air resistance to keep the calculation simple.) Just plug the numbers into the formula, as shown here on the right, and you get t=12.99 seconds. (Use a calculator.)

$$t = \sqrt{\frac{2d}{g}}$$

$$t = \sqrt{\frac{2(828)}{9.8}}$$

$$t = \sqrt{168.98}$$

$$t = 12.99 \text{ sec.}$$

ACTIVITY 8.4: Solve for time or displacement (height)   *(Answers on page 120 and also in teacher's section.)*

1) The Strasbourg Cathedral, built in 1439, was the tallest building in Galileo's day, at 142 meters tall. If Galileo had dropped his balls from the top of this cathedral, how long would it have taken the balls to hit the ground? _____

2) In 2004, the tallest building in the world was Taipei 101 in Taiwan. If a rock dropped from the top of the tower reaches the ground in 10.18 seconds, how tall is the tower? _____

3) As of the writing of this book, the third tallest building in the world is the Shanghai Tower in China, at 632 meters. How long would it take for a dropped ball to reach the ground? _____

Strasbourg cathedral

4) As of the writing of this book, the second tallest building in the world is the Merdeka 118 in Kuala Lumpur in Malaysia. If a piano dropped from the top of this tower takes 11.77 seconds to hit the ground, how tall is the tower? _____

5) The Empire State Building, which is 380 meters tall, was the tallest building in the world when it was built in 1931. In a famous movie, the giant gorilla, King Kong, is shown clinging to the top of the Empire State Building. If King Kong's grip slips and he falls to the ground from the top of the tower, how long will it take until he hits the ground? (Our calculation will have to assume that his body is smaller than it really is, so it can fall from the very top.) _____
Do you think this is enough time for people down below to get out of the way?

6) The tallest building in Australia is the Q1. If a brick dropped from the top of the building hits the ground in 8.12 seconds, how tall is the building? _____

Empire State Building

We're off to visit Galileo again, but this time, he's not dropping things; he is watching cannons being fired.

Aristotle taught that objects that are thrown, or shot, have two or three parts to their motion, depending on how high they go. For a projectile that is shot horizontally, he said that there is a straight line followed by a curve at the end that is a section of a circle. For a projectile that is shot upwards (like most of the cannonballs shown in this illustration from 1629), he said that there is a straight part, then a curved part, then another straight part. Other ancient philosophers taught that projectiles went straight for a while, then dropped straight down. This idea was so obviously false that it fell out of fashion pretty quickly. Aristotle's idea was a little harder to prove false. Even Galileo could not easily prove that Aristotle was wrong. It took a number of years and a lot of experimenting and thinking before Galileo came up with the correct solution. He finally concluded that the shape that all projectiles traced out (in the air) is a parabola.

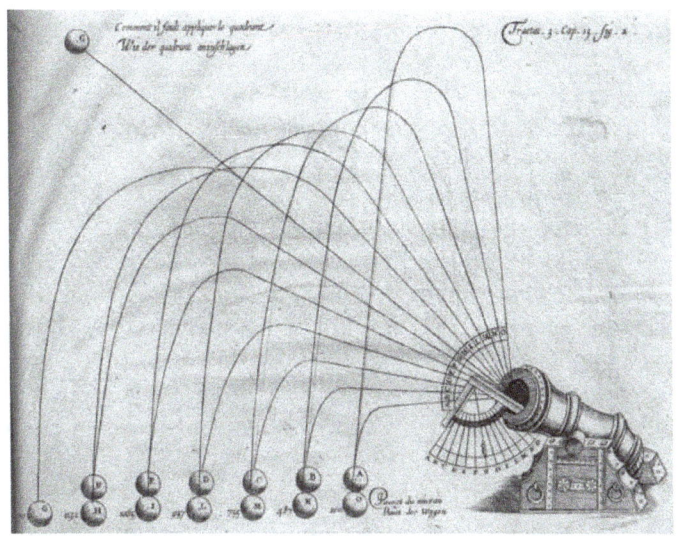

This drawing by Diego Ufino in 1629 shows the prevailing idea of that day, taken from Aristotle.

Galileo had already discovered how to "slow down" gravity so that he could study free fall. Gravity worked the same way for balls on a ramp that it did for balls falling through the air. The ramp just slowed the balls down enough that Galileo could time them. If this worked for objects in free fall, it should also work for studying the motion of objects that have been thrown into the air. All he had to do was "throw" a ball not straight up into the air but up onto a very wide ramp. The motion they made going across the ramp would be mathematically the same as the motion they made going through the air. You can do what Galileo did and observe these motion patterns for yourself.

ACTIVITY 8.5: Observing parabolic motion

You will need: marbles, tape, a few disposable paper bowls (or bowls you don't mind using for washable paint), large sheets of paper (possibly from a standard large sketch pad 18"x24" (45 cm by 60 cm) or a section from a large roll of newsprint, or even some newspapers if that is all you have), a table that can have two of its legs propped up to 12 cm (5 inches), some books that you can use under the table legs, a cereal box (or any box made of thin cardboard), and one of these: water-soluble ink, tempera "poster" paint, or acrylic paint

There are more photos in the teachers' section.

NOTE: A "card table" (square folding table) is ideal for this, but if you don't have one, any table can be made to work. If you don't have a table available, you can also use a large sketch pad instead, or a very large pieces of stiff cardboard. If using a sketch pad or a piece of cardboard, you will still need a large space on a table where you can set the pad or the cardboard.

How to set up:
1) If you are using a table, clear it off. Put a few books under the legs on one side (under two legs) so the table tilts up a bit. Make sure both stacks of books are exactly the same height. A stack that is about 12 cm (5 inches) is plenty high. If you are using a sketch pad or large piece of cardboard, lay it on the table, then prop up one side with the books. Tape the front edge to the table to it doesn't shift around.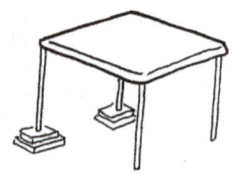
2) Cover the table (or the piece of cardboard if you are using that) with paper. Use minimal tape. Try to make sure that there isn't a huge bump where pages meet if you have to use multiple sheets. Lightly tape the paper to the table (don't have to use tons of tape, just a few pieces will be fine).
TIP: If you are using newspapers, you might want to iron them and try to eliminate the creases as best you can.
3) Cut sections from your cereal box, using the corner edges, so that you have long sections that will make a marble catching tray at the bottom of the table (or pad or board). Tape these in places. This will keep your marbles from rolling onto the floor.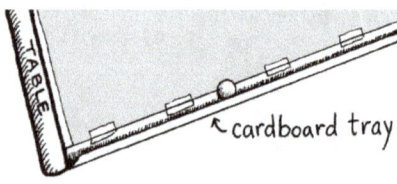
4) Put some paint or ink into a paper bowl and add a little water and mix, until the paint or ink is fairly runny.

What to do:
1) First, roll a marble without any paint on it. Place the marble in one of the lower corners and give it a push upwards so that it rolls up the incline, turns at the top, and comes back down. What shape is it making? You will likely see a distinct shape.
2) Put a marble into the bowl of runny paint or ink. Roll it around until the entire surface is covered.
3) Pick up the marble and hold it in one of the lower corners of the large paper area you have prepared on the table (or pad). Gently give it a push so that it rolls upwards across the paper, then comes back down again. Hopefully the amount of ink or paint on the marble will be enough that it will continue to make a trail all the way up and also all the way down. The trail doesn't have to be extremely visible. Any trail at all will be okay. Anything that records where the marble went.
4) Do this at least a dozen times. (Actually this is so much fun, you'll probably want to do it more than a dozen times!) Try making very low arcs, then very tall and thin ones. Get a wide variety of sizes.
5) Allow these lines to dry. You might want to take a picture of your final results before removing the papers from the table.

The shape that the marbles make is called a **parabola**. Parabolas can be tall and thin, or short and wide. We'll explore the mathematical pattern behind parabolas in the next activity.

Answers to 8.4: 1) 141.82 m. 2) 507.79 m. 3) 11.35 sec. 4) 678.8 m. 5) 8.8 sec. 6) 323 m.

ACTIVITY 8.6: Draw a parabola

A parabola is a geometrical shape that is described by an equation that has a squared term in it, like this: $y=x^2$. You can draw a parabola by filling out the chart below and then graphing each point on this blank graph.

We are going to put all of these equations on this graph. For each equation, you will put a number in for x, then see what you get for y. For example, for $y=x^2$, if you let x=1, this gives you 1^2 (1x1), which is just 1. So when x=1, y=1. If x=2, you get 2^2 (2x2), which is 4. So when x=2, y=4. The points (1,1) and (2,4) have been put on the graph for you. When all the points for one equation are on the graph, then you draw a smooth curve to connect them.

TIP: *When two negative numbers are multiplied, the answer is positive.*

$y = x^2$

x	y
1	1
2	4
3	
4	
0	
.5	
1.5	
2.5	
-1	
-2	
-3	
-4	
-.5	
-1.5	

$y=2x^2$

x	y
1	2
2	8
3	
0	
.5	
1.5	
2.5	
-1	
-2	
-3	
-.5	
-1.5	
.25	
-.25	

$y=x^2 + 3$

x	y
1	4
2	7
3	
4	
0	
.5	
1.5	
-1	
-2	
-3	
-4	
-.5	
-1.5	
2,5	

You might want to make each parabola a different color.

Galileo didn't ever graph a parabola. Descartes had not yet watched that fly on his ceiling when Galileo was experimenting with parabolic motion. All Galileo had was the geometry of Euclid and Archimedes. However, even without the advantage of the Cartesian grid, Galileo was still able to determine that the motion of all projectiles was essentially parabolic. Here is a page from one of his notebooks showing his experiments with objects that were thrown horizontally (parallel to the ground). He carefully measured how far away they landed in comparison to the height of the table (or whatever they were using) and the speed of the object as it left the surface. Using this information, along with careful observation of the arc shape the object made as it fell, he was able to determine that the objects were tracing out a half-parabola shape as they fell.

Galileo made another key discovery about motion during these experiments with objects launched horizontally. But instead of telling you what he discovered, you can discover it for yourself!

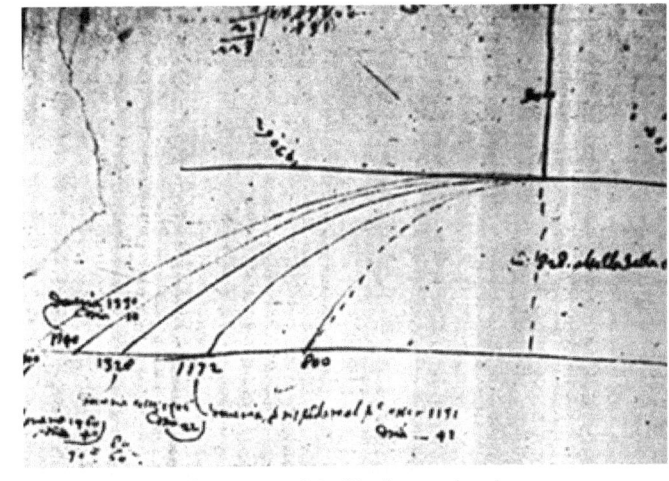

a page from one of Galileo's notebooks

ACTIVITY 8.7: Compare fall rate of launched versus not-launched coin

You will need: a meter stick (or short ruler), two coins, a table

1) Place the meter stick or ruler on the table with the coins placed as shown.
2) Twist the meter stick quickly in the direction shown by the arrow. Coin A should shoot off the table. Coin B should simply fall off and drop straight to the floor.
3) Listen carefully as the coins hit the floor. Do they hit at the same time?
4) Repeat the experiment several times to see if you get the same result.
5) What does this tell you about the downward motion of coin A?

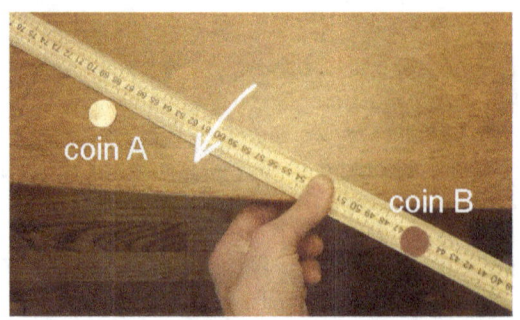

What Galileo (and hopefully you, too) discovered is that the downward motion of an arrow or bullet or cannonball (or launched coin) is exactly the same as an object that is dropped. This means that when we analyze the motion of a projectile, we can separate motion in the x direction from motion in the y direction.

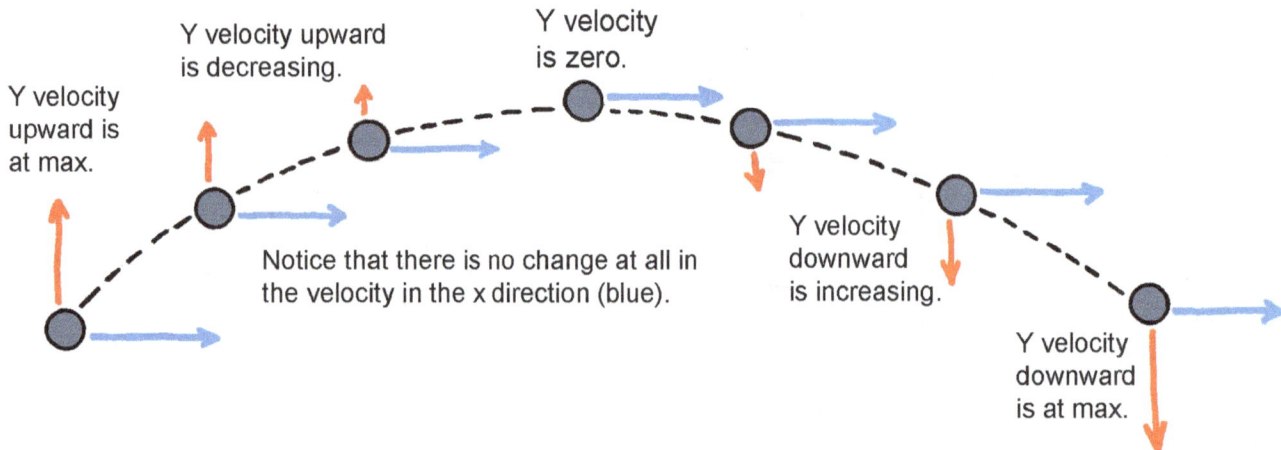

Even though the ball is tracing out a parabolic arc shape, there are two separate motion vectors at each point. There is an x component (blue) and a y component (red). Notice that the blue x vectors do not change throughout the entire flight; the velocity is constant. This is because the ball is not experiencing any acceleration—nothing is pushing on it. It did need a push to get it started moving, but once it left the launching mechanism, that accelerating force was no longer in play. Motion in the y direction (red) has its maximum upward acceleration at the very beginning. As it goes up, however, it is fighting gravity. At the very top, gravity has completely stopped any upward motion and for a split second the ball is going neither up nor down. This occurs at exactly the midpoint of the flight. Then gravity wins the battle, and the ball begins to accelerate downward. You already know how fast this downward acceleration is: 9.8 m/sec². Not surprisingly, the deceleration the ball experiences in the first half of its flight is also 9.8 m/sec².

In experiment 8.7, both coin A and coin B experienced the same velocity in the y direction (red) because gravity was acting equally on both of them. The only difference was velocity in the x direction. Coin B's x velocity was 0, since it fell straight down. Coin A's x velocity was probably at least 1 meter per second.

The horizontal dotted lines show where each coin was at certain time intervals. You can see that despite the fact that coin A was traveling to the right, it was exactly the same distance from the floor compared to coin B. Motion in the x direction had no effect on motion in the y direction.

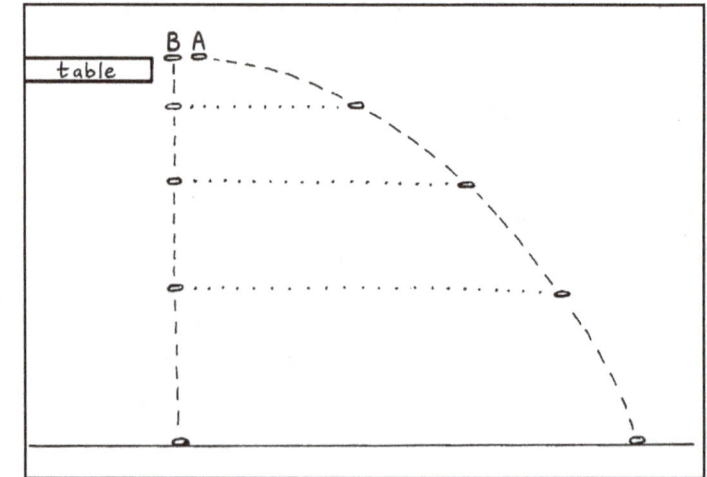

Theoretically, we should be able to do an experiment about what we've just learned. You already know the acceleration due to gravity, and you have a meter stick for measuring the height of the table and the distance at which the projectile lands. The only piece of information missing is the velocity of the projectile right at the moment that it begins soaring through the air. However, measuring the velocity of something that is moving very fast is difficult. You would need a digital timing device capable of measuring in fractions of a second. If you try to use a stopwatch, your reaction time will be too slow to get an accurate measurement. By the time your eye sees the ball pass the line, sends a message to your brain, which then sends a message to your hand muscles to press the button on the stopwatch, the ball will have moved on. Your reaction time is just too slow. So... we'll just torture you with a few math problems to solve, then move on to our final activity about projectile motion (which will be fun).

ACTIVITY 8.8: Using formulas you already know to do something new

You already know the formulas you'll need to solve these problems. Many pages ago we learned **d=rt** (a formula you likely already knew). In this adventure we learned **d= 1/2gt²**. Because there are two formulas that use the letter "d," we are going to label them with subscripts. These subscripts are not necessary, but some students might find them helpful. For vertical distance (down to the ground) we'll add a subscript "v" for "vertical." For horizontal distance, we'll add an "h" or "horizontal."

THINGS WE KNOW: ① $d_h = rt$ ② $d_v = \frac{1}{2}gt^2$ ③ $t = \sqrt{\frac{2d_v}{g}}$

Example 1: Someone kicks a ball off a cliff. At the moment the ball leaves the cliff, its velocity is 5 m/sec. If the cliff is 50 meters tall, how far away from the cliff will the ball land?

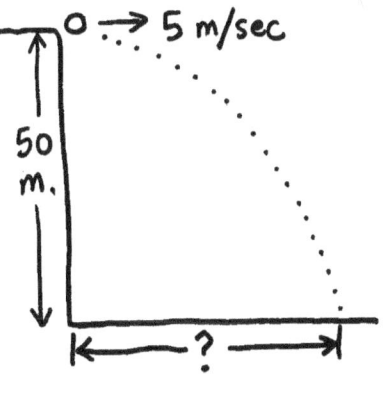

1) Draw a sketch of the situation. We want to find horizontal displacement, d_h, so we want to use **d_h=rt**. But to use d_h=rt, we'll need to find "t" first.
2) We know that the ball will fall at the rate of 9.8 m/sec², and we know the distance that it will fall: 50 meters. Therefore we can use equation #3 to find how long it will take for the ball to hit the ground. Knowing "t" will be helpful.

$$t = \sqrt{\frac{2(50)}{9.8}} = \sqrt{\frac{100}{9.8}} = \sqrt{10.2} = 3.19 \text{ sec.}$$

3) Now that we know how many seconds it will take for the ball to land, we can use equation #1 to find the distance the ball will travel. It doesn't matter that the ball is falling, we can still use **d_h=rt**.
The ball will land 15.95 m. from the cliff. $d_h = (5)(3.19) = 15.95$ m.

Example 2: Someone throws a watermelon off the roof of a building at a velocity of 3 m/sec. The watermelon lands on the ground 5.4 meters away from the building. How tall is the building?

1) Draw a sketch of the situation.
We want to find the height of the building, which is vertical displacement. So we will want to use **d_v=1/2gt²**. But to use this equation, we'll first need to find t.
2) Use **d_h=rt** to find the time till the watermelon hits the ground.

$$d_h = rt \quad 5.4 = 3t \quad t = 1.8 \text{ sec.}$$

3) Now we can use time, 1.8 seconds, to find the displacement in y direction (height).

$$d_v = \frac{1}{2}(9.8)(1.8)^2 \quad d_v = 15.87 \text{ m.}$$

The building is 15.87 meters tall.

Notice that we did not tell you the mass of the watermelon. This is information you don't need because all objects fall at the same rate. In all these problems we are ignoring air resistance. In reality, engineers use formulas that include air resistance, but this adds quite a bit of complexity to the math.

Use these formulas to solve the following problems. Use the space to draw simple diagrams, and make sure to show all your work, like we did in the sample problems.
(Answers on page 126, with extended answers in teacher's section.)

① $d_h = rt$ ② $d_v = \frac{1}{2}gt^2$ ③ $t = \sqrt{\frac{2d_v}{g}}$

1) If someone throws a golf ball horizontally off the top of the Burj Khalifa at a velocity of 15 m/sec, how far away from the tower will the ball land? (Remember, this tower is 828 m.)

2) A secretary working in an office in one of the top floors of the Taipei 101 gets frustrated with her computer and throws it out the window. If the computer is launched at 2 meters per second and it lands 19.68 meters from the bottom of the tower, how high was the window it was thrown out of?

3) Galileo is at the top of the Leaning Tower again. This time, he decides to throw the balls instead of dropping them. He throws them as hard as he can horizontally. They land 13.53 meters from the bottom of the tower. Use 56 meters as the height of the tower. What was the velocity of the balls when they left Galileo's hands?

4) A cliff diver jumps off the top of a 60-meter cliff at a velocity of 3 meters per second. There are rocks at the bottom of the cliff that extend out 8 meters into the water. Will the diver miss the rocks?

In all of the problems we've just done, our projectiles were launched horizontally. But in real life, cannons are usually aimed upwards, not horizontally. The cannon barrel rests on a fulcrum that allows the soldiers to adjust the angle at which the ball is fired. Over the years, soldiers were able to determine by trial and error that when they set the barrel at a 45 degree angle, this would result in the maximum distance for the cannonball. Going either higher or lower than this would reduce the distance the ball traveled. A chart could be kept giving the distance that each angle would produce. If the soldiers were able to determine how far away their target was, they might be able to hit it with reasonable accuracy.

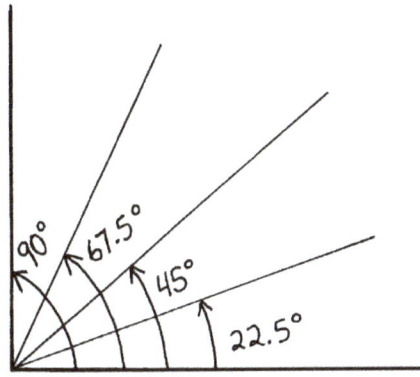

Angles are measured from the ground. Straight up is 90 degrees. (A complete circle is 360 degees.)

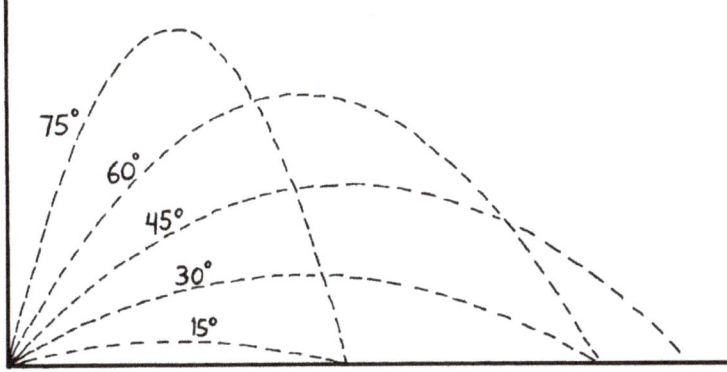

The dashed lines show the parabolic shape that each angle will produce. You can see that the 45 degree angle will result in maximum distance.

Once you raise the barrel and start firing at an angle instead of horizontally, the math becomes more complicated. Due the fact that you are dealing with angles, you have to use a type of math called trigonometry. Since trigonometry is outside the scope of this book, you'll be glad to know that you won't be asked to solve these more difficult problems. Instead of more math problems, we're going to suggest two activities. The first is an online simulation where you can adjust angle and velocity and see how this affects the resulting motion of the cannonball. The second will be more hands-on and will involve you building and operating a small "cannon."

ACTIVITY 8.9:   Online interactive simulation of projectile motion

This link will take you to a simulation where you can experiment with projectile motion. (The simulation does all the math.)

https://phet.colorado.edu/en/simulations/projectile-motion

ACTIVITY 8.10: Experiment with a "cotton ball cannon"

You will need: one long cardboard tubes or two short ones (e.g. one paper towel tube or two toilet paper tubes), a pencil, two medium-sized thin rubber bands, a marker, scissors, a paper punch (if you have one), duct or masking tape, cotton balls (or pom-pom balls found in craft stores—we found that pom-poms worked better than cotton balls), your accelerometer (from activity 6.6), a stack of books, your meter sticks, and some paper bowls or cups to be targets

If you don't have cardboard tubes, you can make some sturdy tubes by rolling up two sheets of card stock and securing with tape.

NOTE: This is a standard STEM project. If you need a video about assembly, search YouTube for "cotton ball cannon."

How to make the cannon:
1) If you are using a paper towel tube, cut it in half to make two tubes. (If rolling your own tubes, make two tubes that are about 12 to 15 centimeters long (5 to 6 inches. Make sure they are very sturdy, as sturdy as a cardboard tube. Make one tube smaller than the other and skip step 2.)
2) Cut one tube lengthwise. Roll it a bit to decrease the diameter, then secure with tape.
3) Punch two holes in one end of this narrower tube.
4) Insert pencil into holes.
5) Use the marker to make lines every centimeter from the end of the tube.
6) Cut slits in the other tube. Insert rubber bands into slits.
7) Place smaller tube inside larger tube and loop rubber bands around the ends of the pencils.
8) Tape your accelerometer to the top of the cannon. (Shown in top photo.)

1

 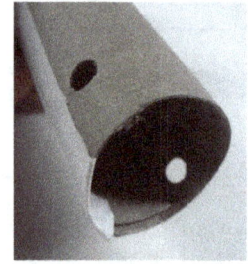 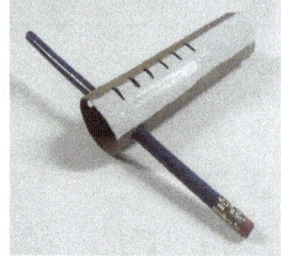 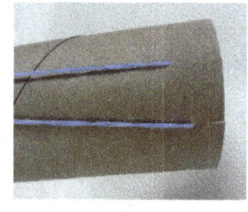 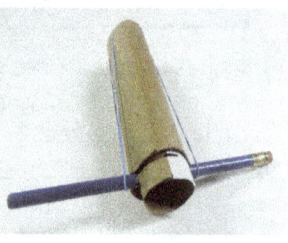

2 **3** **4 and 5** **6** **7**

PART 1: Learn how to use the cannon:
NOTE: Using the cannon is much easier if you have a helper. Sometimes four hands are better than two!
1) Place a stack of books on the floor (or on a table). You will place the center of your cannon on the edge of the stack. This will help you to hold the cannon still so that your shots will be consistent (the same every time).
2) Place the meter sticks, end to end, in front of the stack of books. This will let you measure how far your cannonball goes. If your cannonball goes further than the sticks, just move them farther away from the stack of books.
3) Place a cotton ball (or pom) into the cannon. Make sure it slides down into the cannon as you pull back on the pencils.
4) Decide at what angle you will shoot and tip the cannon until the meter reads that angle. Then decide if you want the acceleration to be small, medium or large. You control this by pulling the inner barrel out until one of the marks is even with the edge of the outer barrel. Pulling it out a lot will create more acceleration. (This is not a very accurate way to control acceleration, but it is the best we can do.)
5) Keeping the barrel steady, let go of the pencils. Notice where the cannonball lands and read the number of centimeters.

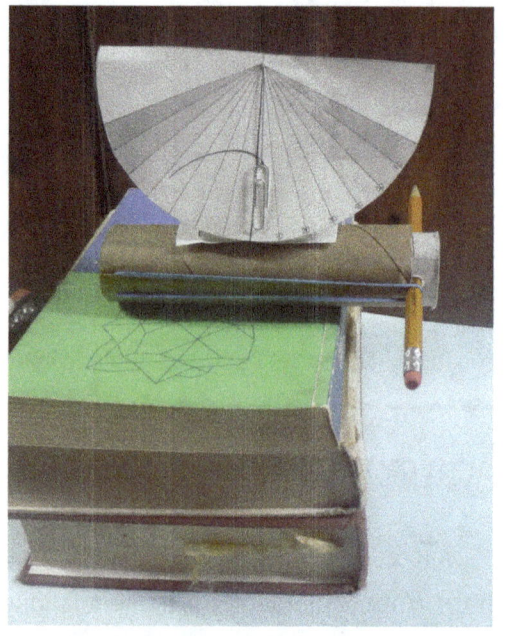

Answers to 8.8: 1) 198.45 m 2) 474.44 m 3) 4 m/sec 4) 10.5 m, so yes he will miss the rocks

PART 2: Make a chart to keep track of how both acceleration and angle of incline affect distance:
You will experiment with 3 accelerations (low, medium and high) and 3 angles (your choice).
1) Shoot a few experimental shots and determine what units to use on your graph below. What will one block represent? (The strength of your rubber bands will be the biggest factor in how far your balls travel. Weak bands will give you results between 1 and 50 cm. Stronger rubber bands will give results from 20 to 200 cm.)
2) Start with your lowest angle. Shoot balls at this angle using low, medium and high force. Make sure to use the same marker line on the barrel each time. (For example, low acceleration might be the line 1, medium at line 3, high at line 5. To get more accurate data, shoot at each setting several times and average the results. Record these results on the graph below using a color code for the accelerations. You will want to watch carefully where the ball lands. The landing spot is what you want to measure. The ball might roll a bit, but we're not going to count the roll in our measure, only the landing spot. (NOTE: You can skip doing the 40 and 50 degrees angles and just do 45 degrees instead. It's up to you.)
4) Do this again for each angle you try. Do at least 3 angles.
5) Draw lines connecting your same-color data points. (Sample graph is shown in answer key.)
6) Analyze your chart. Which combination of angle and accleration gave the greatest distance?

DISTANCE COTTON BALL TRAVELED

PART 3: Hitting a target:
1) Choose a force and an incline angle. Load a cotton ball and shoot. Notice where it lands and place a cup or bowl at that spot.
2) Shoot again, trying to duplicate exactly what you did last time. Did you hit the target? See how many times you can hit the target without missing.
3) Now place a cup or bowl in a random location and see if you can hit it. With all the experience you've gained, your brain might be getting pretty good at this!
3) If you have other classmates or siblings available, try making this into a competition. You can decide on the rules.

ACTIVITY 8.11:   Review questions from this chapter (Answers are in teacher's section.)

1) TRUE or FALSE? By Descartes' day, universities were no longer teaching Aristotle's ideas.
2) TRUE or FALSE? The Cartesian grid has both positive and negative numbers.
3) Descartes died while serving the queen of what country? _____
4) Descartes is most famous for saying this short phrase: "_____." (pg. 58)
5) Which word fits this definition? "The amount something moves in a certain direction"
   a) distance   b) velocity   c) displacement   d) acceleration   e) rate   f) speed
6) TRUE or FALSE? Velocity is a vector.
7) What two things does a vector have? _____ and _____
8) The area under a line on a velocity-time graph equals:
   a) total time   b) distance traveled   c) velocity   d) acceleration
9) TRUE or FALSE? If you multiply both sides of an equation by the same number, the equation doesn't change.
10) What was the tallest building during Galileo's lifetime? _____
11) TRUE or FALSE? A ball dropped straight down from a tower will hit the ground sooner than a ball thrown out (and away) from the tower.
12) What angle will give a cannon the longest range? _____

ACTIVITY 8.12:   Which formula? (Answers are in teacher's section.)

You don't have to solve these problems. All you have to do is determine which formula you would use. Write the letter of the correct formula in the blank after each number.

(A) $F = ma$ (B) $F_g = \dfrac{G(m_1)(m_2)}{r^2}$ (C) $a = \dfrac{v}{t}$ (D) $d = rt$ (E) $p = mv$ (F) $F = \dfrac{\Delta p}{\Delta t}$ (G) $d = \tfrac{1}{2}gt^2$ (H) $t = \sqrt{\dfrac{2d}{g}}$

1) _____ A marble has a mass of 10 grams. When the marble rolls of the end off a ramp is going 1 m/sec. How much momentum does the ball have?

2) _____ If a penny dropped from the top of a tower reaches the ground in .67 seconds, how tall is the tower?

3) _____ What is the strength of the gravitational force between two planets who have the same mass as the Earth?

4) _____ The Shanghai tower is 632 meters tall. If you drop an apple from halfway up the tower, will it hit the ground in less than 5 seconds?

5) _____ A car goes from zero to 50 km per hour in 10 seconds. What is its acceleration?

6) _____ A runner completes a 10 km race in 20 minutes. What was her speed (rate)?

7) _____ A 50-kg satellite is launched with a force of 83 newtons. What is its acceleration?

8) _____ A truck's momentum goes from 10 kg(m)/sec down to zero in 1/5 seconds as it hits a huge pile of gravel. Now many newtons of force was the collision?

9) _____ A tennis ball has a mass of about 55 grams. If the ball is coming towards you with a momentum of 5.5 kg(m)/sec, what was its velocity as it left your opponent's racket?

10) _____ An ant can travel at a rate of 10 cm per minute. If the ant walks for 7.5 minutes, how far does it go?

11) _____ If an egg falls off a 50-cm high kitchen counter. How long does it take for the egg to hit the floor?

12) _____ What is the strength of gravity between you and your dog?

Adventure 9: Round and round we go

To begin our study of circular types of motion, we're going to head back to Galileo's studio again.

A cycloid is the curve that is created as you trace the motion of one point on the outside of a circle as the circle is rolling. The best way to understand how cycloids are made is to watch an animation. If you go to the Wikipedia article on cycloids, you can watch a GIF animation that repeats over and over again so you can watch it as many times as you want to. Here are some snapshots from the animation:

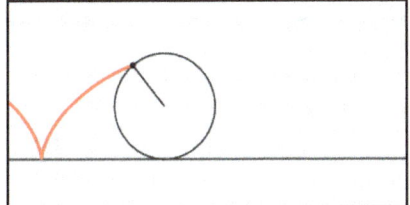 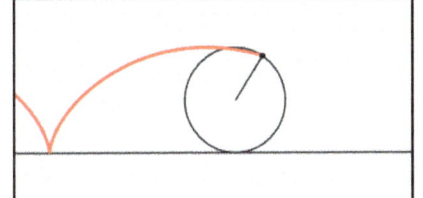 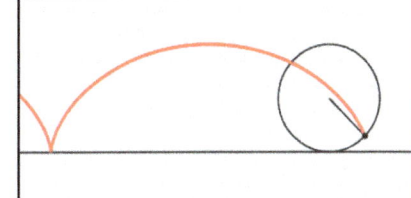

You are welcome to try this yourself, but cycloids are actually a bit tricky to make. It is difficult to keep the pencil or pen point in one place on the outside of the circle while you are also trying to make the circle roll smoothly. You might need someone to roll the circle while you control the pencil/pen. You can try to drill or poke a tiny hole for the pen point but it really has to be at the very edge of the circle. Using a large circle made of corrugated cardboard (or foamcore) might give the best results. A hole is more easily drilled in a thick material.

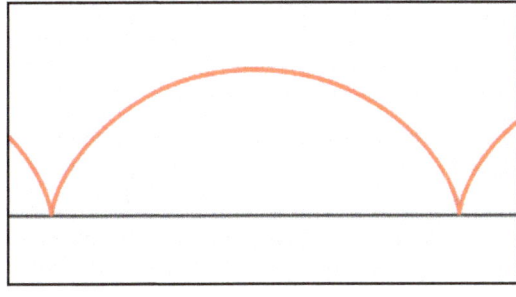

Galileo wanted to find out the relationship between a cycloid and the circle that created it. He guessed that he might find something interesting if he compared the area of the circle to the area under the cycloid curve. In future years, mathematicians would use calculus to tackle investigations like this, but in Galileo's day, calculus had not been discovered yet. (Remember, Newton was one of the inventors of calculus and he was born in the year that Galileo died.) However, Galileo liked to solve problems in very practical ways, so he drew circles and their cycloids on sheet metal and cut out the shapes. Then he weighed the shapes. To his surprise, he found that one cycloid weighed the same as three of its circles. This seemed odd to him. He had not expected to get a nice, even number like "3." This seemed to good to be true, so he assumed that imperfections in his workmanship had created this strange result. It just did not seem possible that the answer could be as simple as "3." However, this result would be confirmed as correct by future mathematicians.

So is a cycloid good for anything? Is it useful in physics or engineering? Amazingly enough, this shape turns out to have some very interesting properties (most of which Galileo never discovered). If you turn a cycloid upside down and cut it in half, you've made a curved ramp that represents the fastest route a rolling object can possibly take. If you make ramps of various shapes (and all the same length) and roll balls down them, the ball on the cycloid ramp will win every time! Even more curious, if you make two cycloid ramps and put them side by side, you can start balls at different points on the ramp and yet they will reach the bottom at the same time. As strange as this sounds, you can start one ball at the very top of one cycloid ramp, and one ball very close to the bottom of the other cycloid ramp, and they'll still hit the bottom at the same time! You can discover this for yourself by watching some videos and by making your own cycloid ramps.

ACTIVITY 9.1: Watch a few short video about cycloids

The YouTube playlist has some short videos about cycloids. A few of the videos show "ball races" with various ramp shapes, as mentioned in the text. If you can't make your own ramps, these videos will let you see these races.

ACTIVITY 9.2: Make your own cycloid ramps (optional)

If you'd like to make some small cycloid ramps out of cardboard or foamcore and do your own ball rolling experiments, there are patterns and instructions in the teacher's section. But you don't have to do this project right now. If you've watched the videos, you can go ahead and continue with the text.

Another interesting property of cycloids is their relationship to pendulums. If you go back and look at activity 3.2 on page 40, you'll remember that although we discovered that pendulums basically maintain the same frequency no matter how high or low their swing is, they do go a bit faster when the swings are very high (meaning when the amplitude is very large). Cycloids to the rescue! If you hang the pendulum between two cycloid curves, when the pendulum swings very high it will bend to follow the curve. Look at the purple and blue lines. Can you see how the string is bending to follow the cycloid curve? This bending slows the bob down just a bit, and the result is that the very high swings end up being exactly the same frequency as the small ones. If all of these pendulums have reached their maximum amplitude and are ready to swing back to the left again, they will all cross the center line at exactly the same moment. (We call them **isochronous** pendulums. "Iso" means "same," and "chronos" means "time.") It's hard to believe that the red and purple pendulums are swinging at the same frequency, but you can see this for yourself if you go to the Wikipedia article on cycloids. You can see these pendulums moving.

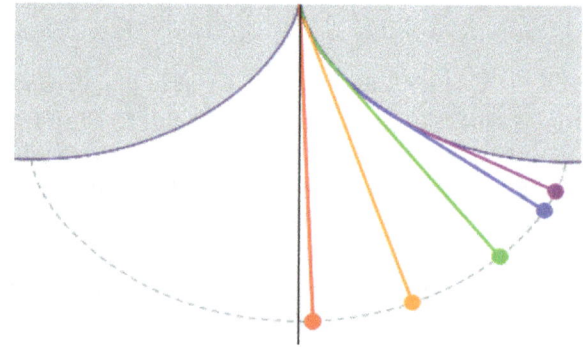

For physicists, the most interesting aspect of the cycloid curve is (as we have mentioned) the fact that it represents the fastest route for a downward rolling object. (As an interesting side note, Galileo did a lot of work with ramps, and we know that he wondered what shape was the fastest, but he incorrectly guessed the fastest shape would be a quarter-circle.) In this diagram, the marble on the red (cycloid) track will win even though it dips down. This snapshot is from the Wikipedia article on the **brachistochrone** curve. *(bra-KIST-o-krone)* You can go to this page if you want to see the animation. The name "brachistochrone" (from a Greek word meaning "shortest time") refers to a cycloid curve that consists of all, or most, of the complete curve. Just half the curve (for example, both curves in the pendulum diagram) are called **tautichrone** curves. ("Tauti" means "same.") If you like math connections, you might be interested in the fact that you can calculate the time it will take for a ball to hit the bottom of the curve: $t = \pi \sqrt{r/g}$ (where r is the radius of the circle that made the curve, and g is 9.8 m/sec²)

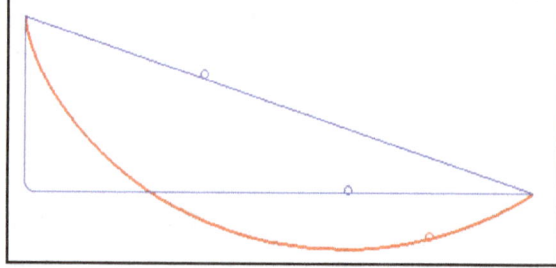

For non-physicists, the most interesting aspect of cycloids is probably the fact that they are part of a larger group of shapes called "trochoids." If you've ever seen Spirograph® designs, you've seen trochoid shapes. A trochoid is any shape that is made by rolling a circle along a surface, whether the surface is straight or curved. In Spirograph®, circles are rolled around other circles, and you can choose where the tracing point is on the circle, as they give you a selection of tiny holes where you can place a pen point. This toy was created in the 1960s, and the set shown here is a "vintage" set that is at least 50

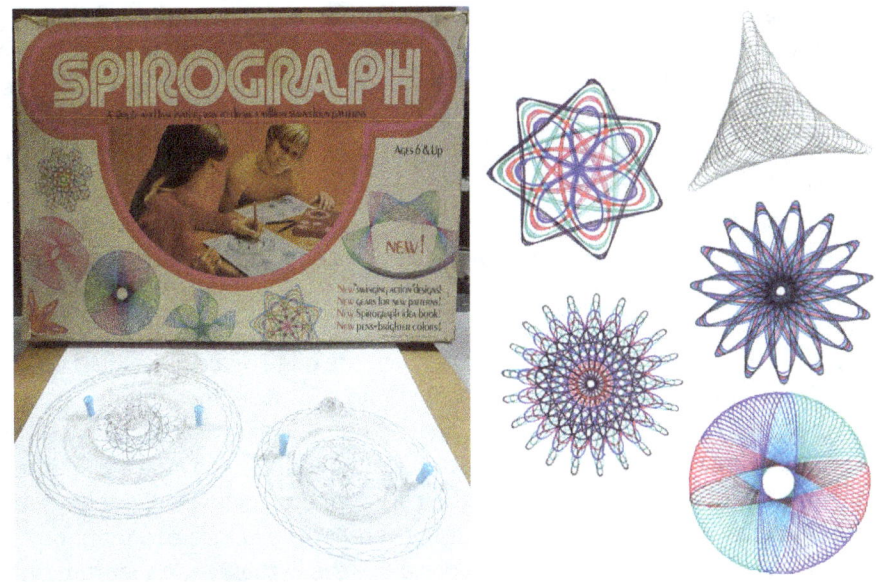

years old. Those blue things are pins holding the rings in place. Yes, back then toy makers trusted kids with pins. Newer Spirograph sets give you sticky dough instead of pins, which is a shame because the pins worked really well.

Now let's take an imaginary field trip to an amusement park. Have you ever ridden on a ride similar to the one shown in this photo? When the ride is stopped, the swings hang straight down so the riders can get onto the swings. Then, as the ride spins around faster and faster, the riders begin to go not only around in a circle but also up and away from the center of the ride. The faster the ride spins, the higher the riders go. Once they get to the maximum height that the ride will take them, the riders experience **uniform circular motion**.

As with the other types of motion we've studied, uniform circular motion is governed by principles (laws) and equations. The equations for circular motion are very similar to those we've already seen for "translational" (non-circular) motion. However, circular motion is different enough from translational motion that physicists have come up with new terms for many of the concepts we've studied. So, in this adventure, we'll be reviewing many of the concepts we've already learned, but putting a new twist on them (pun intended).

Our current understanding of circular motion began with Isaac Newton's thoughts about the moon. He had already realized that the invisible force of gravity is what keeps the moon from flying off, so he isn't

contemplating gravity in this picture. He is wondering if circular motion might <u>always</u> involve an inward force that pulls the object towards the center of the circle. The moon might be giving him a clue about how all circular motion works. In some cases, the source of the force is obvious. If you put a key on a string and twirl it above your head, the string pulls on the key. But what about a horse running on a circular race track, or a bird circling in the air?

Newton then got out his notebooks and began using his knowledge of geometry to try to solve the problem of circular motion. (Newton's proofs for his laws of motion came from mostly from geometry, not from experiments. His book, *Principia*, looks like a geometry book, not a physics text.) His line of reasoning involved imagining the path of a ball inside a metal ring to be a polygon (pentagon, hexagon, octagon, etc.) that produces a pair of forces every time it bumps against the inside of the ring. One of these forces is a push towards the center. He then imagines the polygon having more and more sides until it turns into a circle. But, he reasons, those pushes to the center would still be there! Don't worry if this doesn't make perfect sense—Newton's reasoning isn't easy to follow. So let's skip right to the end and discuss Newton's final conclusion about circular motion.

Newton's conclusion about circular motion involved a type of line called a **tangent line.** Tangent lines are lines that touch the edge of a circle at just one point. Line AB and line segment CD are both tangent to this circle. By definition, tangent lines are perpendicular to the radius of the circle. Perpendicular means that two lines form perfectly "square" corners that are exactly 90 degrees. We can indicate "perpendicular" in a diagram by putting a little square in the corner of the angle. (Tangent lines are helpful to mathematicians because if you connect the center of the circle to a point on the tangent line, you get a right triangle, allowing you to use the Pythagorean theorem ($a^2 + b^2 = c^2$).)

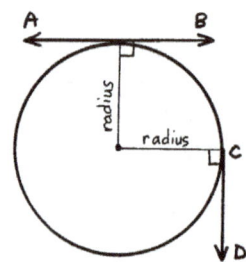

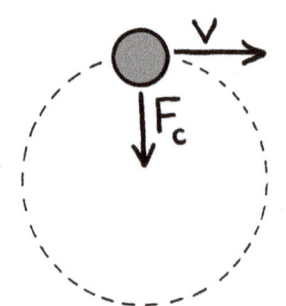

Newton said that objects traveling in a circle always have two forces operating on them. The force that it directed towards the center of the circle is called the **centripetal force.** "Centri" means "circle," and "petal" comes from a Greek word meaning "to seek." The way you pronounce centripetal will depend on where you live. In the U.S., you'll say "sen-TRIP-eh-tull," but in other places you might say "sen-tri-PETE-ull." This force is usually represented by a capital letter F with a small letter "c" as a subscript.

The other force acting on the object is a linear velocity that is always in a direction that is tangential to the circle, and usually represented by "v" for velocity. **It is the combination of these two forces that creates circular motion.** If one of these force suddenly disappears, the remaining force will be the only one acting on the object, causing it to travel in that direction. Thus, Newton answered his question about what would happen if gravity suddenly disappeared. Gravity provides the centripetal force, F_C. If this centripetal force stopped pulling, the moon would go flying off into space, following a straight path (v). Fortunately, it's impossible for gravity to disappear!

ACTIVITY 9.3: Observe what happens when centripetal force disappears suddenly

You will need: *a piece of string at least a meter long, a small object that can be securely tied to the string*
TIP: Choose an object that won't can any damage if it hits a person or a window.

1) Tie the object to one end of the string.
2) Whirl the object over your head so that the string is taught and the object is traveling at a constant velocity.
3) Get ready to make an observation! Let go of the string and watch the path that the object takes. Does it continue in its circular path? Does it fly off in a straight line?

Without the centripetal force acting on the object through the tension on the string, the object will have only its velocity vector left and it will travel in that direction. The direction and magnitude of this velocity vector won't change unless acted upon by an outside force. If you did this experiment in outer space, the object would keep traveling in that direction at that speed until it reached the end of the universe (if it has an end) or until it crashed into a planet or star.

Now for a fun "rabbit trail." (A rabbit trail is when you discuss a side issue that is related to what you are talking about, but it quickly becomes a topic until itself and often presents even more side issues that are interesting. Before you know it, you've wandered far away from your original topic.) This side issue concerns a YouTube video in which a researcher posed this question: Is it possible to create a situation where an object in uniform circular motion <u>doesn't</u> immediately fly off on a tangent?

The YouTube video is titled: "Most Mind-Blowing Aspect of Circular Motion" and was posted by "All Things Physics." The video starts out with some technical stuff, but the main section of the video involves watching Slinkies falling in slow motion. It is well worth searching a video platform for "Slinky falling in slow motion." A Slinky is held up high and allowed to stretch downward to its maximum length. When the top of the Slinky is dropped, the bottom stays right were it is—looking like it is defying gravity—until the shrinking top finally reaches the bottom of the Slinky. Then the Slinky falls to the ground as expected. The researcher then attaches a ball to one end of the Slinky and re-films the falling Slinky with the ball at the bottom end. This makes the effect even more striking. The ball seems to defy gravity as it hovers in the air, waiting until the Slinky finishes recoiling from the top down.

Then the circular motion aspect is incorporated into the demonstration. The researcher ties the ball/Slinky onto a machine that is designed to spin things around and then suddenly release them. He films this experiment using an overhead camera so you can see exactly what happens to the ball. After the Slinky is released, the ball continues on its circular path while the Slinky is recoiling. It is only when all the stretching is gone and the Slinky becomes compact again that the ball begins to go off on a tangent line. While the Slinky is recoiling, the ball continues along in its circular path.

What this demonstration shows, is that the "law" that says that objects always go off on a tangent line when they are released, assumes that the source of the centripetal force, F_c, is rigid, not elastic. The video shows the same demonstration with progressively less stretchy cords and you can see that the ball leaves the circular path more quickly. If the string had no stretch at all, the ball would immediately go off on the tangent.

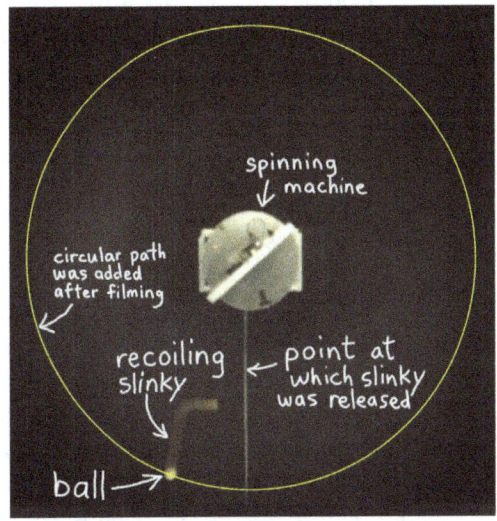

snaphot from the overhead camera view

ACTIVITY 9.4:   Do the falling Slinky demo (or watch slo-mo videos)

Why not try the dropped Slinky demo for yourself? (If not, check out Veritasium's falling Slinky videos on YouTube.)
TIP #1: It is very easy to over-stretch a Slinky. If you try twirling the Slinky in a circle to see if you can replicate the circular path demonstration, be very careful not to go too fast.
TIP #2: Most smart phones today have cameras that can film in slow motion. Why not try filming your dropped Slinky, then watching it on replay?

That's the end of our rabbit trail. Now back to boring physics. We need to talk more about the velocity vector.

As we've mentioned several times in previous adventures, the definition of acceleration is "a change in velocity." Hopefully, you'll also remember that velocity is a **vector** quantity which means it has both magnitude (speed) and direction. "Speed" doesn't have a direction. If we say that a car is going 50 km per hour, that does not tell us in which direction the car is traveling. However, if we say that a car has a velocity of 50 km per hour, we must also specify direction, such as north or southeast. So far, we haven't done much with the direction of velocities. In most of our discussion of velocity we were concentrating on the magnitude. Many of our discussions have focused on the acceleration due to gravity, and the direction was so obvious (down toward the earth) that we did not have to continually mention it; it went without saying. However, direction becomes very important when discussing circular motion because that tangent line is continually changing direction.

If the black dot in this diagram is a tiny moon orbiting a small planet, you can see that the moon's tangent velocity is constantly changing direction as it travels in a circle. We've only drawn six of these tangents, but in reality there would be an infinite number of them.

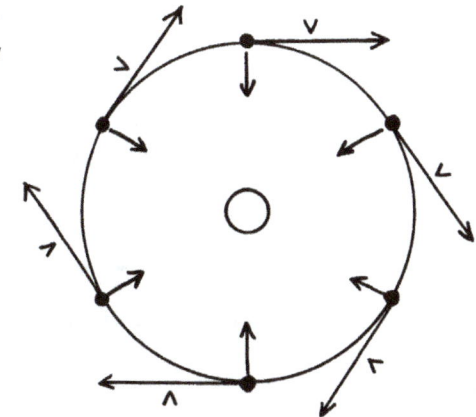

133

In uniform circular motion, we have a change in the direction of the velocity, but not in the magnitude. Only one of these (magnitude or direction) needs to change in order to qualify as acceleration, so in uniform circular motion we have constant acceleration. In linear motion, constant acceleration would mean that the object keeps going faster and faster, with no limit to its acceleration. Technically, this would mean that the object would eventually end up going the speed of light! However, this is a hypothetical situation that we don't see actually happening. Normally, objects accelerate until they reach their maximum velocity, then they maintain this velocity with no more acceleration. When a car pulls onto a highway, the passengers experience acceleration at first, and feel like their backs are being pressed against the seat, but once the car reaches cruising speed, acceleration stops and the passengers no longer have that pressed-back feeling. This is a huge difference between linear and circular acceleration. In circular motion we can experience continuous, never-ending acceleration.

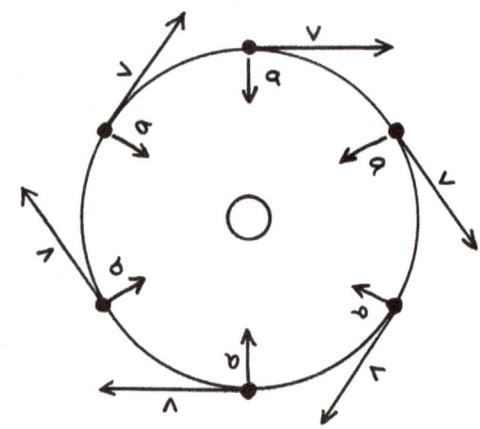

Notice that we've added the letter "a" to the inward centripetal force arrow. We can replace "F_c" with "a" because they are essentially the same thing. You might see either notation in a physics diagram.

Is there a way for a person to experience the feeling of continual acceleration? Yes, certainly! Let's go back to our imaginary field trip to the amusement park. Many parks have a ride that spins you in a circle. Some are called "Gravitron" or "Rotor" but they can go by other names, as well. These rides don't have seats. The riders simply stand against the inside wall of a giant cylinder. As the ride begins, the riders will at first feel sideways acceleration as the cylinder gradually speeds up. Then at maximum velocity, the riders will only be aware of the centripetal force pushing them towards the center of the circle. To the rider, the centripetal force will feel like they are being pressed backwards into the wall behind them. The centripetal force is so strong that the riders can move up on the wall, away from the floor. They look and feel stuck to the wall.

By Saberwyn at the English-language Wikipedia, CC BY-SA 3.0, https://commons.wikimedia.org/w/index.php?curid=4657162

Yes and no. Scientists have been discussing this (even arguing about it) for centuries. In the 1700s, Newton was open to the idea that centrifugal *(sen-TRIF-uh-gul, or sen-tri-FUG-al)* force was the contact force created by centripetal force, forming a force pair. The riders in the Rotor ride certainly feel a contact force, as their bodies apply an equal and opposite force against the inward centripetal force. However, after years of debate, modern physics has finally decided that centrifugal force is a "fictional" force, and the real force is centripetal.

ACTIVITY 9.5: Watch a few videos about "centripetal" versus "centrifugal" force

The YouTube playlist has some videos about this topic.

In common speech, it isn't wrong to talk about centrifugal force. This word works very well for describing how things seem to be pressed outward when spun in a circle. Centrifugal force is used in many machines, for a variety of purposes. For example, old-fashioned steam engines had a part called a "centrifugal governor" that consisted of a pair of metal balls that were connected to a device that controlled the speed of the engine. The engine would make the balls spin around, and, just like the riders on the swing ride, the balls would go not only around, but also upwards. When they reached a certain height, this would trigger a switch to slow the engine down. The balls would begin to fall down as the engine slowed. When they got very low, another switch would turn the engine on again.

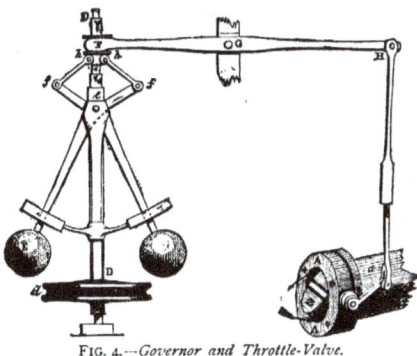

FIG. 4.—*Governor and Throttle-Valve.*

Another machine that uses centrifugal force is a centrifuge. This machine holds liquid samples, usually in test tubes, and spins them at an incredibly high rate. The outward centrifugal force is so great that heavier or more dense particles are pressed to the bottom of the tube and lighter things stay at the top. When the tube is taken out of the machine, the substances in the liquid have been sorted into layers. Some blood tests are performed using a centrifuge. The red and white cells can be separated from the watery plasma.

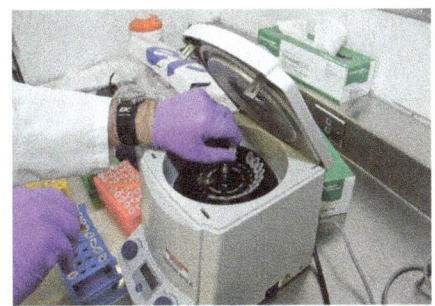

Because circular motion causes constant acceleration, large spinning devices can be used to experiment with how much linear acceleration a human body can withstand inside rockets and jets, or on a roller coaster. Fighter jets can go so fast that the pilot experiences acceleration up to 9 times the force of gravity (9 G's). During take-off, astronauts experience considerable G forces. Just like blood samples in a centrifuge get pressed into the end of the test tube, acceleration can cause your blood to get pushed to one end of your body. If most of your blood ends up in your legs and feet, that deprives your brain of blood and you lose consciousness. Pilots found this out the hard way in the early 1900s, before people knew very much about what high G forces to do bodies. Pilots would black out and then their planes would crash.

This photo shows the 20-G centrifuge at the NASA Ames Research Center. Each end (red and blue) has a tiny "room" where someone can sit. The entire machine spins around, which is why the room is so large. Scientists can hook all sorts of monitoring devices to the volunteers, to keep track of how their bodies are responding to the extreme G forces. Most people can handle 3 to 4 G's, so this is the amount of acceleration you'll meet on a roller coaster ride. However, pilots must be able to handle at least 5 to 6 G's and sometimes more than that. The upper limit that even a trained pilot can handle is about 9 Gs, although black out can occur even at 6-7 Gs. Pilots who fly very fast military planes wear a special G-suit that has inflatable pouches in various places. When the plane is about to do a maneuver that will result in high acceleration, the air pouches inflate and press on the pilot's body in places that will prevent the blood from draining out of his head. However, too much blood is also bad, and this must also be prevented. The eyes are the first thing to be affected by G force. If a pilot notices his vision starting to go, he needs to immediately make adjustments so he doesn't black out.

The big take-away from the previous page is that the actual, "real" force in circular motion is centripetal, not centrifugal, and no matter how it looks or feels, centripetal force is directed <u>inward</u>, toward the center of the circle. The Rotor ride cylinder is pushing the riders towards the center of the ride, and they feel this pressure on their back. Even though the ball on the string that you are whirling around your head feels like it is pulling on your hand, the real force is the tension on the string, pulling the ball towards your hand. Perhaps the hardest examples to understand are liquids, such as the water in your bucket or the blood samples in the centrifuge. We can see the liquids being pressed to the bottom of the bucket or the tube, looking very much like they are experiencing an outward force. However, are just obeying Newton's First Law, trying to keep on doing what they are doing. If left to their own devices, the fluids would keep on traveling along the tangent line of velocity, away from the circle. But the end of the tube or bucket keeps getting in the way.

Just remember that when we are talking informally, not in a scientific situation, it's fine to talk about centrifugal force. People use that word all the time and we understand exactly what they mean. However, if we are in a physics classroom, we need to remember to temporarily delete the word "centrifugal" from our vocabulary.

ACTIVITY 9.7: Try the "penny on a coat hanger" trick (using the fictitious centrifugal force!)

You will need: a penny, a coat hanger (a wooden one might work better than a wire one)

1) Balance the penny on the bottom of the coat hanger, as shown.
2) Hook the hanger part over your finger and carefully give the hanger a twirl around your finger. Can you twirl fast enough to keep the penny on the hanger? (This is similar to the bucket of water trick, where the water stays in the bucket as you whirl the bucket in a circle.)

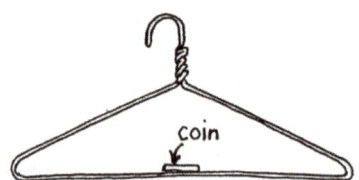

ACTIVITY 9.8: Watch what centripetal force does to water

If you can take the time right now, check out the video on the playlist that shows a clear aquarium rotating on a spinning platform. The water in the aquarium creeps up the walls. Newton observed this happening to water in buckets as he spun them around. What Newton couldn't do was watch the playback in slo-mo. But you can!

You won't be surprised to find out that there is a way to calculate the amount of centripetal force in a given situation. All we need to know is the radius of the circle, the mass of the object, and the velocity at which the object is traveling. It's a fairly simple calculation. As with all calculations of force, the answer will be in newtons. Force is always measured in newtons (kg m/sec²).

The formula for centripetal force is: $$F_c = \frac{mv^2}{r}$$ The lower case "c" after the F stands for "centripetal."

Here is a sample problem that you can solve using this equation:
1) A student is swinging a ball on a string. The length of the string is 50 centimeters and the mass of the ball is 100 grams. The velocity of the ball is 3 meters per second. The string is pulling on the ball with how much force?

First, convert centimeters to meters, and grams to kilograms. 50 cm= .5 m. 100 g = .1 kg
Now plug your information into the equation:

$$F_c = \frac{(.1 \text{ kg})(3 \text{ m/sec})^2}{.5 \text{ m}} = \frac{(.1 \text{ kg})(3 \text{ m/sec})(3 \text{ m/sec})}{.5 \text{ m}} = 1.8 \text{ newtons}$$ (If you want to see what happens to all the units, see the answer key.)

The centripetal force pulling on the ball is 1.8 newtons. Since newtons are "kg per meter per second per second," the "per second per second" part tells us that this is an accelerating force. We know that circular motion produces constant and continuing acceleration, so our answer makes sense.

What would happen if we did not know the velocity of the ball? Could we still do the problem? Well, if we knew a few other facts, we might be able to calculate the velocity of the ball. We know that d=rt, and "r" is "rate," which can also be called velocity. So if we know the distance that the ball has traveled, and the time it took to travel that distance, we can figure out the velocity (rate).

Let's say that the student counts how many seconds it takes for the ball to make 20 circles around their head. We've already said that the length of the string is 50 centimeters, or .5 meter. That makes the diameter of the circle that the ball is traveling in 1 meter wide (because radius is one half of the diameter). We know that the circumference of a circle is the diameter times the number "pi" (3.1415...) so we can calculate the circumference of this circle:

Circumference = πd = (3.14)(1 meter) = 3.14 meters

Every time the balls goes around the circle, it travels a distance of 3.14 meters. If the ball goes around the circle 20 times, it travels a distance of (20)(3.14) = 62.8 m.

If the student uses a stopwatch to time the ball and finds out that it takes 10 seconds for the ball to go around 20 times, this means that the balls travels 62.8 meters in 10 seconds. Using d=rt. we have 62.8=(r)(10). If we divide each side by 10, we have 6.28 = r. The rate (which we are calling velocity) is 6.28 m/sec. Thus, sometimes you are given information that isn't exactly what you need in the formula, but the information can be used to find out what you need.

Before we start our next activity, let's think about this formula for a minute. $F_c = mv^2/r$
We know that when working with regular fractions, if we increase the bottom number, the value of the fraction gets smaller. 1/2 1/3 1/4 1/5 1/6 1/7 1/8 1/10 1/100   If we increase the value of the top number, the value of the fraction gets larger. 1/4 2/4 3/4 4/4 8/4 12/4 40/4

Looking at our centripetal force formula, what happens to the force, F_c, when the radius of the circle gets larger? That means the value of "r" on the bottom gets larger. Will F_c get larger or smaller? Then think about increasing the mass (m). Will this increase or decrease the value of F_c? What about velocity? If we increase the velocity (v) of the spinning object, will Fc increase? What if we increased both the mass AND the velocity?

ACTIVITY 9.9: Experimenting with centripetal force

You will need: a piece of thin string about a meter long, a cheap (non-clicking) ballpoint pen that you can disassemble, one small washer (3/16" size is fine), 5 large washers (5/16" size is best, but 1/4" will also work), two twisty-ties (short, thin wires), a black permanent marker, scissors, a meter stick, a stopwatch, a pencil, four colored pencils

<u>How to set up:</u>
1) Take the "guts" out of the pen so you have a hollow tube. The disassembly process will depend on what type of pen you have. (Some pens will leave a rough edge at the top. You might have to trim and sand it a bit to make it smooth.)

2) Make a slip knot on one end of the string, push the twisty-tie through, tighten, then twist the tie to secure it. Then do this on the other end, also. Now it will be very easy to attach washers using the twisties.

3) Push one of the twisty-ties through the pen tube, so that the pen tube is in the middle of the string.

4) Attach one small washer to the twisty-tie on the end of the string that is coming out of the wide end of the tube. Attach two large washers to the twisty-tie on the other end of the string--the end that is coming out of the pointed end. (We are going to start with two large washers because they will be easier to work with than just one. One is harder to spin.)

(2)   (3)   (4)

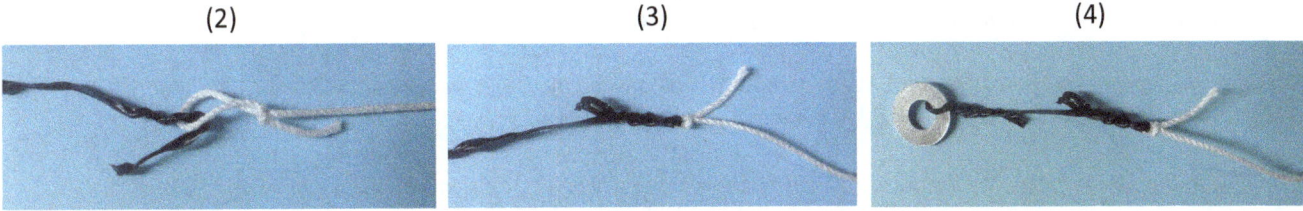

5) Measure from the **center** of the small washer, and use the black marker to make marks on the string at these intervals: 12 cm, 24 cm, 36 cm, 48 cm.
6) Your final set-up should look something like this:

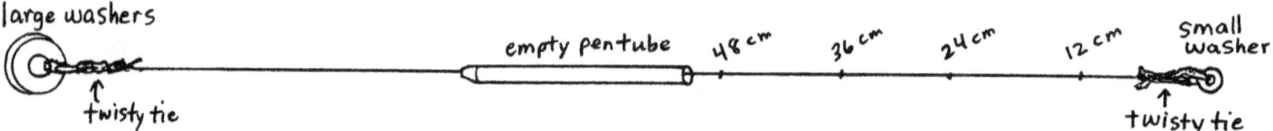

If you have trouble with the string catching on the top of the pen when you spin it, try reversing the orientation of the pen.

<u>What do :</u>

1) First, have some fun experimenting with this device. Hold the pen tube with one hand. Hold it so that the string with the small washer is on top. Keeping the tube up over your head, begin twirling the small washer around in a circle. The two large washers will be hanging down out of the bottom of the tube. Adjust the string so that one of the marks is right at the top of the tube. Twirl the washer at a constant rate so that the mark stays right there at the top of the tube without you touching the string. (You'll need to do this a lot, so get good at it!)

NOTE: We are starting with two washers instead of one, because two washers are actually easier to do the twirling with than just one. One washer gives you a low enough centripetal force that it's harder to keep the small washer spinning at a steady rate. Save the one washer for your last trial; by that time you'll be a washer-swinging pro!

2) Try pulling down on the bottom string. Stop twirling while you pull down. Does the top washer keep going in a circle? Does it appear to speed up or slow down?
3) Try letting the radius increase as you twirl. Do the bottom washers rise?

<u>Now let's start experimenting</u>:   **NOTE: When we use the word velocity in this lab, we'll be referring to just magnitude (speed).**

4) We are going to start with the 48 cm mark (with a two-washer string). You'll need to figure out the velocity of the small washer, and we'll use the method we described at the top of the previous page. The radius of this circle is 48 cm, so the diameter is 96 cm. Convert it to meters: .96 m. So the circumference of the circle is πd = (3.14)(.96m) = **3 m.** That means that every time the small washer goes round in a circle (keeping the 48 cm mark right at the top of the tube) it travels a distance of 3 meters. 3 will be an easy number to work with. The washer will go around the circle 20 times. Therefore, the total distance the washer will travel is 20 x (3m) = **60 m**. (Notice in charts on the next page, the distance of 60 m has been filled in for you.)

5) Operating the stopwatch yourself will be extremely tricky. It's best if you have a friend or lab partner operate the watch. Get the small washer twirling around at a steady pace with the 48 cm mark right at the top of the tube. Say, "Go!" and start the timer. Count 20 revolutions, then say, "Stop!" Write this number of seconds in the first data chart on the next page, under the heading "# seconds."

6) Now calculate the velocity of the washer using d=rt. (We'll call rate "velocity." You can think "d=vt," if that helps.) The distance is 60 meters. For time, "t," you will use the time from the stopwatch which you wrote down under "# seconds." **Divide 60 by your number of seconds to get the velocity (rate)**. Write the velocity in the last column ("v" is for "velocity").

7) Now repeat this process using the 36 cm mark instead of the 48. The new diameter will be 36+36=72. (72cm=.72 m.) The new circumference will be (3.14)(.72m)=2.26m. To get total distance traveled in 20 circles, (20)(2.26m)=**45.2 m**. To make the math easier, we'll round off to **45**. We've written this down for you in the charts. **You will divide 45 by the number of seconds to get your velocity.**

8) Repeat this process for the 24 cm mark. What number will be divided by your time? 24+24=48cm, (3.14)(.48m)=1.5m (20)(1.5m)=**30 m**. Divide 30 by the time you get when you twirl the string at the 24cm mark.

9) Finally, do the 12 cm mark. The distance traveled by the washer will be (3.14)(.24m)=.75m, (20)(.75m)=**15 m** Divide 15 by the number of seconds you get and write this in for the velocity.

DATA CHART for 1 WASHER

radius	total dist.	# seconds	velocity
48 cm	60 m.		
36 cm	45 m.		
24 cm	30 m.		
12 cm	15 m.		

DATA CHART for 2 WASHERS

radius	total dist.	# seconds	velocity
48 cm	60 m.		
36 cm	45 m.		
24 cm	30 m.		
12 cm	15 m.		

DATA CHART for 3 WASHERS

radius	total dist.	# seconds	velocity
48 cm	60 m.		
36 cm	45 m.		
24 cm	30 m.		
12 cm	15 m.		

DATA CHART for 4 WASHERS

radius	total dist.	# seconds	velocity
48 cm	60 m.		
36 cm	45 m.		
24 cm	30 m.		
12 cm	15 m.		

10) Add another large washer, making a total of 3 large washers hanging down. Repeat the previous steps, swinging at each radius and counting seconds it takes to do 20 circles. Record your data in the table that says 3 WASHERS.

11) Now add another washer, making a total of 4 hanging down. Repeat time trials and record data on 4 WASHER chart.

12) Now reduce the number of washers to just one. Repeat time trials and record data on the 1 WASHER chart.

13) Now you are ready to graph some data. We will be making two graphs, looking at the data in two ways. The graph on the left asks you to plot number of washers versus the velocity of the small washer. Choose four colors for the sizes of radii.

14) Now for the graph on the right. Choose **one** of your charts above and plot radius in cm versus velocity.

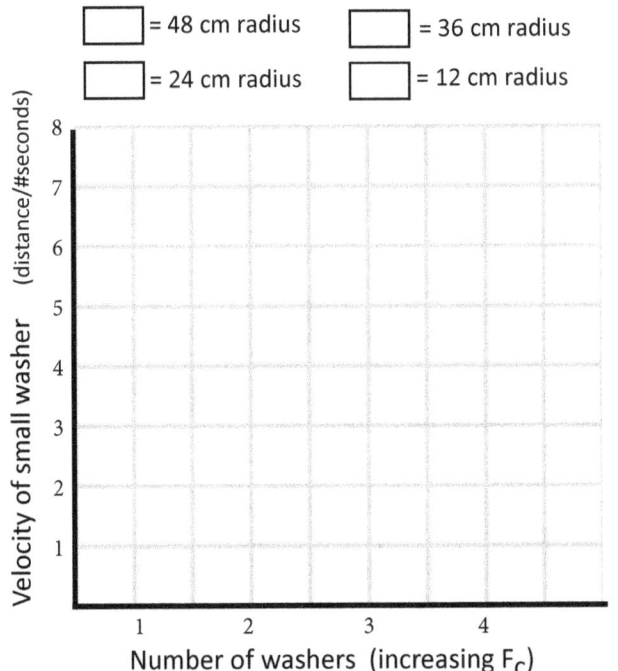

What happens to velocity when F_c increases?

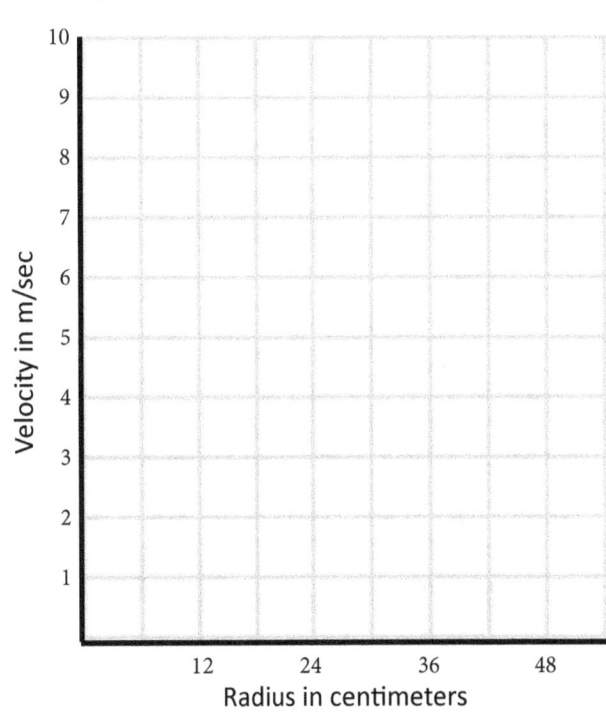

What happens to velocity when F_c stays the same but radius increases?

Questions to answer:
For questions 1-3, use the graph on the **left** on the previous page (centripetal force versus velocity).
Please note that in all of our experiments, we kept mass, "m," the same because we used the same small washer.

1) Look at the data curve for the 48 centimeter radius. Can you extrapolate the velocity the small washer would be moving if you had used 5 washers? _____
(TIP: Make a "best fit curve" by loosely connecting your data points. Then extend the curve a bit, making it continue in the same direction. Look at where it intersects what would be the 5-washer line.)

2) Each of your colored lines represents keeping the radius the same while changing the centripetal force.
In the following statement, circle or underline the correct word inside the parentheses:
When we keep radius the same, the velocity (increases/decreases) as centripetal force (increases/decreases).

3) Look at one of the colored lines, specifically at the 2 and 4 washer data points. Going from 2 to 4 washers means we have doubled the centripetal force. Now look at those velocities. Did the velocity double from 2 to 4 washers? _____

For questions 4-5, use the graph on the **right** on the previous page (radius versus velocity).

4) According to your data, what happens to speed as the radius increases (assuming we keep the centripetal force constant)? _____

5) When you were twirling the small washer at various radius sizes, did it seem like it went slower or faster as the radius got smaller? _____ Is this consistent with your data? _____
Was the small washer completing more revolutions per second at the smallest radius than it was at the largest? _____
However, was the small washer itself traveling through the air more quickly at the smallest radius? _____
What does your data say?

Bonus question:
6) In your warm up activity, when you pulled down on the bottom string were you increasing centripetal force? _____
When you did this, did the small washer speed up? _____

Extension:
The centripetal force, F_C, in this experiment is in the tension of the string. Find the centripetal force, Fc, using 1 gram for the mass of the small washer. Use $F_C = \frac{mv^2}{r}$ (Use a calculator, and don't forget to convert to kg and m.)
For "v" use the velocity from your data charts.

1) 48 cm radius with 2 large washers,

   F_C = _____ newtons (kg m/sec²)

2) 12 cm radius with 3 large washers,

   F_C = _____ newtons (kg m/sec²)

Your answers will depend on your data, but you can check the answer key for an example solution.

3) Try adding more weight to your one small washer (increasing "m"). What happens if you have three washers on top and only two on the bottom?

ACTIVITY 9.10: A few more math problems using $F_c = mv^2/r$ (answers at the bottom of the next page)

TIP: Draw a simple picture of each situation before trying to do the math.

1) A person on a motorcycle is driving around a "roundabout" (a circular path used for making turns) that has a radius of 8 m. The total mass of the vehicle and rider is 120 kg, and they are traveling at the rate of 4 m/sec. What is the centripetal force?

a roundabout

2) If a small car weighing 500 kg is traveling in a circle at 5 meters per second, and the friction between the tires and the road surface is supplying a centripetal force of 25 N, what is the radius of the car's circular path?

3) A 75 kg person gets on this rotating swing ride. When the ride gets to top speed, the riders will be traveling at 7 meters per second. While at top speed, the distance between this person's swing and the center of the ride is 15 meters. What will be the centripetal force experienced by this rider?

3) A child sets a 30-gram wooden block on the edge of the rotating platform of an old-fashioned record player. The platform is designed to play records that require a rotation rate of 33.3 revolutions per minute (rpm). Converting this to a metric velocity, the edge of the platform is moving at 50 cm per second. The platform has a diameter of 30 cm. What centripetal force is the block experiencing? (Be mindful of these units. Do they need to be converted?)

3) The "hammer throw" is a track and field event. The athlete must swing a heavy ball on a rope then release it, causing the ball to fly as far as possible. After it hits the ground, the referees then come and measure the distance the ball traveled. Suppose we are watching an athlete competing with a 10 kg ball, and the rope is 60 meters long. If we know that the tension on the rope is 2,000 newtons, how fast is the ball moving as the athlete swings it around?

(TIP: If you can't figure out how to solve for "v," check the teacher's guide for help.)

141

In the first two math problems on the previous page, we suggested that centripetal force can be supplied by friction on a road surface. Friction is one of the three basic suppliers of centripetal force:
1) **Physical attachment such as a rope or rod**
2) **Friction**
3) **Gravity**

When you ride your bicycle around in a circle, where is the centripetal force? You aren't attached to a string—you are just steering your bike, making it go in a circle. The force holding you in that circular path is the friction between your tires and the road surface. Higher coefficients of friction will let you increase your velocity. Engineers are able to use the centripetal force formula along with friction coefficients to predict the maximum safe velocities that a vehicle can travel around a curve in a road. You've probably seen warning signs for drivers ahead of curved exit ramps on highways, recommending that drivers don't go above a certain speed.

These cyclists are depending on friction for their centripetal force.

Friction is also at work if you have a book sitting on the dashboard of your car and it stays in place as you make a turn. However, if you turn fast enough, you can overcome the frictional force and the book will go sliding. Technically, the book didn't go sliding—the car slid out from underneath the book. The book had inertia and wanted to keep on doing what it was doing. It was traveling in a straight line so it kept going straight while the dashboard followed the curve in the road. From your point of view inside the car, it looked like the book went sliding off.

The third category, gravity, is seen in the orbits of planets and moons. Because the centripetal force is also the gravitational force, we can say that $F_g = F_c$ (force of gravity equals centripetal force) and this helps us to figure out things like how fast a planet is traveling in its orbit around the sun.

Remember that $F_g = \dfrac{G(m_1)(m_2)}{r^2}$

And we just learned that $F_c = \dfrac{mv^2}{r}$

Therefore, if $F_g = F_c$, we can say that:

$$\dfrac{G(m_1)(m_2)}{r^2} = \dfrac{mv^2}{r}$$

Equations like this are very useful if we know all but one of the pieces of information. (Remember, "big G" is the gravitational constant, 6.67×10^{-11}.) For example, if we know the masses of Mars and the sun, and we know the size of Mars' orbit (r), that leaves only velocity, "v," as the unknown.

The math involved in solving for the velocity of Mars is just a bit beyond the scope of this book, but we don't want to leave you in suspense—Mars travels at the amazing speed of 24 km per second! These equations are also the basis for many things that "rocket scientists" do, such as putting satellites into orbit. How important is it to know how much the satellite weighs (its mass) before you launch it into space?

ACTIVITY 9.11: Online interactive lab about gravity and planets (This is optional.)

Here are two online simulations where you can choose the radius and mass of the planets, the size of their orbits, etc. Then hit "Go" and watch what happens. Do your planets stay in orbit? Do they crash? Do they fly off into outer space?

https://phet.colorado.edu/sims/html/gravity-and-orbits/latest/gravity-and-orbits_all.html
https://phet.colorado.edu/sims/html/my-solar-system/latest/my-solar-system_all.html

Answers to 9.10: 1) 240 N 2) 500 meters 3) 245 N 4) .05 N 5) 10.95 m/sec

The scientist's name is Étienne-Jules Marey *(ay-tee-EN zhule mah-RAY)*. He was fascinated with both animals and motion. His first studies were with insects, as he tried to figure out exactly how they moved their wings as they flew. The bugs he studied made a figure 8 with their wings, and Marey designed a working model of the wings so he could demonstrate how they worked. Then he moved on to studying birds and horses. He was the first person to claim that when horses gallop, all their hooves are off the ground for a split second. He even took one of his cameras to an aquarium to film fish swimming! He never got tired of studying motion.

Marey with some of his inventions. He was also interested in sound.

This type of photography is called chronophotography.

One of Marey's cameras might have looked something like the one shown here on the right. It has twelve lenses that will each take a picture. The opening of the lenses was controlled by a mechanism that opened them in quick succession over the course of only one or two seconds. Marey probably had a selection of cameras (including his gun camera) that were designed for particular situations.

So what was he doing with the cat when our volunteers spied on him? Marey knew that scientists had been puzzled for a long time about why cats are able to always land on their feet even if they are dropped from a very short distance. It seems like cats are able to violate Newton's Laws of motion.

According to these laws, the cat should not be able to rotate its body without help from an outside force, such as something to push against. Marey realized that his chronophotography might be able to help solve this riddle. The photos would show the cat's body position every tenth of a second or so. In real time, a cat turns so fast that its motion can't be analyzed. Marey's cat experiment was filmed in 1894 and he was able to publish his results in the prestigious science journal *Nature*.

Falling cats are a fun way to introduce our next topic: **angular momentum**. We've already been introduced to the term "momentum." We learned that an object in motion has momentum, calculated using the formula **p=mv**, where "m" is mass and "v" is velocity. In all types of circular motion, from turning cars to spinning tops, there is momentum. However, when something turns or spins, the momentum is also circular, so we need to specify that it is <u>angular</u> momentum. The word "angular" is used because when working with circles, we are dealing with angles. The Greek letter theta, θ, is used to represent angles in a circle. In this diagram, the angle between the solid line and the dashed line is labeled θ. We can use θ in equations like we use letters such as "m" and "v." Very often, circular motion equations will use "sinθ" or "cosθ." The sine and cosine refer to the ratio of various sides of the triangle created by the angle. In this book we won't be using sine and cosine (though we'll look at cosine in our last adventure), but in any physics course you might take in the future, you will learn how to use sinθ and cosθ.

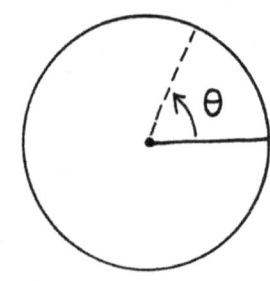

There are many formulas involved in circular motion and angular momentum, and we'll look at several of them in a few minutes, but first let's solve the cat mystery. Angular momentum must be "conserved." We've mentioned this term before, but review never hurts. Conservation means that something has to be preserved or maintained. Many things in science must be conserved, including mass and energy. Mass and energy are never created or destroyed—they just change from one form to another. When angular momentum is conserved, this means that an object can't just create circular motion out of nowhere. Tops don't spin themselves. Something must provide an outside force to get it going, and once it is going it has inertia, so it will take another outside force to stop it. So there's the cat being held upside down by its legs, waiting to drop. It needs circular motion (angular momentum) to flip its body, but there are no outside forces available.

Here is what happens. The cat can "create" angular momentum as long as there is a way to counteract it somehow. An opposite motion will "cancel out" the original motion, satisfying the requirement for conservation of angular momentum. The cat is able to move the front and back parts of its body separately, so the first thing it does is arch its back and pull its front legs close to its body, while spreading out its back legs. The spread-out back legs are very stable and stay put while the compact front body does a 180 degree turn. We now need a counter movement. To do this, the cat now extends its front legs to make them stable, and pulls in its back legs so they are easy to rotate. Having a hard time visualizing all this motion? Let's watch a video about it.

ACTIVITY 9.12: Watch a few videos about falling cats and angular momentum

The playlist has several videos about how cats conserve angular momentum, plus there is a short video about Marey and his photographic legacy. (Don't miss the video where cats are taken into the "Vomit Comet" airplane and experience zero grav!)

The cats were a fun way to introduce the topic of angular momentum (and to meet a new scientist) but they aren't very good for explaining the next part of our discussion. Angular momentum isn't an easy topic to understand. It will take us a few pages to get through it.

We need to look at two separate categories of turning or spinning objects:
1) Point objects: Things that are essentially "points" rotating around a center (planets, balls on a string)
2) Extended objects: Things that are solid such as discs or spheres turning on their axis.

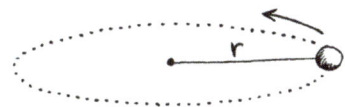

First, let's look at "point objects." Let's imagine a ball on a string. The ball has momentum as it circles around in the air. How much momentum? The formula for finding its momentum is **L= rp**, where "L" represents the angular momentum, "r" is the radius of the circle, and "p" is the ball's momentum. As you know, momentum, "p," is equal to mass times velocity (the velocity at which the ball would travel if it was suddenly set free from the string), so we can replace "p" with "mv" and get: **L= rmv**. This formula will let us calculate the magnitude of angular momentum. Being a vector quantity, L also has direction, not just magnitude, and this is where we meet a bizarre property of circular motion.

The direction of the angular momentum, L, doesn't go along with the ball, as you might expect. The direction of L will be perpendicular to the direction the ball is traveling. To determine if this perpendicular direction is upward or downward, you must use a famous rule of physics called the *right-hand rule*. You curl the fingers of your right hand in the direction that the object is moving, then look at where your thumb is pointing. In this diagram, the circular motion is going counterclockwise (anticlockwise) so the thumb points up. If the motion is going clockwise, the thumb would be pointing down. (Does this mean that there is an invisible force pointing out from the axles of your bike wheels? Yes!)

The best place to look for angular momentum isn't in a point object, however, but in an *extended object*, such as a spinning disc. One of the best examples is a gyroscope, which has a spinning disk in the center. You've probably seen toy gyroscopes and know that when you wind them up, they spin for a long time and are very stable. While they are spinning they resist change. It's almost as if an invisible force is keeping them from falling over. Well, an invisible force IS keeping them up, and this force is the angular momentum. The angular momentum of gyroscopes works so well, that the gyroscope can do amazing feats like the one shown here, balancing on a wire like a tightrope walker. When the spinning stops, the angular momentum will disappear and the gyroscope will fall off the wire. The spinning motion creates the angular momentum (marked L in this photo).

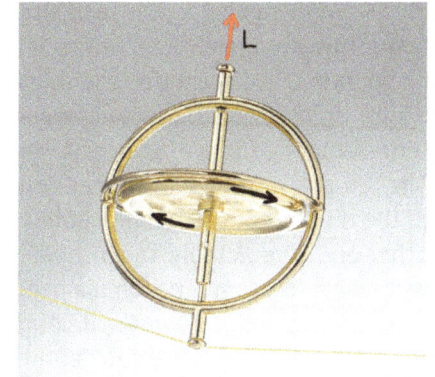

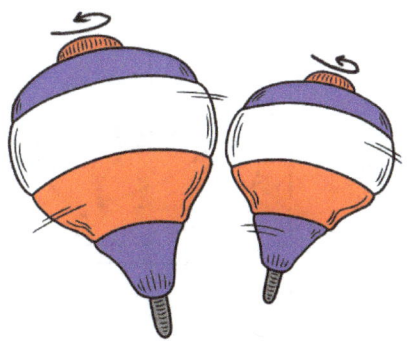

A toy top acts like a very simple gyroscope. While the top is spinning, the spindle in the middle points upwards. If the spinning stops, the angular momentum stops, and the top falls over. If you've played with tops quite a bit, you'll almost certainly have noticed another motion that tops make while they are spinning. The spindle usually moves around in a circle very slowly even while the base of the spindle stays in one place. There is a special name for this slow wobble: it's called *precession*. Precession is one of the natural movements that tops and gyroscopes make. Stability is maintained even while they are precessing. The gyroscope balanced on the wire in the above photo can be precessing while still staying on the wire. Planets also act like giant tops and precess. The earth has a very small amount of precession. It would take 26,000 years for its precession to make a complete circle.

These tops are precessing. The bottom of the spindle stays in place but the top of the spindle moves slowly in a circle.

To continue our discussion of angular momentum and precession, there is another term that you need to know: *torque (tork)*. Torque is a relatively simple concept (compared to angular momentum) and might be a word that you are already somewhat familiar with. However, let's look at a technical definition.

145

In physics, torque is simply a force that is rotational, not linear (not translational). A good example of torque is a wrench (spanner) turning a bolt. The wrench (spanner) is a tool that you use to apply torque to the bolt. The torque force is always perpendicular to the lever arm. In this case, the lever arm is the wrench. Torque is the force (in newtons) multiplied by the length of the lever arm (distance from fulcrum). Length is measured in meters, giving us "newton meters" as the measure of torque force.

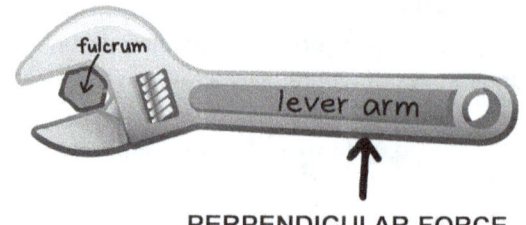

PERPENDICULAR FORCE

Torque is represented by the Greek letter "tau," τ. The formula for torque is $\tau = rF$, where "r" is radius in meters (which is another way of saying the length of the lever arm), and "F" is force in newtons. Looking at this formula for torque, we can see that there are two ways to increase torque. We can increase the force (which is sort of obvious), or we can increase the radius, r. But isn't this what we learned back in our adventure with levers? We learned that as we increase the length of a lever, we gain mechanical advantage. The length of the lever arm is another way to describe "r," therefore, levers create torque.

Another obvious observation is that if we apply the force in a way that is NOT perpendicular to the lever arm, we decrease the torque. Imagine applying a force to the end of the wrench so that the force points towards the bolt. Will the wrench turn at all? No. Thus, we have no torque. If the force is applied sort of diagonally, we'll still get some torque, but only because the diagonal force can be divided into two smaller forces, one of which is acting perpendicularly to the wrench.

Although torque itself is not hard to understand, torque in combination with angular momentum, as we see in tops and gyroscopes, causes effects that are a bit mind boggling. For example, if a spinning disk with a central spindle is suspended by the end of one of the spindles, the entire disk will seem to defy gravity as it remains horizontal, as if an invisible hand was holding the other end of the spindle. This works with spinning bicycle wheels as well as with small gyroscopes. (If you have a gyroscope, you might want to try this right now.)

We're going to pause in just a minute to watch some videos about this strange effect, but first let's go through a written explanation. In diagram 1, we see a disc that is hanging from the end of one of its central spindles. The disc is not spinning in this picture. The blue arrow represents the force of gravity pulling the weight of the disc downward. The red arrow represent the tension in the rope that is holding the weight of the disc. What will happen in this situation, since no one is holding that other end? Predictably, the disc will fall, being pulled down by gravity. It will swing down, and become a pendulum for a few seconds, then finally come to rest in this downward position. This is shown in diagram 2.

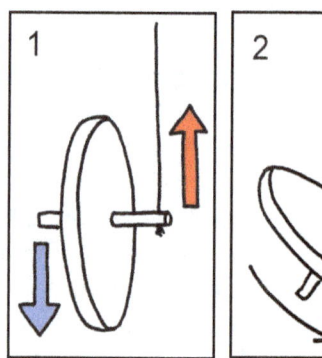

In diagram 3, we look at this same situation again, but this time we focus on the fact that when the disc fell down due to gravity, it was rotating around the point at which the rope is attached. If the downward swing of the disc had been able to continue, it would have made the circle shown by the arrows. This circular motion would technically generate some angular momentum. In what direction? Do the right-hand rule. Hold your fist so that the fingers are curling around in the direction of the arrows. Where is your thumb pointing? Right at you. This is what we are trying to indicate with the green arrow. It is supposed to look like it is coming out of the paper, pointed at you. Torque is being created by red and blue forces in diagram 1, and the result is the green angular momentum arrow. Also, as strange as it might seem, this torque (created by the red and blue forces) is still there even if the disc isn't actually dropping down. This fact will be important in diagram 5.

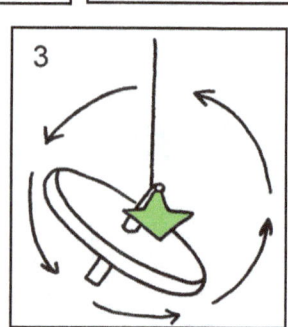

In diagram 4, we finally give the disc a spin. It is spinning in the direction shown by the black arrows. This spinning will create angular momentum in the direction of the orange arrow. Again, use the right-hand rule. Imagine grabbing the spindle of the disc with your right hand. Your fingers curl in the direction of the black arrows, don't they? In diagram 3 we have one source of angular momentum and in diagram 4 we have another source. What will happen if we combine them? In past adventures, we saw two linear forces combine to make a final result that was curved.

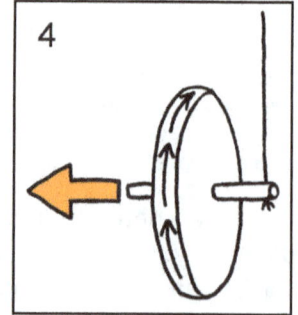

In diagram 5, we have both sources of angular momentum shown, the orange one coming from the spinning of the disc, and the green one coming from the torque created by hanging it from one end of the spindle. When both angular momentums are operating at the same time, the result is motion in the direction of the black arrow.

What would happen if we were to move the rope closer to the disc? That would mean that the lever arm (the distance between the rope and the disc) would become shorter. A shorter lever means less torque. Less torque would mean that the green arrow force would become weaker, and this would lead to less precession (black arrow).

The source of precession in gyroscopes isn't as clear or easy to understand as this spinning disc example we've just looked at. Most physicists say that precession in gyroscopes is partly caused by friction at the place where the spindle touches the surface. That little bit of contact can cause enough friction that a tiny bit of torque is produced. We're not doing to delve into this in depth, as it becomes rather complicated very quickly. If you want to know the details about this, there is a video on the playlist that you can watch. ("How do Gyroscopes Lift Themselves Up?" by The Action Lab channel)

So it's the perpendicular angular momentum force (the orange arrow) pointing out from a spinning disc that makes it stable. This is what enables you to balance on a bicycle. Have you ever noticed how hard it is to balance on a bike that is standing still, or going very slowly? Then you speed up and balancing seems so much easier. That's because your bicycle wheels are acting like gyroscopes. They are resisting any change in direction. They don't want to fall over; they want to keep on rolling along. This is why you can take your hands of the handlebars for a few seconds, just long enough to scare your mother. The moving wheels are relatively stable thanks to angular momentum.

Assuming that the boy is moving forward, where would the force arrows be? Use the right-hand rule.

We're just about ready to take a video break and watch some gyroscopes and spinning discs, but we need to mention one more idea you'll run into in some of these videos about spinning discs. As we mentioned previously when we talked about falling cats, angular momentum is conserved. This means that angular momentum can't be created and it can't disappear. The angular momentum of a system can't just stop. If a change is made, the angular momentum must be transferred to somewhere. This fact is the key to understanding some science demonstrations about spinning discs that you might see at hands-on science museums and in physics classrooms. A very famous demo is to have a volunteer stand on a platform that turns freely. Then they are handed a spinning disc (often a bicycle wheel with handles coming out of the hub). The volunteer holds the wheel so that it is parallel to the ground. Then they flip the wheel over. After they flip it over, their whole body, including the platform they are standing on, begins rotating. If they flip the wheel back over to its original position, they begin rotating in the opposite direction. There is a very nice video about this created by MIT Physics. The video is posted on the playlist. Their channel is MITK12Videos. *(The use of this screen shot from their video is intended to be free advertising for their channel.)*

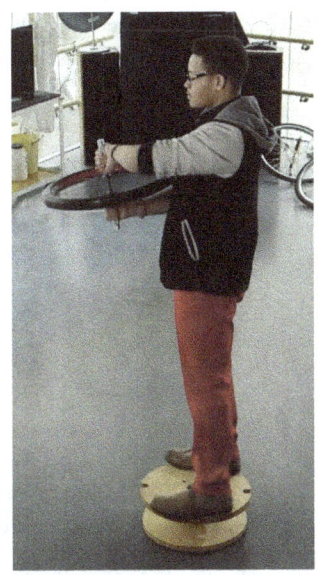

ACTIVITY 9.13: Let's take a video break before we go on

The playlist has quite a few videos about gyroscopes. Watch a few of them, or all of them, whatever you have time for.

The scientist who invented our modern version of the gyroscope is someone we've already met.

Foucault (fu-ko) called his invention "gyroscope" using the Greek words "gyro," meaning "turning," and "scopos," meaning "see." He thought this invention would let you see the turning of the earth. His gyroscope could run for about 10 minutes without losing much energy, which was just long enough to be able to detect a small amount of earth rotation.

Many gyroscopes since Foucault's time have used his method of suspending the spinning disc inside a series of gimbals. A circular gimbal touches whatever is inside it at only two minuscule points, sort of like holding a globe by two pins inserted in the north and south pole. If you use a gimbal inside a gimbal inside a gimbal, the result is the inner disc being free to spin in any direction. Foucault's idea was that his spinning disc would be so stable that the earth's motion would affect the gimbals but not the spinning disc.

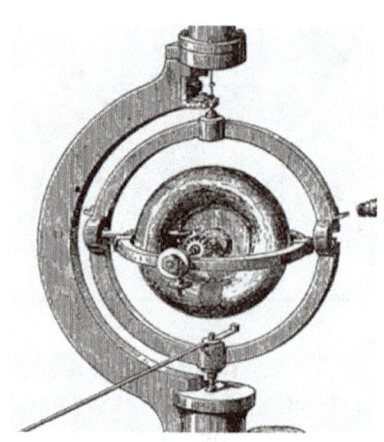

Today, gyroscopes are used in many applications, especially in airplanes, rockets, and submarines. This photo shows a modern gyroscope that uses a motor to keep it running. It won't wind down like a toy gyroscope does. Sensors are placed at certain points around the gyroscope so that an electronic device can record the position of the vehicle with respect to the gyroscope. The gyroscope stays still while the vehicle moves around it. Thus, the gyroscope can let the pilot know how exactly how much the plane is tipping or rolling even at night when the pilot can't see the ground.

Let's exit our gyroscopic rabbit trail and finish the discussion that we started on page 143, about point objects and extended objects. Just as a reminder, these were the definitions:

1) **Point objects:** Things that are essentially "points" rotating around a center (planets, balls on a string)
2) **Extended objects:** Things that are solid such as discs or spheres, which are turning on their central axis.

We learned that calculating angular momentum for point objects can be done using the formula **L=rmv**. We saw how the angular momentum force comes out at a perpendicular angle, the direction of which is determined using the right-hand rule. Before we learn how to calculate angular momentum for extended objects, let's do a few math problems with point objects.

Example problem:
A child is swinging a toy airplane around in a circle. The airplane has a mass of .25 kg and is moving at a velocity of 5 meters per second. The string is 1.5 meters long. What is the angular momentum?

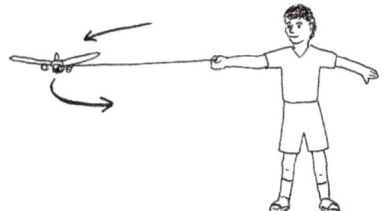

All we need to do is find out what "rmv" is. We know that mass equals .25 kg, velocity is 5 m/sec, and the radius is 1.5 meters.

L = (1.5 m)(.25 kg)(5 m/sec) = 1.87 kg m²/sec

We've determined the magnitude, but angular momentum is a vector, so it also has direction. What direction would it be? We can see by the arrows that the airplane is going counterclockwise if we look from above. Using the right-hand rule, the angular momentum would be up. Physicists decided to define "up" as "positive" angular momentum, and "down" as "negative," so the right answer would be "positive angular momentum." But is there really angular momentum in a situation like this? A child isn't a spinning top. According to the laws of physics, there must be angular momentum even if it is a small amount and isn't visible. In situations like this, however, the direction doesn't play a large role.

ACTIVITY 9.14: Try these angular momentum problems (use L=rmv) (Answers page 150, and teacher's section)

1) A flying insect is circling around a flower. The insect's mass is only half a gram, and it is one tenth of a meter away from the flower. If the insect's velocity is half a meter per second, what is the angular momentum? (Remember to convert grams to kg, and we won't worry about direction in this problem.)

2) A talented performer is doing tricks with a yo-yo. One of her tricks involves spinning the yo-yo around in a circle several times. The yo-yo's mass is 40 grams and it is moving at 3 meters per second. The angular momentum is .12 kg m²/sec. Find the length of the yo-yo's string. (We won't worry about direction of L.)

3) A hammer thrower is spinning around, whirling his hammer in a circle, getting ready to let it go. The ball on the end of the rope weighs 10 kg. The rope is 1.25 meters long. The angular momentum is 100 kg m²/sec. What is the velocity of the ball at this point, just before he lets go?

4) In a very small solar system, a tiny moon is orbiting a planet. The moon's mass is only one million kilograms. Its velocity is 52 km/sec and it is 50,000 km away from the planet it is orbiting. What is the angular momentum?

TIP: When dealing with a lot of zeros, take them all off and set them aside temporarily. (But don't lose them! You'll need to put them back on at the end of the problem!) For example, you can take the six zeros off the number one million so that you have just a 1, with 6 zeros set off to the side. Do this for all the numbers. Then multiply the small numbers (1, 52, 5). After you get the answer, then put ALL the zeros on the end of your answer. After you get this very long answer, you can use another trick to control all the zeros. Take the zeros off and count them. Then write your answer using exponent notation. If your number was 46,000,000 (which isn't the right answer, by the way) you'd write 46 x 10⁶. The tiny 6 after the 10 tells you that there are 6 zeros that need to be added to the 46.

And now for extended objects: discs, cylinders, rings, hoops, solid and hollow spheres. In real life these can be things like rolls of tape, balls, CDs, marbles, cantaloupes, pencils, soup cans, etc.

The formula for calculating the angular momentum for extended objects looks like this: **L = $I\omega$**
This formula looks very strange. We know that "L" is angular momentum, but we haven't met either of those letters yet. "I" stands for something called **moment of inertia**, and ω ("omega") is the **angular velocity**.

Let's explain ω first, since it is a bit easier to understand. Angular velocity is exactly what it sounds like—it's the velocity of something that is traveling in a circular path. The letter omega is used because angular velocity isn't expressed in meters per second like regular velocity is. Physicists express angular velocity in **radians** per second. Radians aren't hard to understand, but using them in equations requires trickier math, so we won't be calculating in radians.

This diagram shows what a radian is. The radius, r, is used as a measure along the outside of the circle. Then we measure the angle that this arc makes. This angle is 1 radian. How many radians are in a circle? Archimedes would have loved this answer. He fascinated with the value π (the circumference of a circle divided by its diameter). A circle has exactly 2π radians. Since π is an "irrational" number that goes on forever (3.1415926535897932384626...) with never-ending decimals, that means that the number of radians in a circle is also a never-ending number. So we can't put a number on how many radians are in a circle. We just have to say "2π" and let it go at that.

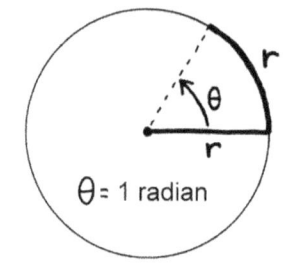

Like it or not, angular velocity is measured in radians per second. A measure of angular velocity would look something like this: ω = 2.5 rad/sec. If a disc is has an angular velocity of 2π rad/sec it would be making one complete turn in one second. (No worries, we won't be doing any calculations with radians!)

And now, what is moment of inertia? This will be a rather lengthy explanation, but it will be the last long explanation of this adventure, so the end is in sight. First, we are going to start with a few very simple experiments. You'll think they are ridiculously simple, and you'll be right, but the principles we'll learn from these experiments will be the foundation for the rest of the discussion.

ACTIVITY 9.15: Discover rotational inertia (also known as "moment of inertia")

You will need: your two meter sticks, a roll of masking tape ("painter's tape," which in the U.S. is often blue, works well), and two identical objects that can be taped to the sticks (to be used as weights)

NOTE ABOUT THE WEIGHTS: *It doesn't really matter what you use as your weights. They don't have to be any specific weight, as long as both objects are the same weight. Aim for about the weight of an apple. We used two disposable juice pouches; juice boxes would also work well. Make sure that whatever you use won't break if it falls to the floor from the height of one meter.*

PART 1:
1) Tape one of your weights to the end of one meter stick. The other stick won't have a weight.
2) Hold the meter sticks upright, as shown in diagram. Lean them forward slightly so that they will both fall in the same direction when you let go. Make sure that they at exactly the same starting point, as this will be a "race" to the ground.
3) Let go of both sticks at the same time and watch them fall. Which one reaches the ground first?
4) Do this several times to see if you get the same result each time.

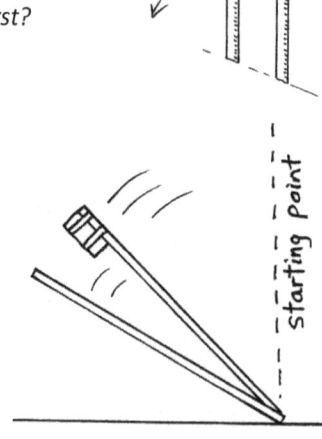

Explanation:
You already know about inertia. Things with more mass have more inertia. (Remember, the definition of mass isn't how heavy something is, but how resistant it is to a change in velocity. We saw how an inertial balance can be used to find the mass of an object by testing how resistant it is to motion.) The stick with the weight had more inertia than the plain stick.

Inertia means that things want to keep on doing what they are doing. The tops of the sticks were standing still before you let them go, so they wanted to keep standing there. However, when you let them go, gravity began pulling on both sticks. It was harder for gravity to get the stick with the weight to rotate because it had more inertia (more mass) and was therefore more able to resist rotation. The unweighted stick was easier to move.

Answers to 9.14:   1) .000125   2) 1 meter   3) 8 m/sec   4) 26 × 10¹⁷

Imagine that your sticks are the radii of circles. The top half of the circle was above the floor and the bottom of the circle was below the floor. Your sticks traveled though one quarter of the circle (90 degrees). Thus, your falling sticks can tell us something about circular motion. This experiment showed us that when mass is located at the outside of a circle it makes it harder for the circle to begin rotating. The principle of inertia also tells us that once the stick with the weight gets moving it will be harder for it to stop.

Rotational inertia works just like linear inertia. Things with more mass will be harder to get moving (or to stop moving), but in a circular path, not a straight path.

FOLLOW-UP EXPERIMENT: What will happen if you simply drop both meter sticks (from a horizontal position)? Will the one with the weight still hit the ground first? What did we (and Galileo) learn about falling bodies? Do heavier things fall faster? This is why we needed to imagine the sticks as being the radius of a circle, as shown in the above diagram. Otherwise, we'd be puzzled as to why the sticks are falling at different rates, seemingly contradicting the law of gravity.

****

PART 2: A balancing act

1) Try balancing the plain stick on the palm of your hand. Notice how easy/hard it is to do this.
2) Now try the same thing with the stick that has the weight tipped on the end. Do you think it will be harder or easier?
3) Notice the difference with the weighted stick.

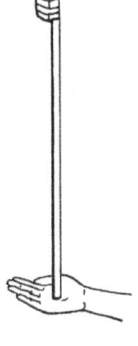

Explanation:

This experiment uses the same principle as part 1 did. The end of the weighted stick has more inertia than the unweighted one. Things with more inertia have more ability to resist a change in motion. The top of the weighted stick is resisting falling, which is to your advantage, because it gives you a few split seconds longer to detect a change in motion and compensate by moving your hand.

****

PART 3:

1) Tape another weight to the end of the weighted stick, so you have a weight on either end.
2) First, hold the middle of the unweighted meter stick with one hand. Keep your fist in the same place while twisting the stick back and forth, so that the top of the stick goes left, right, left, right, left, right, etc. How long does it take you to go back and forth 10 times?
3) Now hold the middle of the stick that has weights on both ends. Do the same thing. Twist the rod back and forth as fast as you can. How long does it take to make the top of the rod go back and forth 10 times? There should be a considerable difference.
4) Now move the weights from the ends of the stick to a position much closer to the middle. Try the motion again. It is easier now than when the weights were at the end?

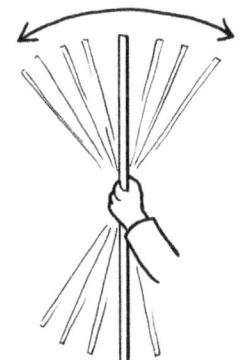

Explanation:

The same principle is at work here, also. Gravity had a hard time getting those weights to start moving and now you are, too! The weights give the ends of that stick more inertia which means a greater resistance to a change in velocity. You can feel that resistance.

When you moved the weights away from the ends and closer to the middle, we learned another important fact about rotational inertia: **the location of the mass is very important.** The further the mass is from the center of the circle, the greater amount of inertia it will provide. The closer it is to the center, the less effect it has on rotational inertia. This allows us to predict that any formula for rotational inertia will contain the letter "r" for radius. We'll have to know how far from the center the mass is, because location is important.

Hopefully, we now have a good understanding of rotational inertia. Next, we are going to look at rotational inertia in extended bodies, using **moment of inertia**. This will be the most confusing term we'll learn in this adventure. We are going to use the word "moment" in a way that has nothing to do with time.

Moment of inertia is all about how the mass of an extended object is distributed. Is most of the mass close to the center, or is it far from the center? In part 3 of activity 9.5, we could feel the results of where the mass was located. When the weights were located at the ends of the stick, it was much harder for us to change the direction of the stick's movement. When the weights were near the center of the stick, movement was easier.

So let's move this idea over to three dimensions. Let's think about a solid ball and a hollow ball. In a hollow ball, all of the mass is far from the center—the center is empty. We can say that a hollow ball has a <u>large</u> moment of inertia, meaning that the amount of mass at a distance from the center is very high. A solid ball would have a <u>small</u> moment of inertia because the mass is as close to the center as it can get.

If you had a rolling race between the hollow ball and the solid ball, which one would win? Let's say that the balls have the same mass and the same diameter, so we don't have to consider those factors. The difference between the balls would be only where the mass is located. Think of your falling sticks in activity 9.15. Which stick was slower to get to the ground? The one with more mass far away from the pivot point. So in the case of these balls, a good guess would be that the one with more mass far away from the center will be slower to start moving. Let's try this and see if we are right.

ACTIVITY 9.16:   Run some races between hollow and solid balls

You will need: *a few hollow balls (ping pong ball, racquetball, plastic practice golf ball, or any other hollow ball (probably not tennis ball because of its fuzzy surface), and a few solid balls (such as marble, baseball, super ball, rubber ball, etc), a long ramp wide enough to race two balls (If you don't have a long board, you can tilt a table like you did in activity 8.5.)*
NOTE: Make sure that the <u>bottom</u> of each ball is touching the start line. *(You might even want to lightly draw a start line or use a piece of masking tape.) If you start them by letting them rest against a ruler, the smaller balls will get a slight head start.*

1) Choose one hollow ball and one solid ball. Make sure your ramp is not too high. You want to be able to watch the race for several second. Make sure the exact bottom of each ball is on the start line, then let them go at the same time.
NOTE: If there is a very, very small difference between two balls at the finish line, repeat the race several times to see if perhaps the small difference was due to a fraction of a second difference in the release of the balls. Any difference caused by moment of inertia should be a very obvious difference.
2) Run a race between two solid balls, then between two hollow balls.

3) When you raced hollow against solid, which type of ball always won the race? _____

4) Did all of the solid ball races end in a tie? _____

5) Did size make a significant difference? _____   *(Teacher's section has a discussion of results.)*

6) Did weight (mass) make a significant difference? _____

We've discovered that when an extended object's mass is located far from the center, this produces a larger moment of inertia, meaning the object will be slower to start moving. This is the only principle you need in order to predict the outcome of a race between different types of rolling objects. Let's test this with cylinders.

ACTIVITY 9.17:   Cylinder races

You will need: *your ramp, two small cans of tomato paste, two medium-sized cans of tomato paste, one can of tomato sauce or stewed tomatoes (or anything else that will slosh around in the can), a round pencil or crayon, a marker, and anything else you want to try—rolls of tape, plastic bottles, cylinders from a wooden block set, etc.*

NOTE: Tomato paste is used because it is so thick that it provides a solid interior, where the mass is equally distributed throughout the can. If you can't use tomato paste, you can use something else as long as it is thick and won't move around in the can.

1) Cut both ends out of one of the small cans of tomato paste and one of the medium-sized cans of tomato paste. Scrape out the paste and rinse the cans.
2) Race various combinations of cans and cylinders:
    both small cans (one full, one empty)
    both medium cans (one full one empty)
    one small hollow can and one medium hollow can
    a small full can versus a large full can
    a can filled with solid paste versus a can with liquid contents that won't roll with the can
    two rolls of tape, one with less tape and one with more tape
    a crayon or marker versus a full can of paste
    a crayon or marker versus an empty can
    other combinations you can think of

Here some charts you can use if you find them helpful. You can write which two things you are racing, then mark the winner.

3) Does size determine the winner of a race? _____

4) Do the hollow cans always lose or always win? _____

5) What about the cans containing sloshy liquids?
   Were they slower or faster than hollow cylinders? _____   Were they slower or faster than solid cylinders? _____

6) Was there a difference between empty/full rolls of tape? _____ Why? _____

7) Did the length of the cylinders play an important role? _____

8) Did any of the results surprise you? Which ones? _____

And now for the championship round—balls versus cylinders!

ACTIVITY 9.18: Racing balls against cylinders

You will need: your ramp and all your balls and cylinders

NOTE: Try to predict the result before you do a race. Think about where the mass of each object is located.
1) Race a solid ball against a solid cylinder. Race a small ball against a large cylinder, and a large ball against a small cylinder.
2) Race a hollow ball against a hollow cylinder.
3) Race a solid ball against a hollow cylinder, and a hollow ball against a solid cylinder.
4) How many results were you able to predict correctly? _____
5) Of all the objects you've raced in the past 3 activities, which object has the largest moment of inertia? _____

The lowest moment of inertia? _____

(Teacher's section has a discussion of results.)

Now we are finally ready to take another look at the angular momentum formula for extended objects.

$$L = I\omega$$

"L" is angular momentum, "I" stands for moment of inertia, and ω ("omega") is the angular velocity. Now that you've done six activities about moment of inertia, this term won't seem as strange as it did before. The "I" is no longer mysterious. But is there a way to calculate a number for the moment of inertia? Equations need numbers.

The numbers associated with the moments of inertia depend on whether you are talking about a hollow sphere, a solid sphere, a hollow cylinder, or a solid cylinder. The math isn't very difficult. It's just multiplication. "M" is mass, and "R" is radius.

Yes, the earth does have moment of inertia and angular momentum! Thanks for pointing that out. We'll discuss this on the next page. About the fractions—this is a tough question to answer without introducing a lot more math. It's enough for us to understand that the fraction is there to represent where the mass is located, how close or how far away from the center. In the hollow cylinder, all of the mass located at the very end of the radius. In the solid cylinder, some of the mass is close to the outer radius, but some of it is close to the center. The fraction 1/2 turns out to be just right for describing the distribution of the mass equally between the outer radius and the center. Notice that in the case of the hollow sphere, 2/3 is more than 1/2 but less than 1. Thus, the moment of inertia for a hollow sphere will be greater than a solid cylinder but less than a hollow cylinder. Does this agree with what you discovered experimentally when you raced spheres and cylinders?

The numbers associated with angular velocity (ω) involve radians, which we explained on page 148. Because we aren't going to take the time to learn how to use radians, we're not going to ask you to do any calculations of angular momentum for extended objects. However, we are going to see how this equation can explain an interesting aspect of circular motion.

Have you ever watched a professional ice skater doing a spin? They'll start the spin with their arms out, then quickly pull them in. As they pull in their arms they start spinning much faster. The formula $L = I\omega$ can tell us why.

We know that angular momentum is conserved. This means that "L" can't change. A spinning skater has a certain amount of angular momentum, and this must stay the same. As they pull their arms closer to their body, they are decreasing their moment of inertia. (becoming more like a cylinder). The value of "I" decreases. In order to keep the value of L the same, the value of ω must increase. For example, let's say that "L" is equal to 12, "I" is 4, and ω is 3. We write: 12=(4)(3). Now, If we decrease "I" to 3, then to keep L=12, we must increase ω to 4. Thus, we've changed it to 12=(3)(4). Now, if we decrease "I" down to 2, then we must increase ω up to 6 in order to keep L=12. The product of "I" and ω must always equal 12 (in this example).

When the skater's moment of inertia goes down, the velocity at which she is spinning goes up. If she pulls in her arms, her spin velocity automatically goes up without any effort on her part. It's just conservation of angular momentum at work.

You can try this for yourself if you have a chair or a swing in which you can spin around. Have someone start you spinning, then pull your arms (and/or legs) in and out and notice how the speed of your spin speeds up or slows down. It really works!

ACTIVITY 9.19: Watch a few short videos showing the conservation of angular momentum in spinning objects, AND/OR try it for yourself

The playlist has a few short videos showing this demonstration. Also, try it yourself if you are able.

On March 11, 2011, a massive earthquake (8.9 on the Richter scale) hit Japan. The epicenter of the quake is shown by the red rings on the map. This is where the quake originated. As a result of the quake, Japan's main island moved about 2.5 meters. Solid ground rarely moves this much! Earthquake scientists around the world recorded the earthquake waves on their seismographs. These machines are able to sense quakes anywhere on the globe. The image below the map shows what the earthquake looked like on one of these seismograph machines. Another piece of data they collected was the angular velocity of the earth after the quake. They found that the earth was spinning a tiny bit faster than it was before the quake. This made our approximately 24-hour day 1.8 microseconds shorter. (One microsecond is one millionth of a second.) This difference wasn't noticeable, of course, but it was significant enough that the scientists could say with confidence that the earth is now rotating faster than it was before the quake.

So what happened? How could the earth just suddenly speed up? There's an obvious solution—the earth's mass shifted in such a way that some of it got closer to the core. The earth was like a giant ice skater who pulled her arms in just a little bit. This gives us a clue about the cause of earthquakes. It is reasonable to conclude that something else is happening besides crustal plates shifting at the surface. What is making those plates shift? Could the answer be found underneath the crust, in the deep mantle rock? Could deep magma be draining downward? In 2006, Japanese researchers at Okayama University in Japan did some experiments with magma, and showed that when magma is heated and compressed it will suddenly become more dense and take up half the space it used to occupy. This happens at about 200-300 km below the surface. Below that depth, magma cannot rise to the surface. If the mantle rock shifted at these depths, the magma produced would go down, not up. (This raises serious doubts about the theory that plumes of magma rise up from the outer core.) If deep magma drained down, mantle rock would move in to fill the gaps left behind, and these sudden shifts would definitely affect the plates above.

Hmm... looks like we took you down a short geology rabbit trail. Must get back to discussing the earth's moment of inertia and angular momentum.

The scientists who discovered that the earth's rotation increased after the earthquake would have also calculated the earth's new moment of inertia, angular velocity, and angular momentum. First, what is the earth's moment of inertia? We use the formula for solid spheres: **$I = 2/5\ MR^2$** Here is what the math looks like:

$$I = \tfrac{2}{5} M R^2 = \tfrac{2}{5}(5.98 \times 10^{24}\ \text{kg})(6.37 \times 10^6\ \text{m})^2$$
$$= 9.71 \times 10^{37}\ \text{kg m}^2$$

If we wrote out the answer instead of using 10^{37}, it would look like this:
$$I = 97{,}100{,}000{,}000{,}000{,}000{,}000{,}000{,}000{,}000{,}000{,}000{,}000\ \text{kg m}^2$$

To find the earth's angular momentum, L, we also need to calculate angular velocity, **ω**. The answer will be radians per second.

$$\omega = \frac{1\ \text{rev}}{24\ \text{hr}} \times \frac{1\ \text{hr}}{3600\ \text{s}} \times \frac{2\pi\ \text{rad}}{1\ \text{rev}} = 7.27 \times 10^{-5}\ \text{rad/sec}$$

Now we can use **L = Iω.** We multiply *I* and **ω**, and the answer is:

L = 2.68 x 10⁴⁸ kg m²/sec

(Imagine how long this answer would be if we wrote out all the zeros!)

So now we know the strength of the (invisible) force vector pointing up from the north pole. (You will notice that the earth and the arrow look tipped. The earth's rotational axis is tipped 23.5 degrees from the imaginary plane created by the earth's orbit around the sun.)

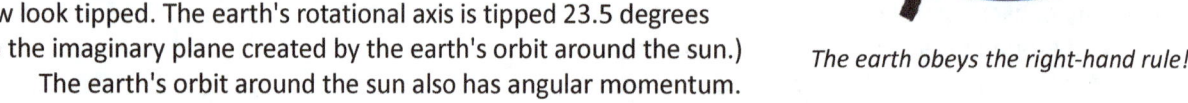

The earth obeys the right-hand rule!

The earth's orbit around the sun also has angular momentum. In this case, we can think of the earth as a point object in a circular path and use L=rmv. But let's skip the math on this and just make a mental note to ourselves that the earth has two types of angular momentum.

Now for the grand finale—perhaps the strangest phenomenon in this adventure. Spinning spheres, including the earth, can do something very odd. They can move <u>through</u> their spin axis without any change to it. In other words, if the mass on the surface of the earth shifted so that there was suddenly more mass on one side than the other, the earth would adjust its spin, trying to re-balance itself, rotating in such a way that the north and south poles would be relocated. This would happen without any change to the size or direction of the angular momentum arrow.

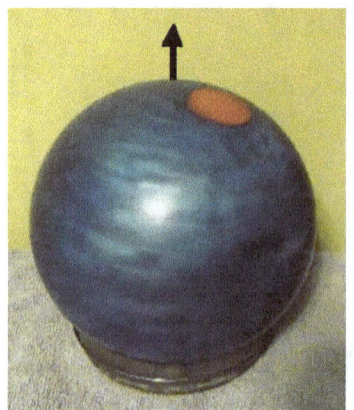 

We can't actually try this with the earth, of course, so the best we can do is to try it with a heavy bowling ball. If we put some lead (which is heavier than the rest of the ball) into the ball at one point (under the red spot in these photos) we can give the ball a spin and see what happens. The axis of rotation (the black arrow) will remain straight up and down the whole time, perpendicular to the floor. You can see that the heavy red spot starts out at the "north pole" but as the ball spins, it turns until the heavy red spot is almost at the "equator."

ACTIVITY 9.20: Watch this bowling ball demonstration

This video is posted on the playlist. If you have a wooden ball, you might be able to try this yourself. Drill a hole the same size as a fishing "sinker" (probably not made of lead but still very heavy) and push the sinker into the hole until it is flush with the surface. Make sure you put the ball onto a very smooth surface when you spin it. (This bowling ball is actually sitting on ice!)

ACTIVITY 9.21: Review questions from this chapter (Answers are in teacher's section.)

1) How did Galileo discover the relationship between the area of a cycloid and the area of the circle that made it?
   a) He used calculus.   b) He measured them with rulers and compasses.
   c) He used logic.   d) He weighed them on a scale.

2) Which of the following is NOT true of cycloids?
   a) They are made using circles.   b) They can be used to draw parabolas.
   c) They represent the fastest path a dropped object can take, other than straight down.
   d) They can be used to make pendulums swing, even high ones, the same frequency.
   e) They are part of a larger group of shapes called trochoids.

3) What would happen to the moon if gravity suddenly disappeared?
   a) Nothing, it would keep on doing what it is doing, because of inertia.
   b) It would go flying off in a straight line.   c) It would fall towards the earth.

4) TRUE or FALSE? A tangent line touches a circle at just one point.

5) TRUE or FALSE? There are two distinct types of circular forces: centripetal and centrifugal.

6) Newton said that orbiting objects (going in a circle) always have two forces operating on them:
   a) centripetal force and centrifugal force   b) angular momentum and centripetal force
   c) centripetal force and linear velocity   d) linear velocity and angular velocity

7) What is it called when an author or speaker goes off-topic? _____ trail

8) A vector has both _____ and _____.

9) Which one of these is NOT a vector quantity?   a) acceleration   b) velocity   c) speed   d) angular momentum

10) Acceleration is defined as a change in either _____ or _____.

11) TRUE or FALSE? An orbiting moon is constantly changing direction.

12) TRUE or FALSE? It isn't possible for a human to experience continual acceleration.

13) What does a centrifugal governor do?
   a) Keep an engine going at the same speed.   b) Keep medical devices spinning around.
   c) Make sure that G forces stay low.   d) Try to get re-elected.

14) What is the maximum number of G forces that military pilots withstand?   a) 3   b) 6   c) 9   d) 12

15) What would be the maximum number of G forces that amusement park rides could create?
   a) only 2   b) 3-4   c) 5-6   d) 9

16) TRUE or FALSE? Centripetal force is a force that is directed towards the center of a circle.

17) Centripetal force is measured in: a) kg   b) kg/m²   c) m/sec   d) newtons

18) In the formula $F_c = mv^2/r$ what happens to the centripetal force when radius, "r," increases? (think about fractions)
   a) it gets smaller   b) it gets larger   c) nothing, it stays the same

19) Which one of these is NOT a supplier of centripetal force?
   a) wheel   b) friction   c) rope   d) gravity   e) stiff rod

20) TRUE or FALSE? When a book slides across the dashboard of your car, it is the car that is moving, not the book.

21) TRUE or FALSE? A bicycle riding in a circle isn't a situation where you find centripetal force because there isn't a string or rope involved, and gravity is pulling down, not towards the center of the circle.

22) What Greek letter is used to represent angles?   a) alpha   b) beta   c) gamma   d) delta   e) theta

23) What rule do you use to find the direction of angular momentum? _____

24) TRUE or FALSE? Only extended objects have angular momentum.

25) The rotational equivalent of "force" is: a) angular momentum   b) torque   c) moment of inertia

26-30) Matching: A) acceleration   B) velocity   C) newtons   D) distance   E) time

26) _____ sec   27) _____ m   28) _____ kg m/sec²   29) _____ m/sec²   30) _____ m/sec

31) TRUE or FALSE? Increasing the length of a lever arm increases torque.
32) TRUE or FALSE? Angular momentum is always perpendicular to the circular motion.
33) A solid cylinder and a hoop had a race. Who won? _____
34) When an ice skater speeds up by pulling her arms in closer to her body, this is an example of what principle? _____ of _____ _____
35) Which has a larger moment of inertia, a hollow cylinder or a solid cylinder? _____

36-40) What does each formula calculate?
36) L=rmv _____
37) Fc= mv2/r _____
38) p=mv _____
39) I = MR² _____
40) L = $I\omega$ _____

ACTIVITY 9.22: Review crossword puzzle (Answers are in teacher's section.)

ACROSS:
5) He was born in the year that Galileo died.
6) The force that works against motion (not inertia)
8) The measure of how fast velocity is increasing
9) Hoops, rings, and hollow cylinders have a large _____ of inertia.
11) The measure of distance in a certain direction (pg 113)
14) The measure of how resistant an object is to a change in velocity
16) How fast something is going, but with direction unspecified
17) The first scientist to record and analyze motion with a camera
18) The shape you get when you follow one point on the edge of a rolling circle
21) The maximum velocity of a falling object
23) Rotational force, represented by the Greek letter "tau"
27) The point (0.0) on a graph
29) The rule that tells you which direction the angular momentum is pointing
31) The force that your feet feel as the ground pushes back against your body's weight
33) What "p" represents in the formula "p=mv"
34) In the expression "3x," 3 is the _____.
35) In a lever, the arm that bears the load is the _____ arm.
36) The gear that has the power behind it
37) This is the physics term for size

DOWN:
1) When Galileo evaluated the lengths and frequencies of pendulums, he discovered the _____ _____ law.
2) The end of a pendulum is called the _____.
3) There are two types of friction: kinetic and _____.
4) A type of force that does not require any contact between surfaces (gravity, for example) (pg 104)
7) The natural frequency of vibration (pg 50)
10) In an _____ collision, both momentum and kinetic energy are conserved (billiard balls, for example) (pg 108)
12) The general term for anything that moves or vibrates with a steady rhythm: _____ oscillator (pg 54)
13) A measure of both speed and direction
15) Pendulums that are found at the top of skyscrapers are tuned _____ _____.
19) In circular motion, the force that is directed towards the center of the circle
20) The tendency of an object to keep on doing what it is doing

21) Half of a brachistochrone curve, name means "same time" referring to its amazing property that balls reach the end at the same time, no matter where they start
22) When the earth sped up after an earthquake, this was an example of _____ of momentum.
24) A line that touches a circle in just one place
25) Foucault named this device; its name means "seeing turning"
26) The formula $L = I\omega$ is used to find the angular momentum of an _____ object.
28) A graph that plots points (x,y) is a _____ graph (named after its inventor)
30) This ancient Greek mathematician wrote a geometry book that was studied by Archimedes, Galileo, and Newton
32) Multiply your mass by the acceleration due to gravity, and you get your _____.

ACTIVITY 9.23: Read about other scientists who dropped cats

In the mid 1800s, before Marey came on the scene with his chronophotography, several notable scientists became very interested in "cat turning." as they called it.

George Stokes was a math professor at Trinity College, Cambridge, holding the same position that Newton had held 200 years previously. Stoke's Law and the Navier-Stokes equations (which are about the movement of fluids), and Stoke's Theorem (in calculus) are used by engineers and mathematicians today. He also made the discovery that fluorescence involves the conversion of invisible ultra-violet light into visible light. However, even brilliant men can be fascinated by falling cats. He is said to have done many cat dropping experiments. However, he never did figure out how they managed to use the conservation of angular momentum to do their split-second swivel. If he had, perhaps we would now have the Stoke's Theorem of Cat Righting. But there was another cat turner at Cambridge who became even more famous than Stokes.

James Clerk Maxwell was a student at Trinity College, Cambridge in the 1850s. He would become the greatest physicist of the 19th century. He made discoveries about light, lenses, gases, fluids, color vision, electricity, and magnetism. He also was an avid reader of classic literature and composed poetry. His most famous discovery was that light is only one part of an entire spectrum of energy called the electromagnetic spectrum. His brilliant math skills allowed him to come up with equations about electricity that are usually known as "Maxwell's equations." But while he was a student at Trinity College, he also studied cats. He did so many cat dropping experiments that rumors began to circulate that he was dropping cats out of windows. Years later, in a letter to his wife, he tried to set the record straight—he had never dropped a cat out of a window. He wrote:

There is a tradition in Trinity that when I was here I discovered a method of throwing a cat so as not to light on its feet, and that I used to throw cats out of windows. I had to explain that the proper object of research was to find how quick the cat would turn round, and that the proper method was to let the cat drop on a table or bed from about two inches, and that even then the cat lights on her feet."

Maxwell during his days at Trinity. He is holding a color wheel.

ACTIVITY 9.24: One last thing that deserves a mention

If we combine circular motion with pendulums we get a ***torsional pendulum***. A typical torsional pendulum in a physics lab consists of a wire with a heavy disc hanging from it. The wire is made of a material that allows it to twist in a "springy" way, so that the disc will rotate one way and then the other. The frequency of a torsional pendulum depends on the moment of inertia of the disc, plus the measure of how stiff the wire is. Disc are used in experiments so that calculation of the moment of inertia is easy: $1/2MR^2$. The other number needed, the torsion constant, is a little bit like the coefficient of friction because it has to be determined experimentally, measuring how stiff the wire is.

The harmonic oscillation of a torsional pendulum is very steady, just like the back and forth motion of a regular pendulum. Even before the official invention of the torsional pendulum in the 1700s, clock makers as early as the 1600s were already beginning to use torsional springs in their clocks, instead of long pendulums. This drawing shows a hairspring design by Christian Huygens, the brilliant Dutch inventor who succeeded Galileo in the 1600s. Eventually, they were able to make these "hairsprings" small enough to fit into a watch. The photo on the right shows a modern German pendulum clock, called the "anniversary clock" because you only wind it once a year. The balls on the bottom are the weights for the torsional pendulum.

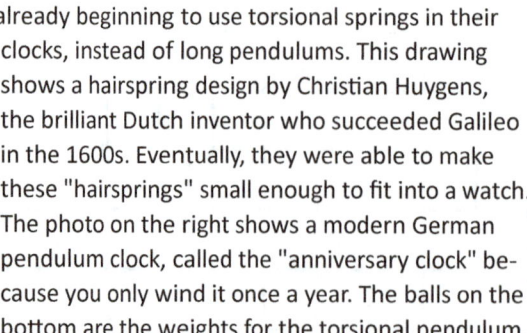

a watch spring design by Christian Huygens

torsional pendulum clock

Adventure 10: Playing and working

Sorry, guys, we've been saving the coasters as a way to introduce **potential and kinetic energy**. Roller coasters are one of the most obvious demonstrations of these terms. Kinetic energy is movement. The word "kinetic" comes from the Greek word for motion, "kinetikos." The modern word "cinema" (movies) comes from this Greek word, and at the time that the first movies were invented, they pronounced it "kinema."

Kinetic energy is present when something is in motion. We've seen a lot of motion in most of our adventures, so we've seen a lot of **kinetic energy**. The roller coaster car rushing down the track has an immense amount of kinetic energy. However, coaster cars have no engine, no motor. The steep track that drags the car up to the starting point does have a very strong motor, but once the car is released at the top, no force except gravity is operating for the rest of the ride. All of the car's energy had to have been present while it was sitting at the top, waiting to start plummeting down. We call this starting energy the **potential energy**.

As you know, the coaster car experiences the acceleration due to gravity, and goes faster every second than it did the second before. Once it reaches the bottom of the dip, it is experiencing maximum kinetic energy. All of the potential energy has been converted into kinetic. Then, it begins to rush up the next hill. Gravity pulls on the car and it experiences deceleration at it climbs. By the time it reaches the top of the next hill, it might be going very slowly. It can be scary for the riders, who might be afraid the car won't make it over the top. But the engineers who built the coasters did the calculations and made sure that the cars would have enough energy to make it over the hill. When the car

This is where ALL of the car's energy comes from—the ride to the top.

is at the top of this second hill, it has converted almost all of its kinetic energy back into potential energy, though some energy was "lost" along the way (converted into heat by friction). As the car begins to plunge downwards again, the potential energy is converted back into kinetic energy, and maximum kinetic energy is realized when it reaches the bottom of the hill. And thus it goes for the rest of the ride, energy being converted back and forth between potential and kinetic.

Motion needs energy. Motion can't happen without some type of energy being applied. Inanimate objects, like rocks, depend on external sources of energy. Living things have internal sources of energy (their own muscles) although this energy originally came from an external source (food).

At the top of a hill, most of the car's energy is potential energy.

The concept of potential energy includes more than just potential kinetic energy. **Anything that stores energy can be a source of potential energy.** Your food contains sugar and fat molecules that your cells can use for energy, so food has potential energy. A battery contains chemicals that can produce a small amount of electrical energy, enough to power a flashlight or a cell phone, so the battery has potential energy in the form of chemical energy. The sun can power solar cells, so sunlight has potential energy. Potential energy can also be stored in springs and elastic bands. Even atoms have potential energy, which is released during the process of fission (splitting).

sources of chemical energy

The type of potential energy seen in roller coasters is called ***gravitational potential energy***. To acquire gravitational potential energy, an object merely needs to be raised up higher than its surroundings so that it can fall back down. We can measure gravitational potential energy fairly easily. All we need to know is the mass of the object and how high it is off the ground. The potential energy equals the mass times "g" times the height.

$$PE = mgh$$

where "m" is mass, "g" is 9.8 m/sec^2, and "h" is height in meters. If the roller coaster car has a mass of 1,000 kg (which includes the weight of the people riding in it) and the height of the first hill is 30 meters, then the potential energy of the car is $(1,000 \text{ kg})(9.8 \text{ m/sec}^2)(30 \text{ m}) = 294,000 \text{ kg m}^2/\text{sec}^2$. Notice that the units in our answer come from all the units in the calculation: the kg weight of the car, the acceleration in m/sec^2, and the meters of height.

Just like we have a special name for units of kg m/sec^2 (newtons), we also have a special name for these units of potential energy, $\text{kg m}^2/\text{sec}^2$. We call them ***Joules (J)***. (Occasionally you might also see them called "newton meters" because they are just newtons plus an extra "m.") As you might guess, they are named after a famous scientist. But before we send our volunteers to meet James Prescott Joule, let's finish our discussion of measuring energy.

Kinetic energy is also measured in Joules. The formula for finding kinetic energy is:

$$KE = 1/2(mv^2)$$

where "m" is the mass of the object and "v" is its velocity in meters per second. Notice that "v" is squared. This means "v" times "v." Because "v" is in "m/sec" this will give us "m/sec" times "m/sec," giving us "m^2/sec^2," which is exactly what we need for Joules.

If our roller coaster car is going 100 km/hour when it reaches the bottom of the hill, what will its kinetic energy be? First, we must convert kilometers to meters, and hours to seconds, because Joules are defined in terms of meters and seconds. When we do that, we find that our coaster car is going 27.7 meters per second. Then we can use our formula: $(1/2)(1,000 \text{ kg})(27.7 \text{ m/sec})^2 = (1/2)(1,000)(27.7)(27.7) = 383,645 \text{ J}$ (Joules) One Joule (1 J) is a small amount of energy, so it's no surprise that a heavy coaster car going very fast has a high number of Joules.

ACTIVITY 10.1:   Try some math problems using these energy formulas   (Answers at the bottom of pg 164)

1) *Someone is holding a 5 kg cat at a height of 1 meter off the ground (getting ready to drop it). How much potential energy does the cat have?*
NOTE: Notice that in your calculation, you will use "mg," which we noted in a past adventure was how you find your weight in newtons. Therefore, we could substitute "N" for "mg" and write: PE= N(h). Since height will be measured in meters, we can say that the answer is in "newton meters" which is another name for Joules.

2) *A skydiver is just about to jump out of an airplane. The airplane is flying at an altitude of 3,000 meters (3 km). The skydiver weighs 80 kg. What is the diver's potential energy?*

3) A very small meteor with a mass of only one pound (2 kg) hits the moon while going 10 km/sec. (Meteors travel very fast.) How much energy was expended onto the surface of the moon? Make sure you convert kilometers to meters!
NOTE: Notice here that we don't need to account for the different gravity of the moon. The formula only asks us for the mass of the object and the velocity of the object. We don't need "g."

4) A car is going 30 km/hr. The car weighs 1,500 kg. How much kinetic energy does it have? (Don't forget to convert km/hr to m/sec. Round your answer to just one decimal point.)

5) The same car (1,500 kg) is now going 60 km/hr. How much kinetic energy does it now have? If you want to save time, you can avoid converting km/hr to m/sec by simply doubling the figure you got in the previous question, because 60 km/hr is double 30 km/hr.)
NOTE: Notice the increase in Joules when the velocity is doubled. Are the Joules doubled? More than doubled?

Okay, let's go meet James Prescott Joule. The S.N.O.O.P. machine should be set for England in 1845.

Joule wasn't a full-time scientist. He was the son of a brewer (someone who makes beer) and early in his adult life he inherited the family brewery business. He had to be at the brewery from 9 am until 6 pm five days a week, so science was more like a hobby. However, he did manage to attend Manchester University for a few years, and it was here that he met James Dalton, who would become known as the "father" of the atomic theory—the idea that matter is made of individual pieces called atoms. In the early 1800s, the atomic theory was not yet generally accepted, so having Dalton as one of his teachers gave Joule an advantage over other young scientists. Joule was also able to read widely and he eagerly read anything he found about the new science of electricity. He and his friends rigged up simple batteries and used them to shock each other (and other people, too, if they could get away with it).

Joule's fascination with electricity and magnetism continued into his adulthood and he made sure he stayed current with all the latest discoveries. People did not know exactly what electricity was at this point, but they had discovered how to use it to make simple motors. Joule wondered if perhaps someday his brewery might be able to switch from coal-powered equipment to electrical equipment. He eventually accomplished this.

Joule spent a lot time thinking about energy because for him it was a very practical subject. He was running a business that used a lot of energy. The only powerful engine of that day was the steam engine. A steam engine could run on anything that could make water boil—usually wood or coal.

This is an electric motor from the 1840s which Joule gave to his friend, the future Lord Kelvin.

In Joule's day, scientists did not yet understand energy and heat. The most popular theory about heat was that it was a self-repellent fluid called "caloric." This mysterious fluid always went from hotter things to colder things. This wasn't the first time that scientists had invented a mysterious substance to explain their observations. Attempting to explain combustion, scientists in the 1600s invented a mysterious substance called "phlogiston." Even the highly respected and brilliant scientist, Robert Boyle, was an advocate of the phlogiston theory. Eventually, oxygen was discovered in 1774, and soon the famous French chemist Anthony Lavoisier was able to provide a better explanation for combustion. Phlogiston theory was abandoned a few years after that.

In the 19th century, "caloric" was thought to be a hot substance that was transfered from object to object. Because caloric was considered to be a substance, it was widely believed that it could be created or destroyed. After doing many experiments, and discovering ways to make very precise measurements of temperature, Joule began to doubt the caloric theory. Dalton's theory about atoms and molecules might have influenced Joule's thinking. What if heat wasn't a substance at all, but merely the motion of atoms and molecules? Joule thought perhaps the motion of the molecules might be rotational motion. (In the future, it would be discovered that atomic and molecular motion occurs in straight lines, as Newton predicted for all matter.)

The actual equipment used for the experiment that our S.N.O.O.P. machine was watching is shown below.

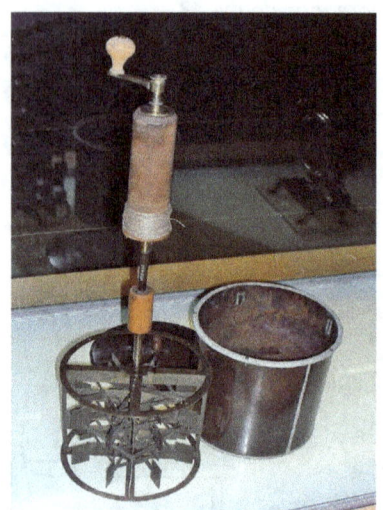 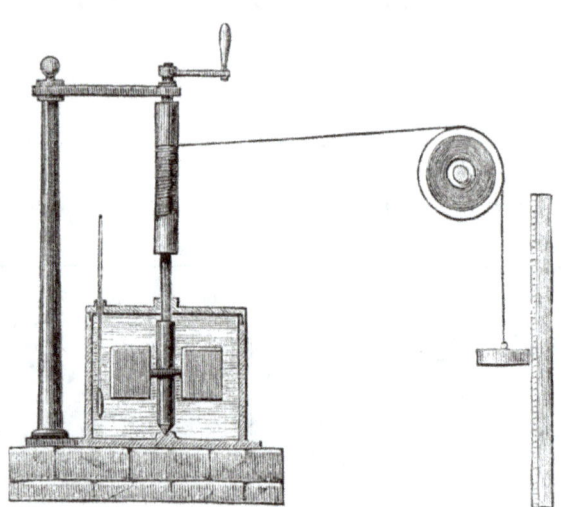 

James Prescott Joule

The metal bucket contained water. The paddles were inserted into the bucket so they could agitate the water. The stirring of the water would increase its heat because the stirring would increase the motion of the molecules. The diagram in the middle shows how the spindle and crank were attached by a string to a weight. The pulley is shown just hanging in midair, but it would have been attached to something, of course. The ruler could measure how far the weight traveled downward. Joule combined all the measurements—the mass of the weight, how far it traveled, how fast the spindle turned the paddles, and the temperature of the water before and after the experiment, and from this he concluded that almost all of the energy of the falling weight had been put into raising the temperature of the water. This was the beginning of the theory of the **conservation of energy**.

Joule designed other experiments to test his idea and they led to the same conclusion. Finally, he was able to publish a few papers about his work, and in 1847 Joule presented some of his work to the British Association for the Advancement of Science. **George Stokes** (who we met on page 158) and **Michael Faraday** (the biggest name at that time in electricity and magnetism studies) attended this lecture. Stokes and Faraday immediately accepted this new idea about conservation of energy. **William Thomson**, (who would later become the very famous Lord Kelvin after whom the Kelvin temperature scale is named), wasn't so convinced. However, Thomson struck up a friendship with Joule and even helped with more experiments. After several years of experimenting and thinking, Thomson finally decided that Joule was right, and in his later years, Thomson was able to provide further proofs that the theory of conservation of energy was, indeed, a physical law of the universe.

Joule's influence on science in the 19th century was profound. Today, he is remembered for: the First Law of Thermodynamics, the Joule cycle, the Joule effect, the Joule expansion, Joule's First Law, Joule's Second Law, and the Joule-Thomson effect. (These laws and effects, though very important in physics and chemistry, are outside the scope of this book, so we'll let you investigate on your own if you want to.) The "joule" was officially adopted as the international unit of energy in 1889, a few weeks before Joule's death.

Answers to 10.1: 1) 49 Joules 2) 2,352,000 J 3) 10,000 Joules 4) 51,667.5 J 5) 206,670 J

ACTIVITY 10.2: Online lab: "Energy Skate Park"

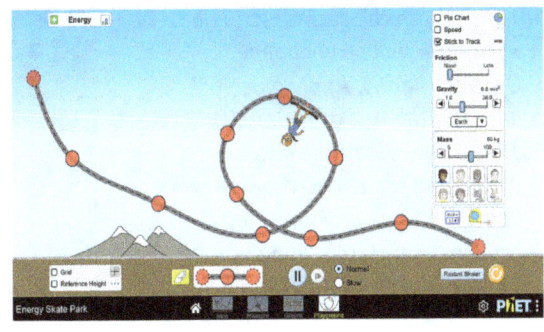

Take a few minutes to explore an online simulation lab at:
https://phet.colorado.edu/en/simulations/energy-skate-park-basics
This is the closest we could get to a roller coaster lab. In this lab you'll be able to place a cartoon skateboarder onto a track you've designed, and adjust the mass of the rider, the amount of friction, and the strength of gravity, and then release the rider and see if he/she makes it to the end of the track, having navigated up or over any twists, bends, or loops you put into the track. Hopefully you'll provide enough potential energy at the beginning of the track so that your rider won't fall off during the loop?

The simulation will provide a meter that constantly check the rider's potential and kinetic energy, as well as thermal energy generated. Friction between the wheels and the track will generate heat energy. The simulation uses Joule's principle of conservation of energy, so the potential energy at the beginning will always be equal to the combined kinetic and thermal energy during the ride.

There are two levels for this lab. If you like this one and want to do more investigation with more measure, click on the word SIMULATIONS on the top menu bar. Choose PHYSICS from the drop-down menu. Then look for "Energy Skate Park" without the word "basics" after it.

ACTIVITY 10.3: Watch a few roller coaster videos

The playlist has some videos about the physics of roller coasters.

ACTIVITY 10.4: A "loopy" lab that combines potential and kinetic energy with centripetal force

You will need*: copies of the following pattern pages printed onto cardstock, scissors, glue or glue stick, clear tape, a few marbles, two meter sticks, your cup from activity 7.2 (or make a new one)*

<u>How to make the apparatus</u>:
1) Print pattern pages onto cardstock. Cut out pieces. You should have 6 long pieces and one "closed C" shape (circle with slit).
2) Fold carefully on dashed lines to make the long pieces into "troughs" through which marbles can roll. Make sure that the pieces with the "ladder lines" get folded so that the short lines are on the inside of the trough. For the other long pieces, it doesn't matter whether the dashed lines are on the inside or outside of the trough. (The folding process might be easier if the dashed lines are on the outside because you can look at the dashed l lines while folding.) Fold carefully, keeping the fold right on those dashed lines.
3) Cut slits in the sides of two of the plain troughs (without ladder lines). The spacing between the slits should be about 2 cm.
4) Apply glue stick to one half of the circle. Start right at the slit, and bend and stick one of the long pieces shown in (3). The slit flaps will overlap a bit.

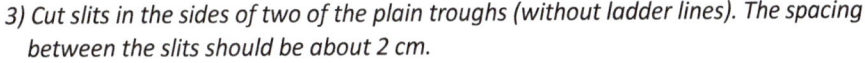
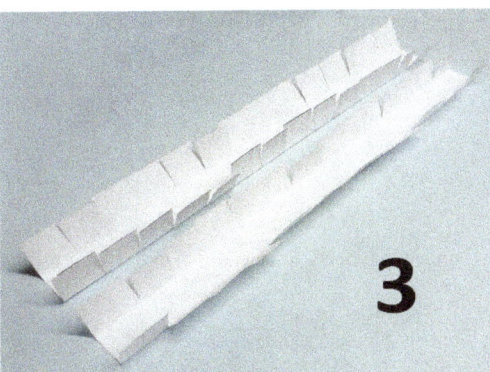

5a) Begin with the second strip right where the first strip ended. Overlap slightly. Apply glue stick all along the rest of the circle.

5b) Press the rest of the second strip along the circle. You will end up with a tiny bit of circle left. You'll use this remaining piece of the circle in just a minute.

6) Cut a few slits in the end of another long trough piece.

7) Apply glue stick to the remaining part of the circle and then stick the slit end of the piece from (6) to it. You'll be left with the rest of this third piece sticking off. This will be where your marble exits after going over the loop.

8) Use clear tape to secure the side of the circle that wasn't glued, as shown in 8 (close up). Next, secure the loop itself, as noted in (8).

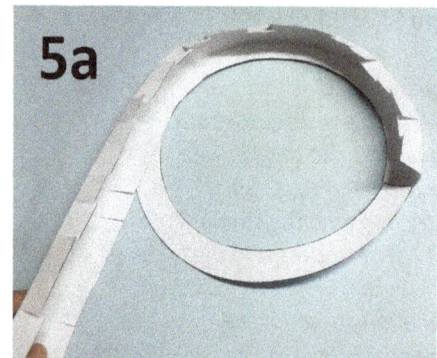

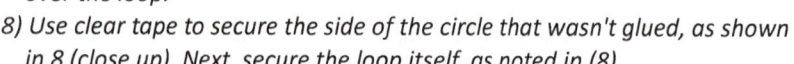
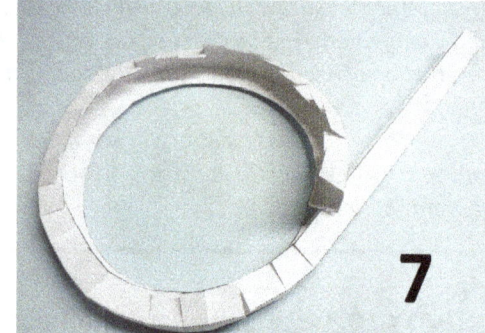
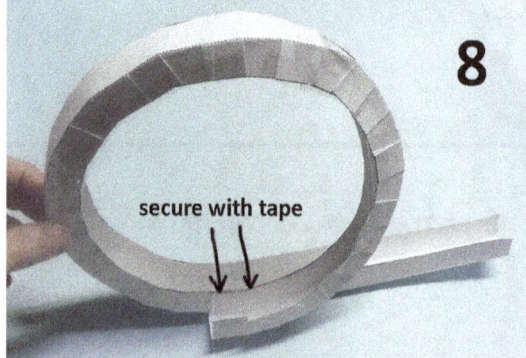
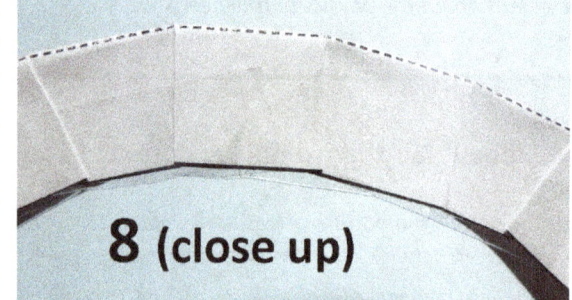
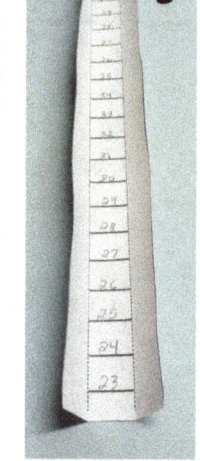

9) Take one of the "ladder" trough pieces and write the numbers 23 to 40, one number per line.

10) Glue the other "ladder" piece to the bottom of the number piece. You can trim one of the pieces just a bit if you want the ladder stripes to match up perfectly.

11) Use your last trough piece to connect the latter trough to the loop. You'll have to make a few snips in this last piece just at one end, so that it curves up a bit. To see the ideal amount of curvature between the straight ladder piece and the loop, see the diagram that is inside the circle pattern.

12) The last step is to secure your tiny roller coaster using chairs, boxes, table legs—whatever works for you. The coaster needs to be firmly in place so it does not juggle much when marbles roll through it.

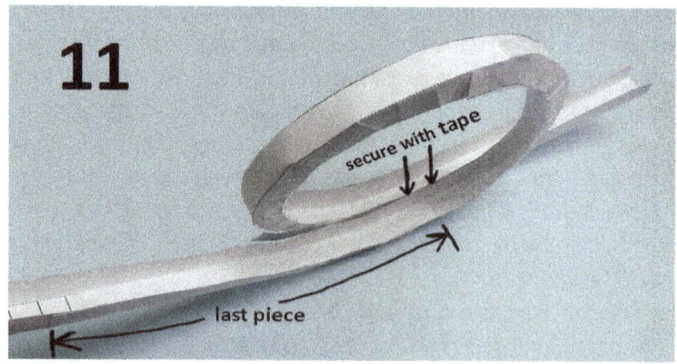
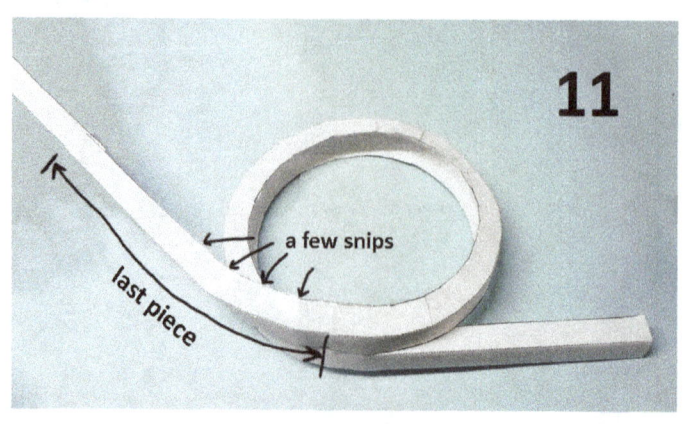
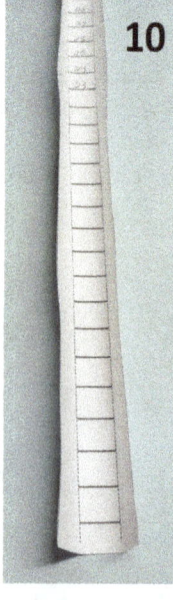

MAKE 2 COPIES OF THIS PAGE ON CARD STOCK
Cut on the long, solid lines. (Don't cut the short lines.)
Dashed lines are for folding.

MAKE 1 COPY OF THIS PAGE ON CARD STOCK
Cut out ring. Also cut the slit on the bottom of the ring.
You won't need the center circle piece, so we used it for the diagram of set-up.

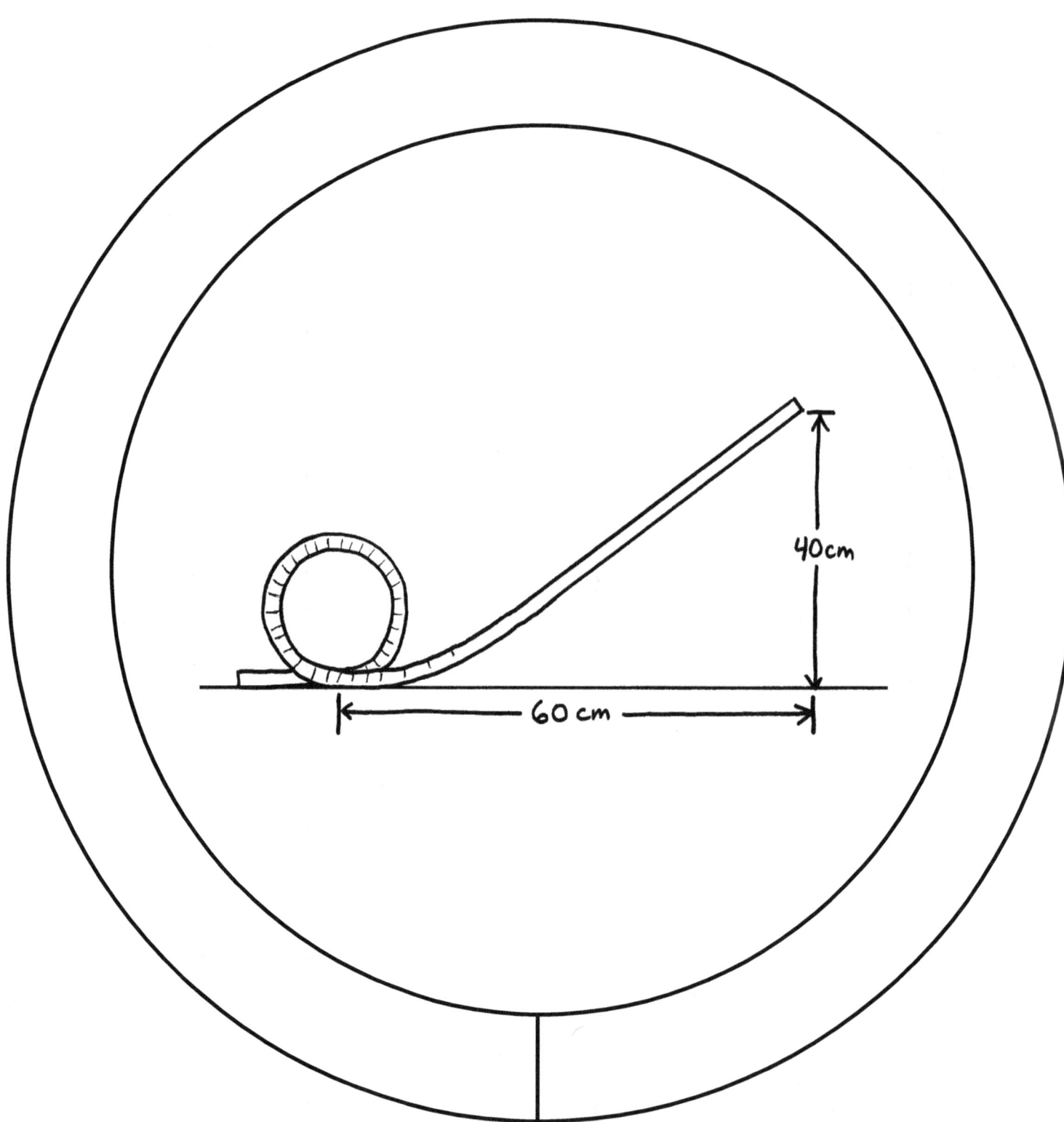

12) continued: We can't give you exact instructions for setting up your paper roller coaster because each person or class will have a unique selection of objects to choose from. Chairs are a good choice for securing the first straight part, but you can use anything you have at hand, boxes, stacks of book, table legs, etc. As shown in the diagram (inside the circle), make sure that the end of the track (the "40" line) is exactly 40 centimeters above the floor. Use your meter stick to check placement. Tape the track to the floor if necessary.

This photo shows one solution that someone came up with. Your solution won't be exactly the same as this one, but this gives you an idea of what we have in mind.

Collecting your data:

1) First, you need to experiment a bit with your set-up. Spend at least 5 minutes rolling marbles through the loop. Does your loop shake as the marble goes through it? If so, some of the marble's kinetic energy is going into this shaking, making the marble have less kinetic energy as it comes out, so you'll want to find a way to make sure the loop is stable and doesn't move. When you are sure your track is running well, start collecting data.

2) Now start collecting data. Hold a marble right at the 22 centimeter line. Let it go and observe what happens. Do this 4 more times. If you get the same result 5 times in a row, record the results in the chart below. Put a check mark in the appropriate box under 22 cm. The reason we're going to do 5 trials at each line is because tiny imperfections in the set-up might cause a different result once in a while. For example, just when you are very sure that the marble will make it over the loop every time at a certain height, you'll get one trial where, for some reason you can't determine, it won't make it over. We are going to have to ignore these occasional abnormal results and just go with what usually happens. We're establishing "5 in a row" as the criteria for a normal result.

3) Now start a marble at the 24 cm line. When you get 5 in a row of the same result, check the appropriate box under 24 cm.

4) Continue on up the "ladder." Eventually, you'll reach a point where you need to observe carefully what the marble is doing inside the track. The difference between the marble staying on the track the whole time and the marble completing the circular motion but losing contact with the track, is very important to roller coaster engineers.

WHAT HAPPENS TO THE MARBLE?	22 cm	24 cm	25 cm	26 cm	27 cm	28 cm	29 cm	30 cm	31 cm	32 cm	34 cm	36 cm	38 cm	40 cm
Doesn't make it to the top														
Makes it to the top, then drops														
Makes it around the loop but loses contact with the track														
Makes it around the loop and stays in contact with the track														

We will analyze these results in a minute, but first let's generate some numerical data.

Calculating potential energy:
How much potential energy does each marble have at each height?
We learned that **PE=mgh**. We can use this formula to calculate the potential energy of our marbles. We won't do every number, just a few at key points. The mass of a marble is about 5 grams.

1) EX: Find the potential energy of a marble that is at a height of 22 cm.
   Remember that to use physics formulas, you units must be in meters and kilograms. After we convert to this we have:
   PE= (.005 kg)(9.8 m/sec^2)(.22 m) = 0.01078 J (the answer is in joules)

2) Find the PE of a marble at height 26 cm: _____   At 27 cm _____

At 28 cm: _____   At 30 cm: _____   At 40 cm: _____

Don't forget to put a "J" for "joules" after each number!

Do your values for potential energy make sense? You should have found that the marbles have much less than a joule of energy. A joule is the amount of energy expended when the force of 1 newton is applied over one meter. Remember the 1 N (102 gram) candy bar on page 100? If you drop that candy bar from a height of one meter, it will hit the floor with one joule of energy. It also takes one joule of energy to lift it back to its original height of one meter. (You might also think of a small apple being dropped, instead of a candy bar. Apples make us think of Isaac Newton, and a joule can also be called a newton-meter.) If a tennis ball is moving at 6 meters per second (the speed it would travel if you gave it a gentle toss), it has a joule of energy. As long as the ball is in motion, it still has that one joule of energy. A marble is much less massive than an apple or a tennis ball, so it isn't surprising that our calculations gave us a fraction of a joule.

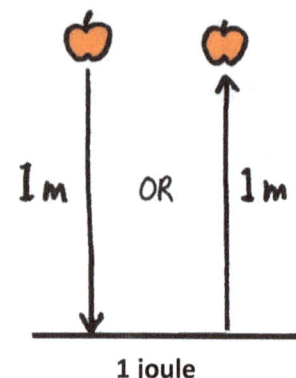

1 joule

What about kinetic energy? The marble's maximum kinetic energy occurs as it reaches the bottom of the track, just before it starts upward into the loop. Would it be easy to calculate its kinetic energy?

The formula says that KE= $1/2(mv^2)$. We need to know mass and velocity. The mass is 5 grams. How would we calculate the velocity of the marble? One way to do this would be to use a stopwatch to time how long it takes for the ball to go down the ramp. However, our reaction times (how fast we can press start and stop on the stopwatch) are too slow to give accurate results. The marble moves too fast. It's possible to figure it out using just math (no stopwatch) but the process is very complicated. Therefore, we're not going to make you calculate the kinetic energy.

Now let's talk about your data chart. At what height did the marble finally make it over the loop without losing contact with the track? Was it close to 27 centimeters? If your experiment was absolutely ideal, with a perfect track and perfect data collection, and your loop was a perfect circle with a radius of exactly 10 centimeters, then a starting height of 27 centimeters would prove to be the critical point where any starting height below that would not maintain contact with the track, and anything above that would stay on the track. We know this because physicists used math to prove that the starting height of the ball must be at least 2.7 times the radius of the loop before a ball will travel all the way around the inside. For our loop, 10 cm x 2.7= 27 cm.

ACTIVITY 10.5: Watch this experiment on a perfect track

Find the video called "Loop the loop" by TSG Physics on the YouTube playlist. You'll see a physics teacher (or perhaps a graduate student) performing the same experiment that you just did. The track probably isn't perfect, but it is close enough. They'll show you each run in slow motion so you can see exactly what the ball is doing. They use three heights: 2 times the radius, 2.5 times the radius, 2.7 times the radius. They don't tell you what the radius is, but that doesn't matter. You only need to know how much taller the initial point is than the radius. You will see "ho" used for the initial height. A zero is often used to represent the word "initial" because zero is the ultimate starting point. Makes sense. Scientists use the British word for zero, "naught." So "ho" is pronounced "h naught."

At this height the ball does not maintain contact with the track

The second part of the video explains how the numbers 2.5 and 2.7 were determined. If you like math, you can try to follow the logic they are using. They use equations we have learned about, including the formula for a rolling solid sphere ($2/5MR^2$ on page 152), but the calculations go by pretty fast.. If the math is beyond you, don't worry about it; just skip it.

Since engineers don't want coaster cars falling of the tracks at the top of the loops, they need to know the minimum height for the loop. No worries, though—coaster cars that go around loops have wheels underneath the tracks so that it is impossible for them to fall off! However, engineers still do these calculations because there are other factors involved, too, such as the comfort of the riders. The engineers need to take into consideration how many G's the riders are experiencing at every point on the ride. There were some mistakes made on the very first "Loop the Loop" roller coasters.

Oops! The men who designed and built this first Loop the Loop used the same design you did—an almost perfect circle. They didn't do calculations ahead of time to figure out what the G forces would be as the car went into the loop. Many people came out of the ride unconscious; some had injured necks and backs. It turned out that as the car started into the loop, the sudden change in direction caused a G force of 12-14. Even military pilots black out when the G force goes above 9. The amazing thing is that the ride was not shut down immediately. Perhaps the time that this G force was applied was short enough that

some people's bodies were able to immediately recover, and enough people recovered that they decided to keep the ride open. Soon after this, engineers began designing loops that were shaped like an upside down teardrop, making the approach into the loop a much gentler curve. Just by making this small adjustment in the shape, they were able to bring the G forces down to a safe but exciting 4 or 5.

The man who started Sea Lion Park was quite a showman and did not mind danger. If you want an interesting person for a history report, check him out. His name was Capt. Paul Boyton, an Irish immigrant who lived out his "American dream." He got his start by being an aquatic stunt man. He had someone build him a rubber suit that turned him into a "human boat." He would float down major rivers, even going over waterfalls and perilous rapids. If the journey was a long one, he would tow a waterproof suitcase behind him, containing jars of fresh water and some emergency food. One of his more dangerous stunts almost killed him. He tried to get on board a steam liner that was headed across the Atlantic Ocean, intending to jump off the boat when it was several hundred miles out, and swim or float his way back to shore. He was open about his intent ahead of time, trying to get publicity. So when the captains found out about his crazy idea, none of them would take him on board. Finally, he had to sneak on board, disguised as a cargo loader. When he got out his suit and tried to jump overboard, the ship hands stopped him. Finally, he managed to convince them to let him jump as they got close to their destination, Ireland. Right after he jumped, a hurricane storm hit, and this almost ended his life. Miraculously, he made it to shore, and what a site he was for the people who saw him emerge from the ocean wearing a rubber suit and covered in seaweed! His story hit the press the next day and he received many invitations to speak about his adventures.

After several decades of crazy aquatic stunts all over Europe and the Americas, he decided to settle down in New York and open the world's first amusement park. Until then, no one had ever thought of the idea of having a huge park with many rides inside it. His park also featured a sea lion show, of course, since it was called Sea Lion Park.

View of Sea Lion Park from the top of the Shoot the Chutes ride.

Sea Lion Park would close in 1902, after only 7 years in operation, but in its place would be several other parks. Boyton's popular Shoot the Chute ride (perhaps based on his love of riding down rivers) was purchased by the next park owner and survived for 40 more years. The Flip Flap Railway was replaced by safer Loop the Loops.

The amusement parks of Coney Island began to deteriorate by the 1960s and were never restored to their former glory. Today, there are still some rides along the beach, but nothing like what was there during the island's heyday in the early 1900s.

Coney Island today

ACTIVITY 10.5: The Roller Coaster Song

Need some help trying to remember all these formulas? Music to the rescue! The adults in your life have their heads stuffed full of silly songs and TV commercials they heard when they were kids. The human brain is great a taking in information in musical form. So why not put this genius ability of your brain to good use and learn this science song? If you take more physics in the future, this song will give you a head start on learning the many basic formulas you meet in physics. The sillier the song, the more likely it is to get stuck in your head, and the more stuck it is, the better.

The soundtrack for this song is posted at **www.ellenjmchenry.com/music**

HISTORICAL NOTE:

The tune was borrowed from a very famous song around the turn of the 20th century: "The Good Old Summertime." This song was one of the "top ten" during the years that Sea Lion Park and Luna Park were open. In 1903, the famous march composer, John Philip Sousa, recorded this song with his band. You can hear it if you search YouTube for "John Philip Sousa Good Old Summertime." Calliopes (automated music machines) at circuses and amusement parks were often programmed to play this tune.

The song stayed popular until the 1930s and was recorded by famous singing stars such as the Andrews Sisters. In 1949, Judy Garland (of "Wizard of Oz" fame) and Van Johnson made a movie called "The Good Old Summertime," and opened and closed the movie with this song. (The movie is a freebie on YouTube.)

This is how people dressed at the time the song "Good Old Summertime" was written. It was called the Edwardian Era.

The Roller Coaster Song

Tune based on "The Good Old Summertime" (a very popular song in America from 1902 through the 1930s)

I love the days of summer, but one thing do I fear,
I hate the roller coasters during this time of year.
My friend said, "Not to worry—just think of math instead.
And concentrate on formulas and science that you've read."

So off to the roller coasters I went with my new plan,
And as the car got near the top, my strategy I began.
I lost my hat atop the hill, but loudly I declared:
"The distance that my hat will fall is $1/2\ gt^2$!"

The height of that roller coaster was a bit too much for me,
But I knew height's important for potential energy.
"I know PE is "mgh,"— converted it will be,
As gravity turns it all into kinetic energy."

"It's only math," I told myself, why should I be so scared?
I know KE is nothing more than $1/2\ mv^2$;
And if you multiply "ma," it's Force that you will find;
A's acceleration—velocity over time.

The scariest part was coming-- I hoped that I'd survive;
As we went racing through the Loop, the G-force went to 5.
I kept on thinking math-y thoughts and gripped the safety bar,
"Centripetal force is 'mv' squared, divided by radius, r!"

Along the straight parts of the track, momentum is "mv."
It's odd that for momentum, we use the letter "p."
And at the next loop, I yelled out, "L is mvr!"
As angular momentum was experienced by the car.

The forces that I'd been feeling during that frightful ride
Are properly measured in newtons—"m" and "g" are multiplied.
Sir Isaac and his motion laws are dear to me, and I'm
Thankful they will help me though the good old summertime!

Well, that was a fun little rabbit trail, but now that we've had some play time with roller coasters, we need to get to work and talk about— **work**.

In science, the definition of work doesn't have anything to do with household chores or going to an office. Here are several scientific definitions of work. They all say basically the same thing, though some go into more detail than others.

WORK: Force multiplied by distance. **W = F(d)**
WORK: A force applied through a distance.
WORK: When an external force moves an object through a distance.
WORK: The energy transferred to or from an object by the application of force along a displacement.
WORK: The measure of energy transfer that occurs when an object is moved over a distance by an external force.

Which one do you like best? Notice that the more technical definitions use "displacement" instead of the word "distance." As we learned on page 113, displacement is the movement of an object in a particular direction. Displacement is like velocity in that it has both magnitude and direction.

Let's think about scenarios that might, or might not, involve work:

1) Imagine someone trying to push a car. They work up quite a sweat as they try to budge it. After 30 minutes, they are a sweaty mess and finally give up. The car never moved. Did they do any work? Their muscles are tired and they are covered in sweat, but the car did not move, so, technically, no work was done.

2) Imagine a penny lying on the table in front of you. You have a penny collection, so you pick up this penny and put it into your pocket. Did you just do some work? Yes, because an external force (you) moved an object (the penny) through a distance (from the table to your pocket). It didn't seem to you like work, but according to the physics definition, that was work.

3) You are sitting still in your chair, thinking about your homework. You are memorizing some formulas and you go over and over them in your mind. You think of some tricks that might help you remember them. Then you recite each formula ten times. Have you done any work? If you were perfectly still while doing this, then no work was done.

In examples (1) and (3), an element of work was missing. **Work requires motion**. That's why work is in this book about motion. All of the people in the pictures below are participating in mechanical work, though only the boy taking out the trash would call what he is doing "work." (He seems to have paused his work here in the photo, but we can assume that just a few seconds ago he was rolling the can down the sidewalk.)

Someone's paying attention! Yes, joules are a very handy unit that can be used in a number of ways, including measuring work. It actually isn't surprising that we would use the same unit for both energy and work because the two are related. **Energy is the ability to do work**. "Energy" is the potential, and "work" is the action. Just as we use joules for both potential and kinetic energy, we also use joules for both work and energy. In fact, joules can be used to measure many types of energy. One joule is the amount of energy it takes to run a 1-watt device for 1 second. A joule is the amount of energy it takes to raise the temperature of 1/4 gram of water one degree Celsius. A food calorie is a tiny fraction of a joule (.00024 J). Your body releases about 60 joules of heat energy every second. But instead of just throwing more facts and definitions at you, let's do something more creative. How about a very silly story that makes our scientific points, AND entertains you? We'll have to go a little bit off-topic here, straying away from our theme of "discovering motion," but as long as we are discussing joules of energy, it's worthwhile to learn a little more about this energy unit, and about energy in general.

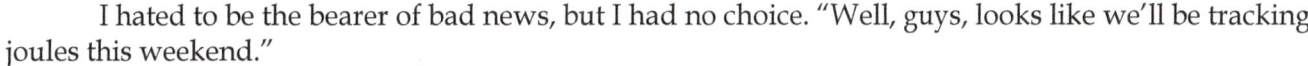

"THE CASE OF THE MISSING JOULES"

It was Friday afternoon at the detective agency, and we were all looking forward to the weekend. If the phone stayed quiet and no one came through that door, we'd be out into the sunshine in less than an hour. But it wasn't meant to be. Not this weekend."

The phone rang and I reluctantly picked it up. "Hello," I said very flatly. "You've reached the Decent Detective Agency. No one is here right now to take your call. If you leave your name and—"

"Smith!" came the voice at the other end. "This is Sergeant Argent down at headquarters. I know you're there! I've got an emergency at the power plant. There's a million joules missing. Probably stolen. Get down here on the double!" Click.

I hated to be the bearer of bad news, but I had no choice. "Well, guys, looks like we'll be tracking joules this weekend."

"Don't tell me the crown jewels are missing again," groaned Lefty.

"She can get along without them for a while," said Righty.

"No," I said, "this time it's joules, not jewels. You know, joules, as in energy units. Make sure we've got conversion tables in our gear. Let's go."

As we arrived at the power plant we couldn't see anything unusual. The supervisor met us at the door. "Thanks for coming, gentlemen," he said.

"Can you show us to the scene of the crime?" I asked.

"Not really," he said. "You see, energy is invisible. We don't really know what happened. All we know is that a million joules are missing."

"Any clues at all?" asked Lefty.

"Our only lead," said the supervisor, "is that one of our employees quit at exactly 3:00, the same time that we realized the joules were missing." "So you think he took them?" asked Righty.

"That's a possibility," he said.

"What was the employee's name?" I asked.

"John Doe."

That clue was all I needed to start the chase. "That's one of the aliases used by the notorious energy thief, Jon Dough. Come on, guys. We've got a criminal to catch."

The supervisor had one warning for us. "You know, of course, how hard it will be to find the evidence. Energy can change from one form to another."

"That's okay," I called back as we peeled out of the parking lot. "I've got conversion tables!"

The tires screeched and we were off. We had dealt with Jon Dough before. He had robbed a bakery last year, and a bank the year before. We knew right where to start our search: the county fair. Dough's favorite trick was to get lost in a crowd.

As soon as we pulled into the fairgrounds, Lefty spotted him. "There's that ring-head! Standing in front of the cotton candy booth!"

"That'll be his first conversion," I wagered. "He'll try to throw us off by converting the joules into food calories."

"Maybe he's just in line," said Righty.

"Get out the conversion tables, Lefty," I said. "Tell me how many kilocalories a million joules will make."

Lefty glanced at the tables. "To convert joules to kilocalories, you multiply by .0002389. That's one million times .0002389. Hmm...that's a big calculation."

"It's easy," I said. "Just think of it the other way around. It's .0002389 times a million. Just slide the decimal point over six places. That makes 238.9 food calories. So we're looking for about 240 calories.

"How much does a bag of cotton candy weigh?" I asked Righty, knowing he had extensive experience with cotton candy.

"65 grams," he said as he wistfully watched the cotton candy being made.

"That's it!" I announced. "If you multiply the weight of the sugar times the amount of calories in each gram, you get 240! That's exactly the amount of energy we are looking for. Undoubtedly, Dough is converting those joules into food calories to throw us off the trail!"

Just then, Dough left the vending stand. Sure enough, he was holding a big wad of cotton candy.

"Look, he's not even eating it," said Lefty.

"We'll follow him closely, before we nab him," I said. "We have to get more evidence."

Our suspect headed for the animal judging tent. We followed him in and watched his every move. Righty slapped me on the arm. "Look! Look!" he said. "He's feeding the cotton candy to a horse!"

Sure enough, there was Dough, holding the candy while the horse munched away.

"There goes our evidence!" moaned Lefty.

"Not so fast," I said. "Get out the conversion tables again. Convert kilocalories into horsepower."

Lefty looked it up and Righty did the math.

"240 times .00156. The answer is .374 horsepower per hour."

"Hmm..." I mused.

Our suspect was talking to the owner of the horse. The owner untied the horse and began leading it to the competition area. The pulling contest was about to begin. The horse was hitched to a sled with a weight sitting on it. A voice over the loud-speaker began to announce, "Ladies and gentlemen, the pulling contest will now begin."

I thought fast. "Using the conversion of one horsepower equal to 33,000 foot-pounds per minute, that means if this horse pulls that 1000-pound weight for 11.45 feet, we may have enough evidence to at least start an investigation." The horse began pulling. Its muscles were busy converting food energy to kinetic energy. It pulled the half-ton weight exactly 11.45 feet, then stopped.

"We've got him!" cried Lefty.

"Get out the handcuffs!" yelled Righty.

"Not so fast, guys," I said. "What do we have for evidence? All the energy has now been converted into motion, plus a little heat from friction. How are we going to take that to court? We can say we saw it, but we don't have any hard evidence."

"You mean he's going to get away with it?" asked Lefty.

"The visible evidence is gone," I said.

"That dastardly Dough is a mastermind of crime!" said Righty.

"That's the advantage of doing energy crimes," I explained. "You trade the energy around into different forms, but it eventually ends up as non-recoverable things such as movement or heat. Motion and heat are happening one minute, and are gone forever the next."

"So he got away with it?" said Lefty.

"Yeah, this time," I said. "Maybe next time he'll make a mistake and we can catch him with potential energy on his person. We'll call into the station tonight and report the situation to the Sergeant."

"Hey! It's after five o'clock!" said Righty. "Since we are already here, how about a Friday night out at the county fair?"

"Sounds like a plan," I replied.

Hopefully, the science behind this story came through loud and clear. Energy is always conserved, meaning that it can't be created or destroyed, though its form can change. In this story, we were not told what the original joules were, but we saw the energy going from chemical potential energy (from food) to kinetic energy as the horse used that chemical energy to pull the sled.

Many scientists have done experiments similar to the one that James Joule did, finding ways to measure various types of energy. This painting show James Watt, who invented ways to turn steam energy into kinetic energy. A unit of energy, the Watt, is named after him. (We'll say more about Watt a few pages from now.) Because of the ability of these scientists to do precision measurements, we now have conversion tables that allow us to figure out exactly how energy goes from one form to another.

Here is the conversion table that the Decent Detective Agency was using. To convert from the unit in the first column to the unit in the second column, multiply by the conversion factor in the third column.

The Decent Detective Agency's Energy Conversion Tables

To convert:	To:	Multiply by:
BTU (British Thermal Units)	Kilowatt hours	.0002928
BTU	Horsepower	.0003931
BTU	Kcal (food calories)	.252
BTU	Joules	1,054.8
Horsepower hours	Kilowatt hours	.7457
Horsepower	BTU	2544
Horsepower	Kcal (food calories)	641.1
Horsepower	Joules	2,684,000
Joules	Kcal (food calories)	.0002389
Joules	BTU	.000948
Kcal (food calories)	Horsepower	.00156
Kcal (food calories)	BTU	3.968
Kcal (food calories)	Joules	4,186
Kilowatt	Horsepower	1.34
Kilowatt hours	Joules	3,600,000

ACTIVITY 10.6: Try some energy conversions (Answers are at the bottom of page 180.)

You will need: *a calculator*

Help the detectives get a head start on their week's work by using table above to make these energy conversions. Thanks!

1) On Monday, they will need to know how many joules of energy are in a slice of pizza that contains 300 food calories. (Food calories are also known as kilocalories.)   300 kcal = _____ joules

2) On Tuesday, they will be flying to England and will need know how many British Thermal Units (BTU) are equivalent to 15 horsepower.   15 horsepower = _____ BTU

3) On Wednesday, they will be back from England feeling very jet-lagged and needing to finish the case of the missing BTU. They will need to know how many BTU it takes to make 10,000 joules. 10,000 joules = _____ BTU

4) On Thursday, they will be investigating a cafeteria food fight and will need to know how many calories are equivalent to 12 horsepower hours. (Don't ask about the horse in the cafeteria. It's a long story.)   12 horsepower hours = _____ kcal

5) On Friday, they will be once again looking forward to an uninterrupted weekend and will need to know how many kilowatt hours they can expect from their horse-powered lighting system, which provides up to 5 horsepower hours from a single horse. (The horse-powered light is another long story.)
5 horsepower hours = _____ kilowatt hours

6) On Saturday, they will be putting in overtime, solving the case of the missing detectives. They will need to know how many kilowatt hours a million BTU will make.
1,000,000 BTU = _____ kilowatt hours*

*We'll learn about watts at the end of this adventure.

Let's stay in math mode and talk about calculating work using W=Fd. In the story, we read about a competition where horses pulled heavy weights. The weights they use in these events are extremely heavy, so the horses have to work very hard to pull them. It's easy to see that work is being done by the horses as they strain and sweat.

We can make the pulling horse into a free body diagram. We saw free body diagrams at the end of our adventure with friction. Anyone can draw a free body diagram—no art skills are required. All objects are represented as boxes with a dot in the center (at the center of mass). You don't have to draw the source of the force, just an arrow. (This means we don't have to draw a horse.) Forces acting on the box would include friction, gravity, the normal force, and pushes or pulls called "applied" forces. What forces are in play as the horse pulls on the weight?

You should recognize F_f, F_g, and F_n. These are the forces of friction (f), gravity (g), and the normal force (n). The force F_h is the force of the horse (h). Look at the sizes of the arrows. The length of the arrow shows you the

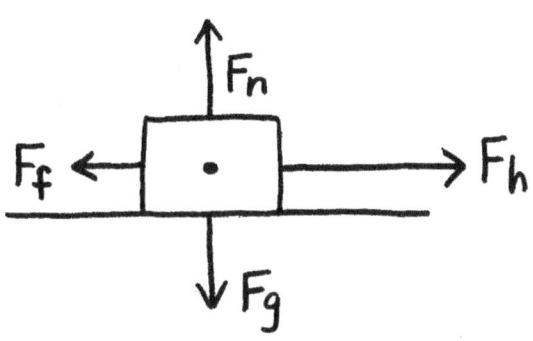

magnitude of the force. The arrows for gravity and the normal force are the same size. Even though we think of the normal force as coming up from the ground, pushing back against the gravitational force that is pulling things down, we can show it above the block. It doesn't matter whether the F_n arrow is drawn above or below the box. Because the F_n and F_g arrows are the same size, this indicates that they are balanced. If we removed the normal force arrow, we would have an unbalanced situation and the block would begin moving down. If we removed the gravity arrow, the block would then be moving up. As it is, these forces are equal and opposite. The friction arrow is smaller than the horse arrow, indicating that at this moment the horse is winning the battle against friction and the block is accelerating to the right.

In a real-world scenario, we'd have to find out what the coefficient of friction is between the block and the surface of the ground, and factor that into our calculations. If the block is resting on ice, the horse can apply a smaller force and still move the block. If the surface is concrete, the horse's work will increase dramatically. Friction greatly complicates the math. We are going to find work-arounds so we don't have to calculate F_f.

In all of these examples, we are going to assume that $F_n = F_g$, and because these are equal and opposite, they will "cancel out," so to speak. Thus, we can ignore these forces and just look at F_f and F_h.

1) If the horse is pulling with a force of 500 newtons across a frictionless surface, and it pulls the box a distance of 10 meters, how much work was done?

We use W=Fd, which tells us that work is equal to Force times distance. Force is always in newtons and distance should always be converted to meters. Fortunately, we were give newtons and meters, so we don't have to do any conversions. Since the surface has no friction, we don't have to subtract anything away from the pulling force. W= (500 N)(10 m)= **5,000 joules** (Don't forget, joules are also known as newton-meters, N-m.)

2) The horse is pulling with a force of 500 N, the force of friction is 50 N, and the horse can only manage to pull the weight a distance of 90 cm.

First, we need to do something called "summing the forces." This means that we add up opposing forces and find the final result. We have the horse pulling to the right at 500 N, and friction applying a force to the left of 50 N. We need to subtract off the force of friction from the horse's pulling force. 500 - 50= 450. Second, we need to convert centimeters into meters. 90 cm= .9 m

Now we can use W=Fd. W= (450 N)(.9 m)= **405 joules**. This number is smaller mostly due to the fact that the horse pulled the weight for less than a meter, whereas in the first example it pulled for 10 meters.

3) If the horse does 1,500 joules of work, and it pulled the block with a force of 300 N, how far did the block move?

W=Fd, but we can rearrange the formula by dividing each side by F, giving us d= W/F. Now we just plug in the numbers. 1500 J/300 N = 5m.
The horse pulled for 5 meters.

$$\frac{W}{F} = \frac{\cancel{F}d}{\cancel{F}} \Rightarrow \frac{W}{F} = d$$

Now we must be more realistic, and consider what is actually happening when a horse pulls a weight. Notice in this photo that the lines attached to the sled are going up at an angle. Usually the weight is fairly close to the ground and the horse's harness is much higher than the weight. The horse isn't pulling parallel to the ground. This means that not all of the pulling force can be represented by a horizontal arrow. Some of the force will be directed upward (but not enough to lift the sled off the ground). What do we do in this case? Is there a way to find out how much of the pull is contributing to the horizontal motion of the sled?

By Cgoodwin - Own work, CC BY-SA 3.0, https://commons.wikimedia.org/w/index.php?curid=4976124

The way we deal with this situation is to imagine a triangle under the force line. We'll call the angle between the ground and the force arrow "θ" (*thay-tah*), the Greek letter used for angles. On the right we see just this triangle. We've labeled the sides of the triangle according to their relationship to the angle. "O" is the opposite side, "A" is the adjacent side (adjacent means "next to"), and "H" is the hypotenuse (the longest side).

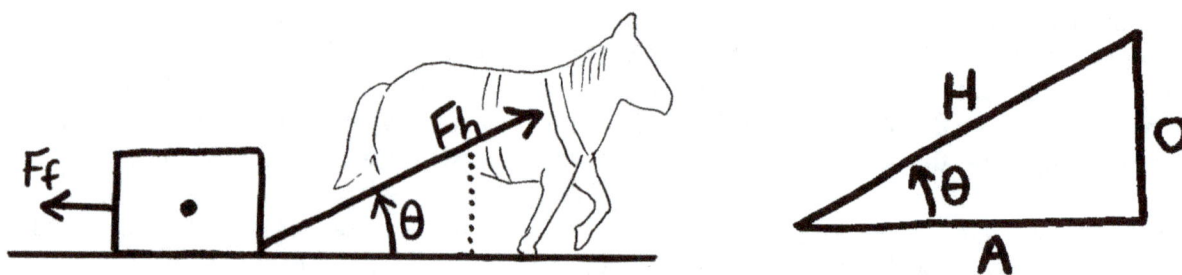

If we divide the length of A by the length of H, (A/H), we get a fractional number that happens to be the number we need in order to calculate our force. For example, if A=7 and H=8, we divide 7/8, which is .875. We simply multiply "Fd" by this number. If F=500 N, and d=10 m, our calculation will be **(500 N)(10m)(.875)= 4375 J**.

A question immediately arises: Where, exactly, should we drop the line for the O side? Where should that dotted line go? Close to the box? Far from the box? Is there just one right place, or can you put it anywhere? It turns out that no matter where you drop the line, the fraction you get when you divide A/H is always the same. The numbers might get larger as you move the O line farther from the box, but the fraction you get when you calculate A/H stays the same. If we moved the O line twice as far away, and made A=14, H would also increase such that A/H would still be .875. It doesn't matter where you draw the line. As long as you keep the angle the same, you will always get .875 when you divide A by H. (Of course, .875 is what we get in this example. If we change the angle, A/H will change.)

Doing this division of A over H is called **finding the cosine of the angle**. The word "cosine" means "A/H." Now we are going to show how to "cheat" and avoid doing the division. It turns out that you don't need to drop an O line at all! You can find the cosine (A/H) by using the search box on any Internet browser. Type in: "cosine ___ degrees =" Fill in the blank with the angle that you are given in the problem. Hit "enter," and the Internet will return a list of links to answers, with many of them showing the cosine number you need right there in the preview line of text. You probably won't even need to actually click on any of those links. You can also use a calculator that has trig functions on it (sin, cos, tan). If you take a lot of higher math, you'll need one of these calculators, but for now, you can just ask the Internet.

In our horse diagram, the angle is close to 29 degrees. If we made the angle of F_h steeper, making it 60 degrees, then we'd have cos(60)= .5. H would always be double the value of A. If A was 6, H would be 12. (6/12= 1/2= .5) If A was 100, H would be 200. (100/200= 1/2= .5) Cos(60°) is always .5 no matter how long the sides are.

One last thought on this before we try some problems. Imagine reducing the angle of F_h. As we make that angle smaller, the triangle gets narrower. Notice that the sides A and H become very close to the same length, while the opposite side, O, becomes smaller and smaller. Eventually, when the angle reaches zero, the triangle goes flat and sides A and H will be the same length; they'll be the same line.

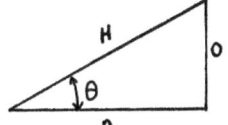 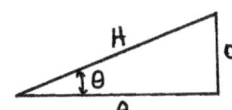 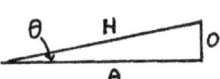 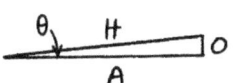 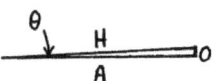

Once the angle and the opposite side reach zero, and A=H, then A/H will be equal to 1 because any number divided by itself is 1. We will get the same result if we consult with a calculator or the Internet to find $\cos(0°)$. The cosine of zero is 1. When something is multiplied by 1 it doesn't change, so we might as well just ignore $\cos\theta$, and use $W=Fd$ instead of $W=Fd(\cos\theta)$. But it is good to know where the vanishing $\cos\theta$ went.

ACTIVITY 10.7: Try some W=Fd and W=Fd(cosθ) problems   (Answers are at the bottom of page 178.)

You will need: a calculator or the Internet (Note: When you look up cosines, you might get a very long string of numbers, such as .639292720. Two decimal points is plenty for us. In this case, round to .64.)

1) A repairman had to push a refrigerator away from a wall. He slowly moved the fridge a distance of .5 m. The force he used was equal to 38 N. How much work did he do? _____
(NOTE: It doesn't matter whether F is a push or a pull. W=Fd still applies.)

An important follow-up thought: Why did we not need the mass of the refrigerator? Isn't that pretty important? We did not need to know this because the mass was already taken into consideration when we told you that the man pushed with a force of 50 N. In order to calculate the force, we had to use F=ma. This is where the mass, m, of the fridge was used.

2) A tug boat is pulling a barge out to sea. If the tug pulls (horizontally) with a force of 900 N for a distance of 250 meters, how much work did the tug boat do? _____

3) A child pulled her wagon down the sidewalk. The angle (θ) between the handle and the ground was 20 degrees. She pulled with a force of 3 N for 2.5 meters. How much work did she do? _____

4) The child's mother came out and began pulling the wagon so that her daughter could ride in it. The mother is taller than her daughter, so the angle (θ) between the handle and the ground increased to 40 degrees. The mother pulled the wagon with a force of 5 N for 25 meters. How much work did she do? _____

5) Three people pushed a broken-down car to a nearby repair shop. If each person pushed with a force of 30 N, and the total amount work that was applied to the car was 7200 joules, how far away was the repair shop? _____
Note: You will need to rearrange the formula from W=Fd, to d=W/F, as shown at the bottom of page 177.)

6) In Alaska, a team of dogs pulled a sled for a distance of 5 km. The combined weight of the sled and the driver is 115 kg, the acceleration of the sled averaged 1.5 m/sec². How much work did the dogs do? _____
(You'll have to use F=ma first, to find F. Also, make sure to convert to meters.)

We're almost finished with our study of motion, but before we call it quits, let's briefly explore how we measure power. Power is directly related to work and energy. We'll be able to keep our horse theme going, too, as we find out why we measure power in units of "horsepower."

Our volunteers will take us to S.N.O.O.P. on a scientist who discovered how to convert steam energy into kinetic energy. English mining engineer Thomas Newcomen made this discovery around the year 1712. A steam device was already being used to suck water out of coal mines, but Newcomen saw a better way to use the steam. But wait—you can witness the power of steam for yourself! This demonstration will show you what Newcomen knew about cooling steam quickly.

ACTIVITY 10.8: Do an experiment with steam (safely!)

You will need: an empty soda pop can (or several cans if you want to see this demo several times, as the can will be ruined after the first demo), a shallow pan (or bowl) filled with ice water, hot pads to protect your hands, a stove top burner

NOTE: If you can't do this experiment, you can watch a video of it. Just type "air pressure soda can crush experiment" into the search box of YouTube. Many videos will pop up. as this is a very popular experiment.

1) Put some ice water into the shallow pan or bowl. It doesn't need to be deep. An inch or two is plenty.
2) Put a little water in the empty can. An inch (2 cm) in the bottom of the can is plenty.
3) Put the can onto the stove burner until you see some steam coming out of the can.
4) Using your hot pads (and being VERY careful), take the can off the burner and quickly turn it upside down and place it into the ice water.
5) In just a second or two the can should collapse towards its center.

<u>What is happening</u>:
   As the can heats, the water in the bottom turns to steam. When you put the steam-filled can into the ice water, the steam cools very quickly. As the steam cools, the steam droplets turn to water and either fall to the bottom or stick to the sides of the can. This is called **condensation**. (You see condensation on the outside of a glass of ice water on a hot day.) The condensation causes the air pressure inside the can to drop very quickly. The air pressure outside the can is suddenly much higher than the pressure inside the can. This higher pressure outside the can presses in on the can and causes it to collapse. This experiment works with 55 gallon steel drums, too, not just small aluminum cans!

ACTIVITY 10.9: Watch the can crush experiment being done with a 55-gallon steel drum

The playlist has videos of this experiment, but it is also easy to find using a key word search. The demonstration is quite impressive! It is amazing how powerful a vacuum can be.

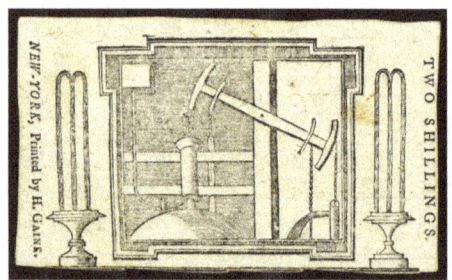

Thomas Newcomen wasn't the first person to think of using the power of condensed steam. Coal miners were already using simple devices that used vacuum suction to suck water out of the bottom of their mines. Water in the mines was a continual problem that kept the coal mines from being profitable. Newcomen analyzed the situation and saw that there was a better way to use the steam.

The image shown here is a monetary note printed in New York in 1775, celebrating the success of what Newcomen is about to invent.

Answers to 10.6   1) W=(38N)(.5m)=19 J   2) W=(900N)(250m)=225,000 J   3) W=(3 N)(2.5m)(.94)=7.05 J   4) W=(5N)(25m)(.76)=95 J   5) (7200 J)/(90N)=80 m   6) F=(115kg)(1.5m/sec²)=172.5N, W=(172.5N)(5,000m)=862,500 J

Newcomen's engine used the vacuum power you witnessed in activity 10.8 to pull down one end of a giant wooden beam (side D in this diagram). Side F is attached to a pump that is under the ground. E is the pivot point of the beam. (Yes, it is a first class lever.) Dome A is where the steam is generated. The brick structure underneath the dome is where wood or coal is used to heat the water. A valve allows the steam to escape up into the cylinder labeled B. Once the steam enters, another valve opens and cold water from tank C is squirted into the B cylinder. This cools the steam almost instantly, making the vacuum that you witnessed in your soda can experiment. The pressure of the atmosphere was now much greater than the pressure inside the cylinder, so it pushed down the gray piston (gray thing that looks like an upside-down T), which is attached to the lever arm, D. Once the steam has condensed into liquid water, it runs out the bottom of the cylinder at point E. Then the cycle begins again.

The very first atmospheric engines had valves that were operated by humans, but it soon become obvious that it was much more efficient to have valves that were controlled by metal pieces constructed to automatically open and close at the right time. (These are not shown in this diagram.) Even with automatic valves, this engine could only do about 12 strokes (up and down) per minute. A faster engine was needed.

This is where James Watt comes into the story. If you ask someone, "Who invented the steam engine?" they will likely say, "James Watt," unless they have studied this part of history more carefully. Watt became so famous that his name is the one we associate with steam engines. Watt wasn't primarily a scientist; his specialty was repairing machines and scientific instruments. One day, he was asked to repair a table-top sized model of a Newcomen engine. Even after the repairs, the model engine still didn't function very well. Watt studied the model carefully and realized that heating and cooling the steam inside the same cylinder wasted a lot of heat energy. He figured out a way to cool the steam in a separate cylinder, allowing the hot steam to go on doing its job. Because less heat was wasted, less fuel was required to run the engine. Needing less fuel would mean that coal mines could save a lot of money on their expenses.

Watt's success was not overnight. He did not have the large sums of money required to pay someone to make the massive metal parts. When Watt was about to give up, along came Matthew Boulton, himself an inventor and engineer, but also a business man with investment money. The two men worked together to start producing steam engines not just for coal mines, but for other businesses, as well.

A Watt steam engine on display today

183

Another innovation that Watt introduced came about in response to a problem he ran into. Watt had realized that his engines would be much more useful to many businesses if they could make circular motion, not just up-and-down motion. After designing a way to make circular motion, he found that someone else had come up with the same idea and had already received a patent on it. Patents are like copyrights; they are legal documents that allow the owner of the patent to have exclusive rights to an invention. This meant that Watt could not use his idea—he had to come up with something else. Watt worked with one of his employees to use gears in an arrangement they called "sun and planet." The "sun" gear in the center was stationary, while the "planet" gear traveled around it. The shaft attached to the planet gear went up and down, causing the orbital motion of the planet gear. (If this

By Newtown grafitti from Sydney, Australia - flywheel of the Boulton-Watt steam engine, CC BY 2.0, https://commons.wikimedia.org/w/index.php?curid=38852277

motion looks familiar, it's similar to the motion you see on train wheels.) This circular motion was then transferred to a very heavy flywheel that used the principle of inertia. It was hard to get the flywheel going, but once it was going it was hard to stop it. The flywheel continued to go around even if the power went down for a short time.

And... last but not least, Watt was also the person who invented the centrifugal governor, a device we met in our adventure with circular motion. The addition of the governor added to the efficiency of his steam engine design, keeping the engine running at a constant pace.

It wasn't long until steam engines were adapted to make trains, tractors, and sawmills. Steam energy would power the Industrial Revolution of the 1800s. The chemical energy found in wood and coal was used to heat water to make steam, and the steam engine then turned the heat energy into motion.

ACTIVITY 10.10: Turn heat into motion: a quick and easy demonstration

NOTE: This activity uses open fire (a candle). Please take precautions. Work safely!

You will need: *heavy-duty aluminum foil, scissors (a pair that can be used for rough jobs--don't use your best pair!), a long bamboo skewer (the kind used for shish kabobs), at least two small candles (the short "votive" candles, or "tea lights" are ideal),*

<u>What to do:</u>
1) Cut a spiral out of heavy-duty foil. (20 cm dia.?)
2) Push the blunt end of the skewer into the edge of one of the candles.
3) Make the center of the foil spiral into a tent shape (mold around a pencil point?) and balance it on the tip of the skewer.
4) You might need to trim the bottom of the spiral just a bit to make sure it doesn't touch the table.
5) Put this structure onto a piece of foil so you don't get wax on the table.
6) Set the other candle next to the one with the skewer.
7) Light both candles. They will create warm convection currents in the air, which will turn the foil spiral.
8) If it doesn't go very fast, add another candle.

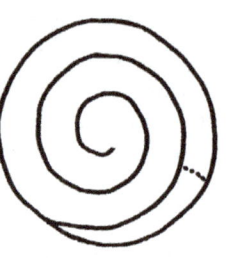

If you want to try a more complicated project, you can search YouTube or the Internet with key words "Turn heat into motion" or "steam engine craft experiment"

ACTIVITY 10.11: Optional video break

The playlist has a few short videos summarizing what we just learned. Or, search for "Steam Engine-- How does it work?" by the Real Engineering channel, and "James Watt and the Transition from Horses to Steam" by the History Guy.

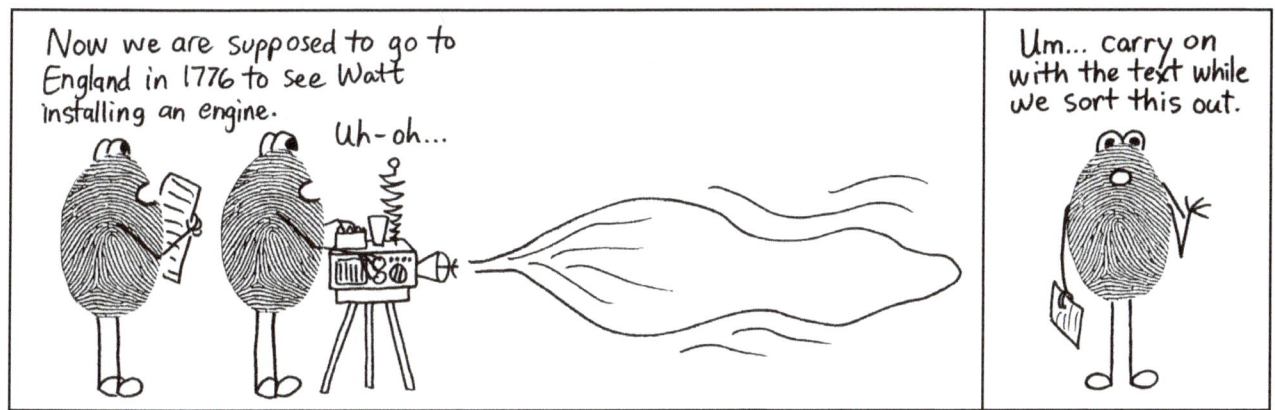

Okay, will do. Good luck, guys.

When Watt was advertising his steam engines to prospective buyers, they often wanted to know exactly how much energy the engine could produce. Would these new engines be more powerful than the horses they'd been using for decades? Watt was able to estimate how many horses could be replaced if these businesses switched to steam. He started using the term "horsepower" because it was easily understood by everyone in the late 1790s. Watt calculated that a horse harnessed to a 24-foot-diameter circular mill could turn the mill 144 times in an hour. He used that figure to calculate that the horse was pulling with a force of 32,929 foot-pounds. He rounded this off to 33,000 foot-pounds, and this became the standard that everyone used.

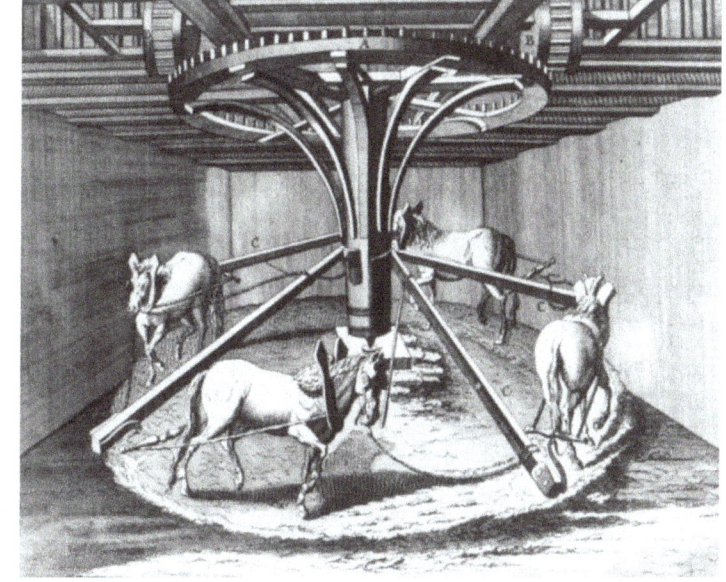

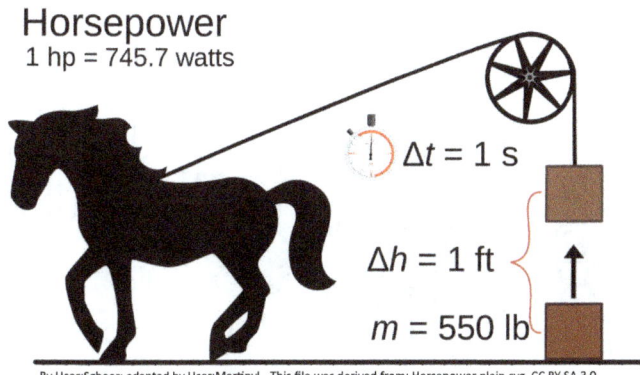

By User:Sgbeer; adapted by User:Martinvl - This file was derived from: Horsepower plain.svg, CC BY-SA 3.0, https://commons.wikimedia.org/w/index.php?curid=57359285

Today, horsepower (hp) is defined as the amount of energy required to lift 550 pounds to a height of 1 foot in 1 second, as shown in this diagram. The metric version of horsepower is lifting 75 kg to a height of 1 meter in 1 second.

Horsepower is still a useful measure for things that use a lot of energy, like cars, tractors, mowers, or motors (things that have replaced horses). It is not useful at all for things that require only a tiny amount of energy. Scientists needed a small unit— something that could be used in laboratories.

In the late 1800s, an international group of scientists met together for the purpose of establishing new units of measure. We've already learned about one of these new units, the Joule, which became the standard unit for energy. It wasn't difficult to choose a name for the new unit of power. "Watt" was the obvious choice.

One watt is defined as "one joule per second." The newton, the joule and the watt are all related.

$$1\,N = \frac{1\,kg \cdot m}{sec^2} \quad \Rightarrow \quad 1\,J = \frac{1\,kg \cdot m^2}{sec^2} \quad \Rightarrow \quad 1\,W = \frac{1\,kg \cdot m^2}{sec^3}$$

FORCE

One newton is the force it takes to accelerate a 1 kg object at the rate of one meter per second per second. The "2" in (sec²) tells us this is acceleration.

ENERGY/ WORK

One joule is the amount of work done when one newton of force is applied to an object for the distance of one meter. We add another "m," making "m²."

POWER

One watt is one joule of work per second. More watts per second means more power. Watts can be used to measure many kinds of power.

Watts can be used to measure any type of power, from electricity to physical exercise. Watts are commonly used as a measure for heaters, light bulbs, lasers, medical machines, motors of many kinds, and radio transmitters. Prefixes such as "micro" and "mega" can be added to the word "watt" to make smaller or larger units. One **kilowatt** (a thousand watts) equals about 1.34 horsepower (hp). The average house in America consumes energy at about the rate of one kilowatt. Every square meter of the earth's surface receives about one kilowatt of sunlight. **Megawatts** are used for planes, trains, ships, and large buildings. **Gigawatts** are used for power plants and power grids.

Laborers maintain an output of about 75 watts during their work day.

Well, everyone, looks like we are at the end of our adventures. Thanks to our volunteers for taking us so many places! Join us for more adventures in other books.

ACTIVITY 10.12:   Review questions from this chapter (Answers are in teacher's section.)

1) When a roller coaster reaches the top of the first hill is has its maximum amount of _____ _____

2) Motion energy is called _____ energy.

3) TRUE or FALSE? Motion can't happen without some type of energy being applied.

4) A roller coaster car has a mass of 500 kg. The first hill is 100 meters tall. How much potential energy does the car have as it sits on top of the first hill? (Use 10 m/sec² for "g.") _____ joules

5) This coaster car reaches the bottom of the first hill (the first valley). How much potential energy does it have now? _____

6) When this coaster cars gets to the bottom of the first hill, its velocity is 10 meters per second. How much kinetic energy does this coaster car have in this first valley? _____ joules

7) Which one of these is NOT a standard unit of measure that you find in physics formulas?
    a) meters   b) kilometers   c) kilograms   d) seconds

8) What was James Prescott Joule's primary occupation? _____

9) Which one of these is NOT a source of potential energy?
    a) food   b) water   c) sunlight   d) batteries   e) rubber band

10) A book is sitting on the edge of a table. Does it have any potential energy? _____

11) What scientific principle was the result of James Joule's very precise measuring?
    a) the existence of atoms   b) the creation of the Kelvin temperature scale
    c) the use of electricity to heat water   d) the conservation of energy

12) Before Joule's experiments, what did scientists think heat was?
    a) a mysterious substance called phlogiston   b) a mysterious substance called caloric
    c) a mysterious substance with no name   d) the motion of molecules

13) TRUE or FALSE   Loops the loops in today's amusement park are perfect circles.

14) One definition of work says that it is a _____ applied through a _____.

15) Which one of these is NOT a vector, meaning that it does not have both magnitude and direction?
    a) displacement   b) velocity   c) distance   d) acceleration

16) Which one of these people is NOT working, according to the scientific definition of work?
    a) Someone who swinging a golf club   b) Someone who holding a book
    c) Someone who is stacking books   d) Someone who is sleep-walking

17) TRUE or FALSE? It is possible to find how many horsepower hours are in a sandwich you are eating.

18) TRUE or FALSE? Energy can't be destroyed; it only changes its form.

19) TRUE or FALSE? The normal force is usually the same magnitude as the force of gravity, so often both of these can be excluded from a free body diagram.

20) Who found a faster and more efficient way to cool steam?   a) Watt   b) Newcomen   c) Joule

21) Who should receive credit for the invention of the steam engine?   a) Watt   b) Newcomen   c) Joule

22) TRUE or FALSE? Watts can only be used to measure the output of light bulbs.

23) What were the gears called in Watt's improved steam engine? _____ and _____ gears

ACTIVITY 10.13: America's tallest roller coaster

Write the letters of each missing word. When you are finished, if you read the boxed letters going down, they will spell the name of the tallest coaster in the U.S. (at the time this book was written). It is located in the state of New Jersey at the Six Flags amusement park.

The equation ($1/2mv^2$) is used to find ☐ ___ ___ ___ ___ ___ ___ energy.

To find the ___ ___ ___ ☐ ___ ___ of an angle, you divide adjacent side by hypotenuse.

Use "mgh" to find ___ ___ ___ ___ ☐ ___ ___ ___ ___ energy.

Motion needs ___ ___ ___ ___ ☐ ___.

Work equals force times ☐ ___ ___ ___ ___ ___ ___ ___.

One joule per second is a ___ ☐ ___ ___.

Energy is the ability to do ___ ___ ___ ☐.

___ ___ ___ ___ ___ ___ ☐ ___ ___ ___ ___ means preserving or maintaining.

ACTIVITY 10.14: More physics art—no talent needed! (A review of free body diagrams)

Free body diagrams are fun to draw! Yeah, sure they are! No matter what you are asked to draw—a car, a dog, an ice skater, a polar bear—all you have to do is draw a box (with or without a dot to represent center of mass).

First, some reminders about a few strange characteristics of free body diagrams:

1) If a surface is frictionless, drawing the surface is optional.
Therefore, diagram A shows an object that it is accelerating to the right.
The object could be on a friction-free surface, or it could be floating in space.
(However, most students prefer to go ahead and draw surfaces.)

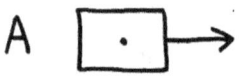

2) Arrows represent FORCES, not motion. If something is moving at a constant velocity, this means no acceleration is happening and therefore no force is being applied. So in diagram B, we don't know whether the box is moving at a constant velocity or sitting still. It could be either!

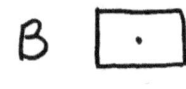

3) The arrows can start from the central point or from either side of the box. In diagram C, all of these arrows show the normal force. Draw what you think is best for your diagram.

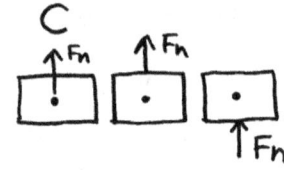

4) The sizes of the arrows are important. The bigger the arrow, the greater the force. Equal forces are indicated by arrows that are the same length. So in diagram D, we have the force of static friction equal to the force being applied to the box. Since the forces are equal, the box is not moving.

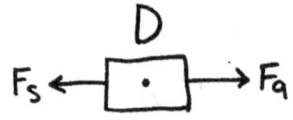

5) The force of gravity always points down (diagram E).
The normal force is always perpendicular to the surface that the object is sitting on. (diagram E)

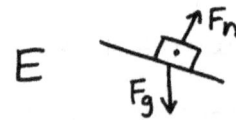

MOTION OCCURS WHEN THERE ARE UNBALANCED FORCES (unequal arrows).

Draw a free body diagram for each of these situations. (Answers in teacher's section.)

1) A car is sitting in a parking lot and is not moving. What forces are acting on the car?

2) A rightward force is applied to a box across a rough surface. The rope is parallel to the ground.

3) A hockey puck is gliding across the ice at a constant velocity. (We'll ignore air resistance and the tiny amount of friction from the ice.)

4) An elevator is accelerating upwards.

5) A skydiver has opened her parachute and is now descending slowly, moving at a constant velocity. Indicate air resistance with $F_{(air)}$.

5) A box is sitting on a ramp, but there is enough friction to keep it from sliding.

7) A ball was thrown into the air. It is slowing down as it approaches the peak of its parabolic flight. (Is its horizontal velocity constant? Review on page 122.)

8) A refrigerator magnet is stuck to the fridge. It is not moving. Show a side view of the magnet and the forces acting on it. Use F_m for the magnetic force. Don't forget the friction between the magnet and the fridge surface!

9) A horse is pulling a sled across a rough surface, but the sled is accelerating. The angle of the rope is 45°.

10) A boy is trying to drag his stubborn dog across the kitchen floor, but the dog weighs too much, and try as he might, he can't make the dog budge. The leash is parallel to the floor.

11) Two people are lifting a log, one at each end. They are pulling with equal force. Their two pulling forces must be greater than the force of gravity in order to lift the log. (Remember, the size of the arrows indicates strength of force.)

12) A child is hanging from a horizontal bar in a playground. Show each arm as a force arrow. How long should these arrows be in comparison to the F_g arrow? (Use F_t for tension force in arms.)

13) A sky diver has not yet opened his parachute. He has not yet reached terminal velocity. Include air resistance.

14) A satellite is traveling through the solar system. It has instruments on board that can keep its velocity at a constant 100 m/sec.

Add some magnitudes (numbers) to the force arrows in these problems:

15) A sailboat is moving to the right with a wind force of 40 N. Water currents are working in the opposite direction with a force of 5 N. (We know what Fg is, and that Fn is equal and opposite so you don't have to label these with numbers.)

16) A penguin is walking into an Antarctic wind that is blowing with a force of 8 N. He is walking with a force of 9 N. (What will be the total force of the penguin's forward motion?)

17) A satellite was launched with a force of 2 N. Since it is in outer space, it is not experiencing any air resistance, and therefore continues to accelerate.

18) A crane is lifting a box with a force of 12 N. At the same time, a gust of wind blows the box to the right with a force of 12 N. Use a dotted line to indicate what the combined motion of the crane and the wind will look like.

EPILOGUE

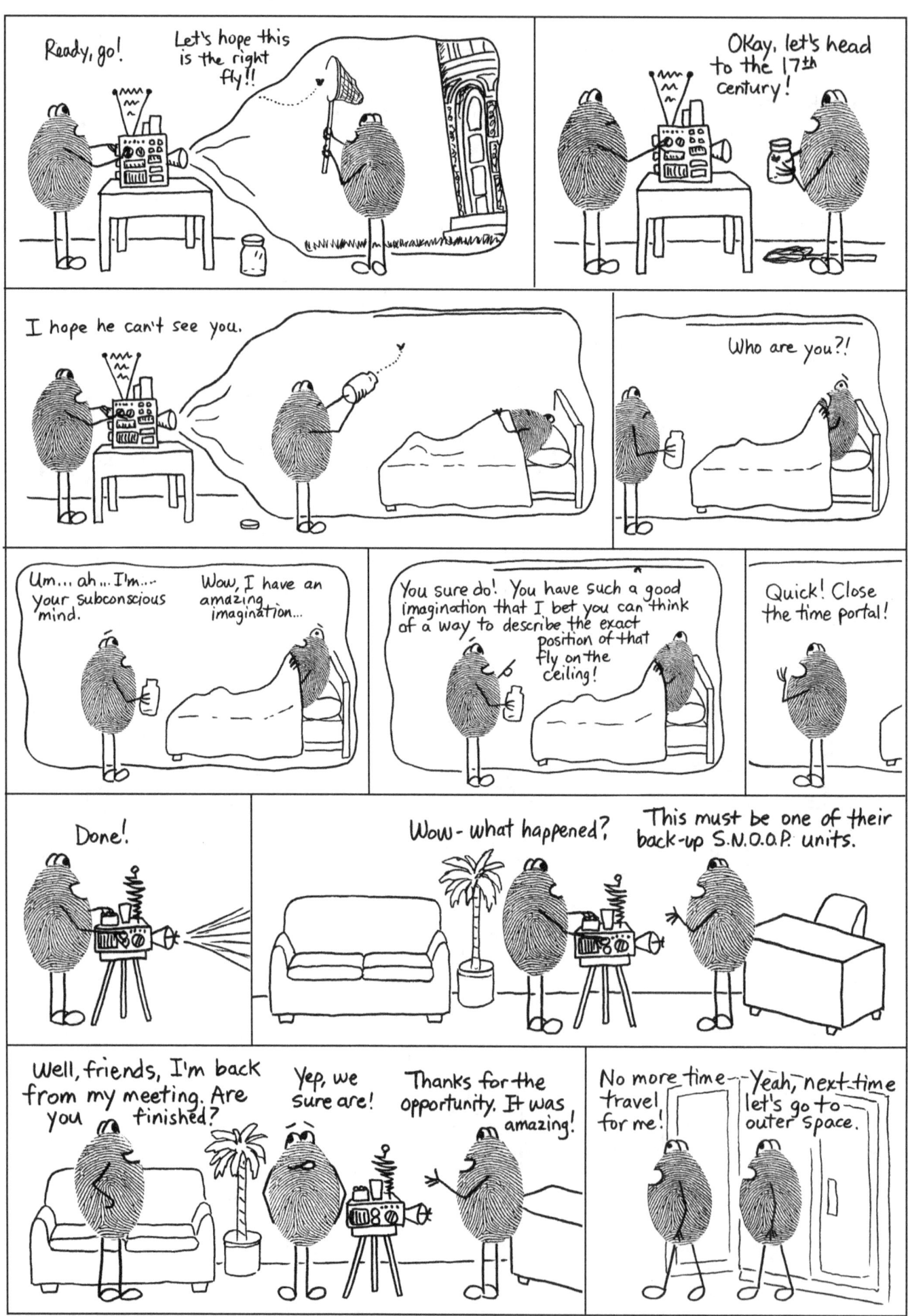

TEACHER'S SECTION

ANSWERS TO REVIEW QUESTIONS and SUPPLEMENTAL ACTIVITIES

ANSWERS and SUPPLEMENTAL ACTIVITES

NOTE: The numbering of these extra activites picks up where the numbering stopped in the chapters. This will avoid confusion between chapter activities and supplemental activiities.

CHAPTER 1:

ACTIVITY 1.13: Immobilize someone with your pinky finger

You will need: a chair and a volunteer for this activity.

1) Tell your volunteer to sit in the chair, with his/her back against the back of the chair and hands in lap.
2) Place your pinky finger on his/her forehead. (You can use any finger-- it does not have to be the pinky.)
3) Tell your volunteer to stand up. Use your finger to keep his/her head from tilting forward.
4) Your volunteer will probably not be able to stand up.

    When you are standing, your center of mass (somewhere in your abdomen) is directly over your feet. When you are seated, your center of mass is above the seat of the chair, not over your feet. In order to stand up you need to move your center of mass from over the chair to over your feet. To accomplish this, you need to lean forward. Strangely enough, the amount of force needed to keep someone from leaning forward is not all that much. You can exert this force with one finger.

ACTIVITY 1.14: The "girls always win" chair lifting challenge

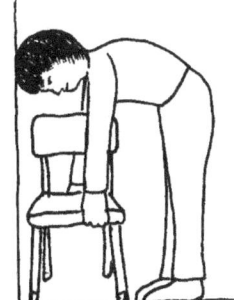

You will need: two volunteers (one male and one female), a chair, and a wall against which you can set the chair. The older the volunteers are, the better. The difference between the center of mass for boys and girls (under the age of 12 or so) might not be enough to make this trick work.

1) Place the chair against the wall.
2) Have one volunteer bend over the chair so that his/her head touches the wall and his/her upper body is parallel to the floor, as shown in the diagram. Tell the volunteer to try to lift the chair and then stand up.

    Men and women have their center of mass in different places. Men tend to have broad shoulders and narrow waists, giving them a higher center of mass. In this case, a higher center of mass is a disadvantage because adding the weight of the chair makes the center of mass even higher, creating an impossible situation where there just isn't enough counterbalancing weight in the lower half of the body. Girls and women have their center of mass closer to their hips, which in this case is an advantage because even with the added weight of the chair, the overall center of mass is close enough to being over the feet that the body is able to right itself.
    (NOTE: In my classroom, we found it hard to get the "correct" results, probably due to the young age of the boys. Our boys could lift the chair. For young students, try a relatively small chair with a book or two sitting on it.)

ACTIVITY 1.15: Balancing a coin on a dollar bill

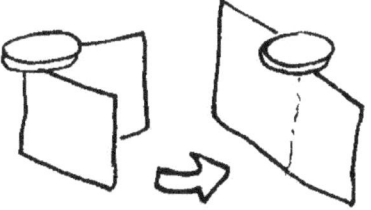

    Did you know you can balance a coin on a paper bill? The trick is to fold the dollar a bit, rest the coin on the folded corner, then slowly open the bill. If your hands are steady and you pull the ends of the bill very slowly, the coin will balance right on the edge!

ACTIVITY 1.16: Balance a ruler with a hammer

You will need: a hammer, a one-foot ruler (30 cm), and some string or a strong rubber band

This balancing act looks like it should not be possible. The hammer hangs precariously underneath the ruler, looking like it should make the ruler fall off the table. But there it hangs, perfectly balanced!

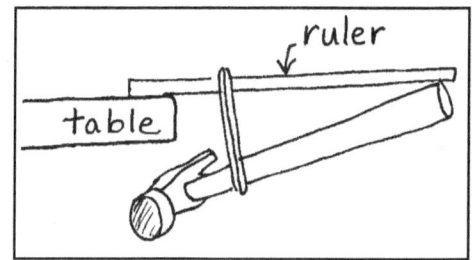

1) Take the rubber band or string and make a loose loop around the hammer and ruler, as shown in the picture.
2) Make sure the end of the hammer is touching the ruler, and then position the ruler at the edge of a table, as shown. (You might have to reposition the string/rubber band a few times to get it just right.)
3) Why does this trick work? Analyze where the center of mass might be. Where is the balance point? What is the heaviest part of a hammer?

ACTIVITY 1.17: Make a "balancing bird"

For each bird, you will need: a copy of the following pattern page printed onto heavy card stock, scissors, tape, two pennies, a toothpick and a glue stick (Optional: colored pencils or markers to color the bird)

NOTE: You can avoid having to tape a toothpick under the wings by printing the pattern onto heavier cardstock, or gluing the bird to thin posterboard (or cereal box cardboard). If the bird is thick enough, it won't need the toothpick to help support the wings.

Follow the directions on the pattern page (skipping the part about the toothpicks if your bird is thick enough).

Someone unrelated to me made a video of how to construct this project and they posted it on YouTube.
The poster is Declan Holmes and the video is titled "Balancing Bird Two."

Answers to Activity 1.12 Review questions:

1) Syracuse, Sicily 2) T 3) F 4) mid-point or center. intersect 5) T
6) T 7) T 8) F 9) F 10) bell tower
11-13)
car-- keep the center of mass low to make it stable around turns
plane-- center of mass affects how it flies
athletes-- shift their mass to make efficient jumps
If students have other answers that make sense, give credit.
14) T 15) librarian of Alexandria 16) estimating the circumference of the earth
17) no 18) yes 19) circumference
20) radius 21) F 22) F
23) F 24) grains of sand 25) parchment was very expensive.

BALANCING BIRD TOY

You will need:
• two pennies, a toothpick, clear tape, glue stick (optional: coloring supplies)

1) Do any coloring you want to do. 2) Cut out bird and tail. Make sure to cut along the wing lines that go into the body area. 3) Fold the bird in half. 4) Apply glue stick to inside of the forward half of head (eye and beak area) and stick halves together. (Note: Beak can be reinforced with clear tape if it seems too flimsy.) 5) Make a slight crease along the lengths of the wings, to stiffen them. 6) Tape toothpick to underside of wings, across the center, (like the cross bar of a kite). 7) Insert tail piece and secure with tape on the underside. 8) Roll two pieces of tape and apply one to each penny. Stick pennies on the undersides of the ends of the wings and then check balance. Adjust the pennies if necessary to make the bird balance well. Once pennies are in the right place, secure them with a little more tape.

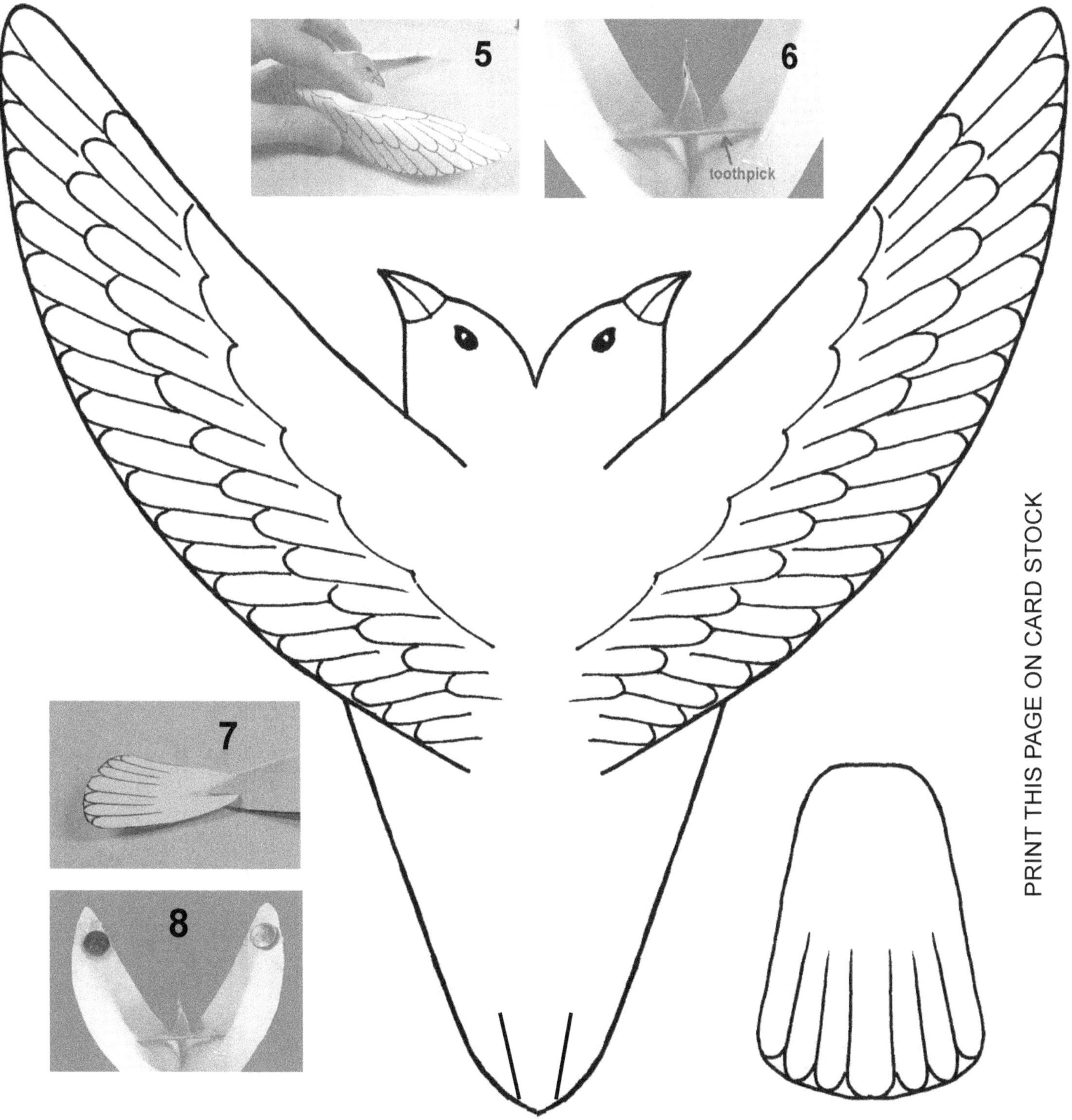

PRINT THIS PAGE ON CARD STOCK

CHAPTER 2:

ACTIVITY 2.11   Identifying levers (or other simple machines) in ordinary household objects

You will need: a variety of objects found around the house that use leverage. Here are some ideas: pliers, hammer, scissors, fly swatter, nail clippers, stapler, chopsticks, racket, table knife, fork, spoon, bottle opener, manual can opener, tweezers, tongs, wrench (spanner), meat grinder, old fashioned (manual) egg beater, etc.

Do this activity in a way that suits your student(s). If you are working with a large group, you might need to pass the objects around. If you have a small group, they can just gather around one table. You can decide how you want to manage responses, as well. You can call on students one at a time and ask them to identify a simple machine, or you allow open and informal comments. Classroom management is at your discretion. The goal of the activity is to encourage critical thinking and analyzing, and any way you can accomplish this is fine.

ACTIVITY 2.12   Identifying levers in parts of the body

This activity will give your student(s) a chance to move around, which is always a wonderful thing in a classroom setting. Demonstrate each type of leverage and then have the student(s) analyze their own body movements.

SUGGESTION: You might want to ask for ideas first, before giving the answers below. For each category, give the student(s) a few minutes to think, and see if they can come up with their own answers. The first and third class levers will be easier that the second class lever.

First class levers in the body: (There aren't very many of these.)

1) Tilting the head, either side to side or back to front.
The top of your spine, right under your skull, acts as the fulcrum. Your head is the seesaw, even though it is round, not flat. The muscles in your neck (and going up the back and sides of your head) provide the force. Imagine your head to be a 3-dimensional seesaw. You can make your ears be the riders on the seesaw by tilting your head to the left, then to the right. Your ears go up and down. When you tighten one side, making it the force, you automatically relax the opposite side, allowing it to be pulled. When you tilt your skull forward then backward, it again acts as a seesaw, with your face and the back of your head as riders.
Challenge: Can you tilt your head in an oblique ("catty-corner") way so that your head tilts down not really side to side but also not really back to front? Your two muscles group will have to cooperate, so it will take some concentration.

2) Bending your torso at the waist.
This is similar to tilting your head. The vertebrae at the bottom of your spine act as the fulcrum. Your torso is the seesaw, even though it is not long and flat. The muscles on each side of your body take turns supplying the force. The muscles not supplying the force must relax and allow themselves to be stretched. Muscles only contract; they can only pull, not push.

Second class levers in the body: (There aren't many of these, either.)

1) The calf muscle, or gastrocnemius *(gas-tro-NEE-me-us)*, lifting the body to a tip-toe position
Look at this diagram to see how this can be a second class lever. Notice the similarity to a wheelbarrow. The ball of your foot acts as the fulcrum, the load is the weight of your body, and the force is supplied by your calf muscle. It is very important that the calf muscle is attached to the back of the heel bone. The force needs to act on the <u>end</u> of the lever, keeping the load in the middle. If the load was on the end, it would no longer be a second class lever. The way to exert force is to contract your calf muscle, making it shorter.

Challenge: How many times (in a row) can you go from standing to tip-toe? The first 10-20 times will be easy.
Challenge: Try move your foot side to side without moving your leg.

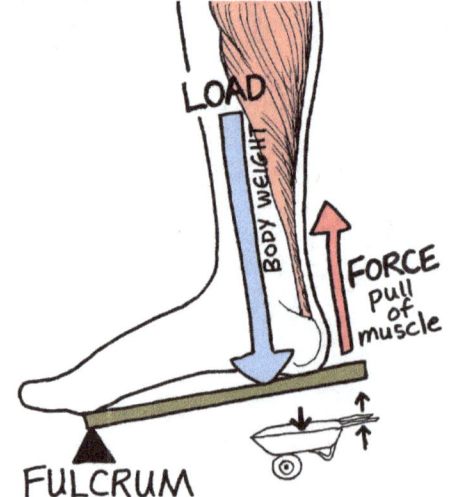

2) The jaw, while we are chewing food
Think of a nutcracker. While using our molars to chew, the load is right under the force. Notice how the back molars in this diagram are right under the masseter muscle. This isn't a perfect second class lever (a nutcracker is a much better example) but it is close enough to qualify. (For a description of how the jaw can be a third class lever, see the next section.)

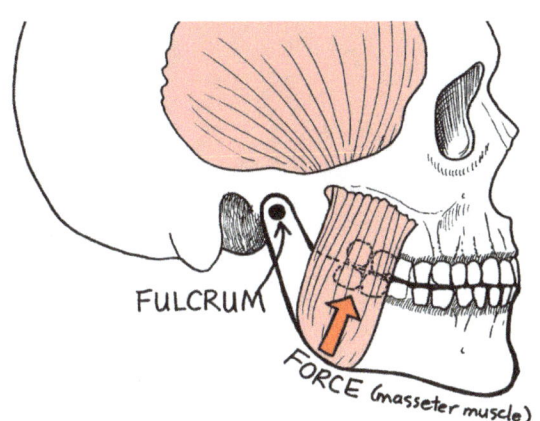

Challenge: Put your fingers on your jaw and also on the side of your head, where your temples are. Open and close your jaw. Can you feel the muscle that is shaded in pink in this diagram? It goes quite a ways up the side of your head. Keep sliding your fingers up your scalp until you can no longer feel any motion when moving your jaw.

Third class levers in the body:
There are far more third class levers in the body than there are first or second. This is not a complete list.

1) Lifting the forearm (from elbow to wrist)
This is shown in the diagram. The force is provided by your biceps. This muscle attaches to the bones in your forearm, but close to the elbow. The fulcrum is at a point inside your elbow joint. If the attachment point of the muscle on the forearm was closer to the hand, you would have a better mechanical advantage, but you would sacrifice range of motion. The placement is at a point where you get adequate range of motion but still have enough power in the muscle to be able to lift most of the loads we need to lift during our daily routines. A good design!

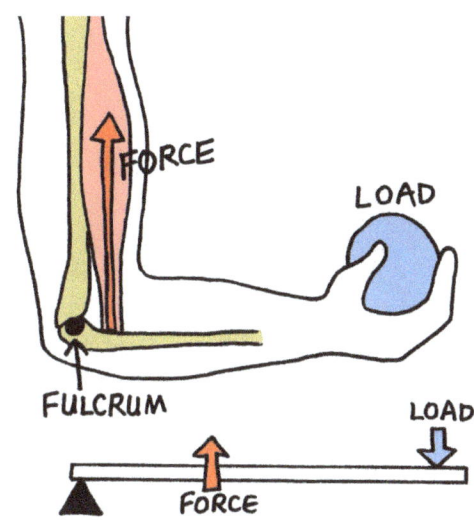

2) Lifting the hand (using wrist as fulcrum)
If your wrist is the fulcrum and the load is something you are moving with your fingertips, the force must be between the two of them. Feel your palm and the back of your hand as you bend you hand. Can you feel the muscles and tendons pulling?

3) Lifting your arm from the shoulder, your leg from the hip, your lower leg from the knee.

4) Sitting up from a lying down position
You don't have a whole lot of mechanical advantage for this, so you often need to use your arms to get you started.

5) Lifting each finger, with the bottom knuckle as the fulcrum.

6) Moving each finger bone (each phalange)
Since the finger bones are linked together with tendons and muscles, it is very difficult to move the tip of a finger without moving any other segments.

7) Your jaw when you are biting into something, like an apple.
You can see in this diagram that the load (whatever you are biting) is at the end of the lever (your jaw bone). This arrangement makes the jaw a third class lever.

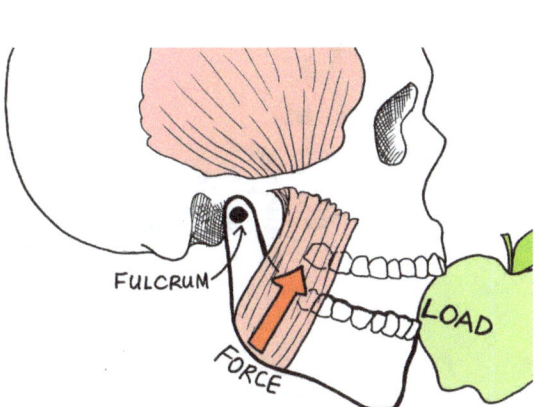

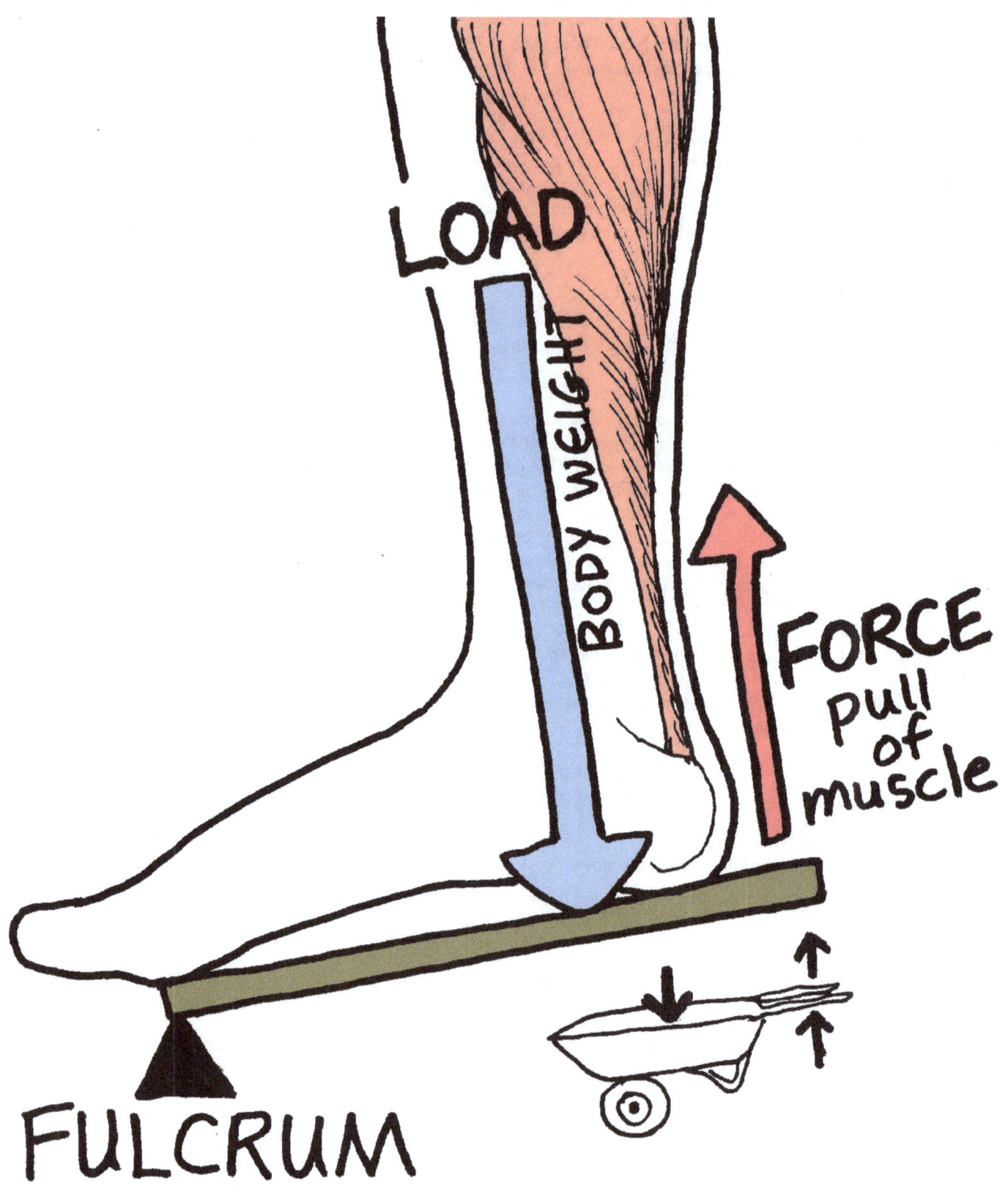

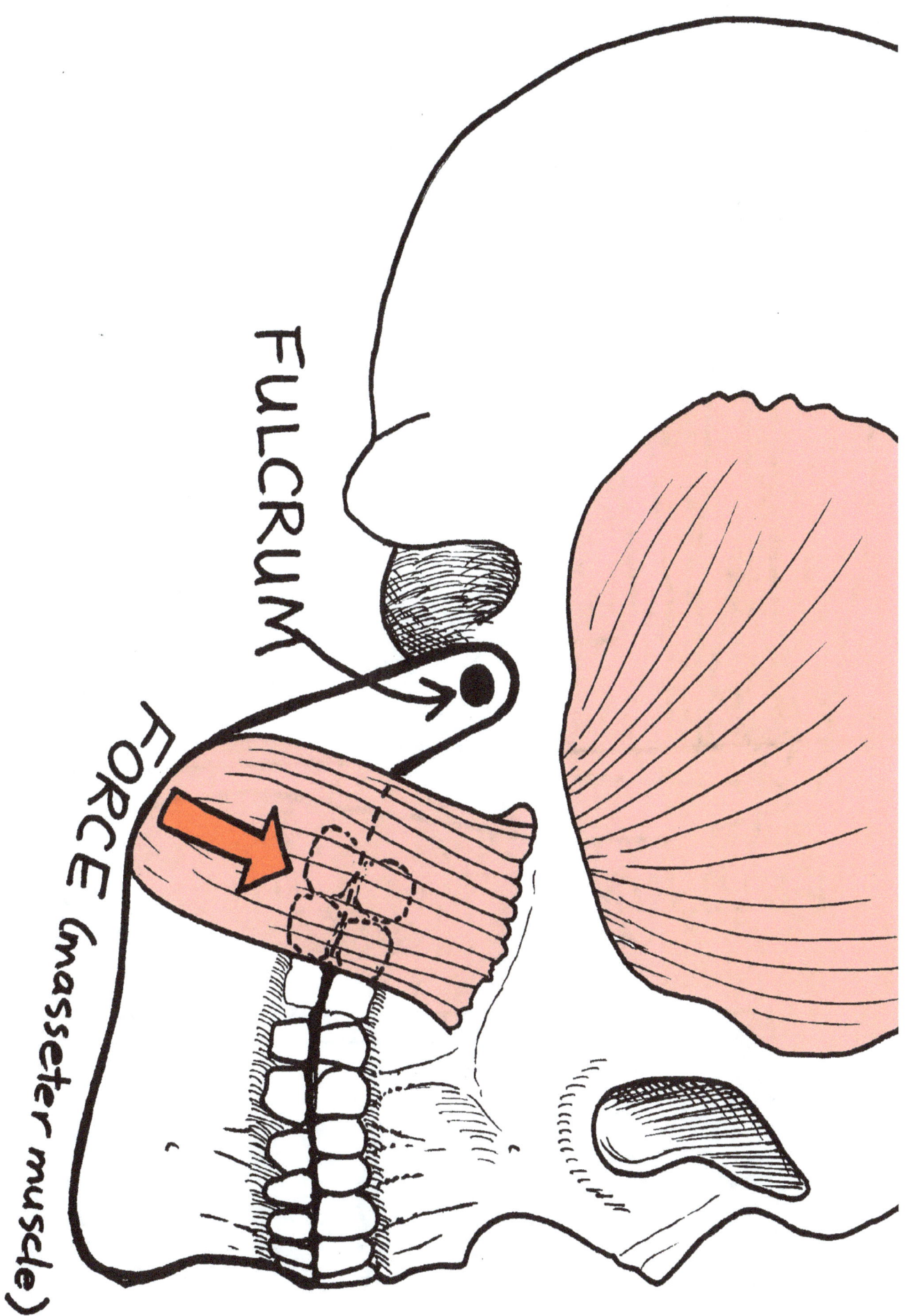

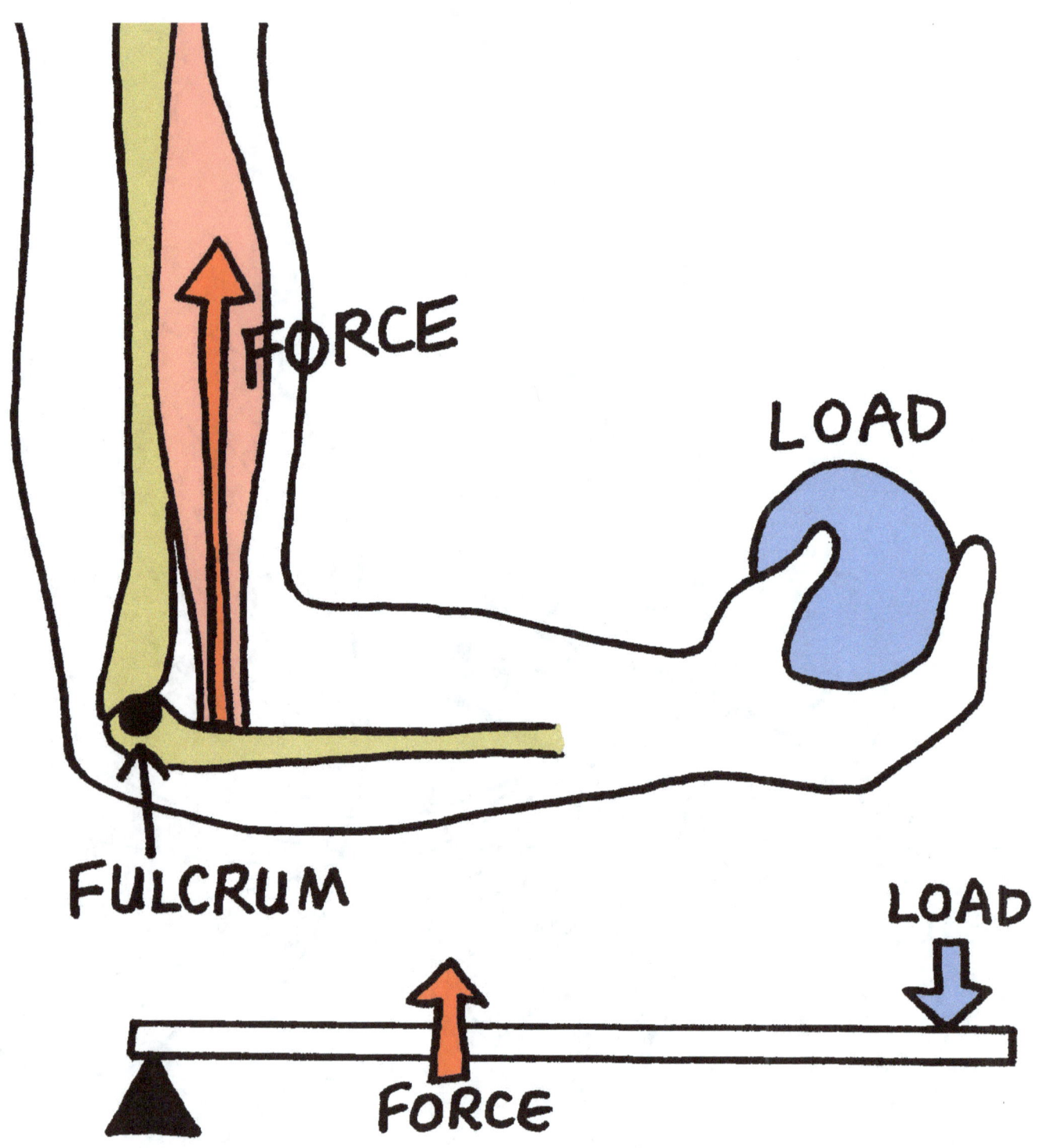

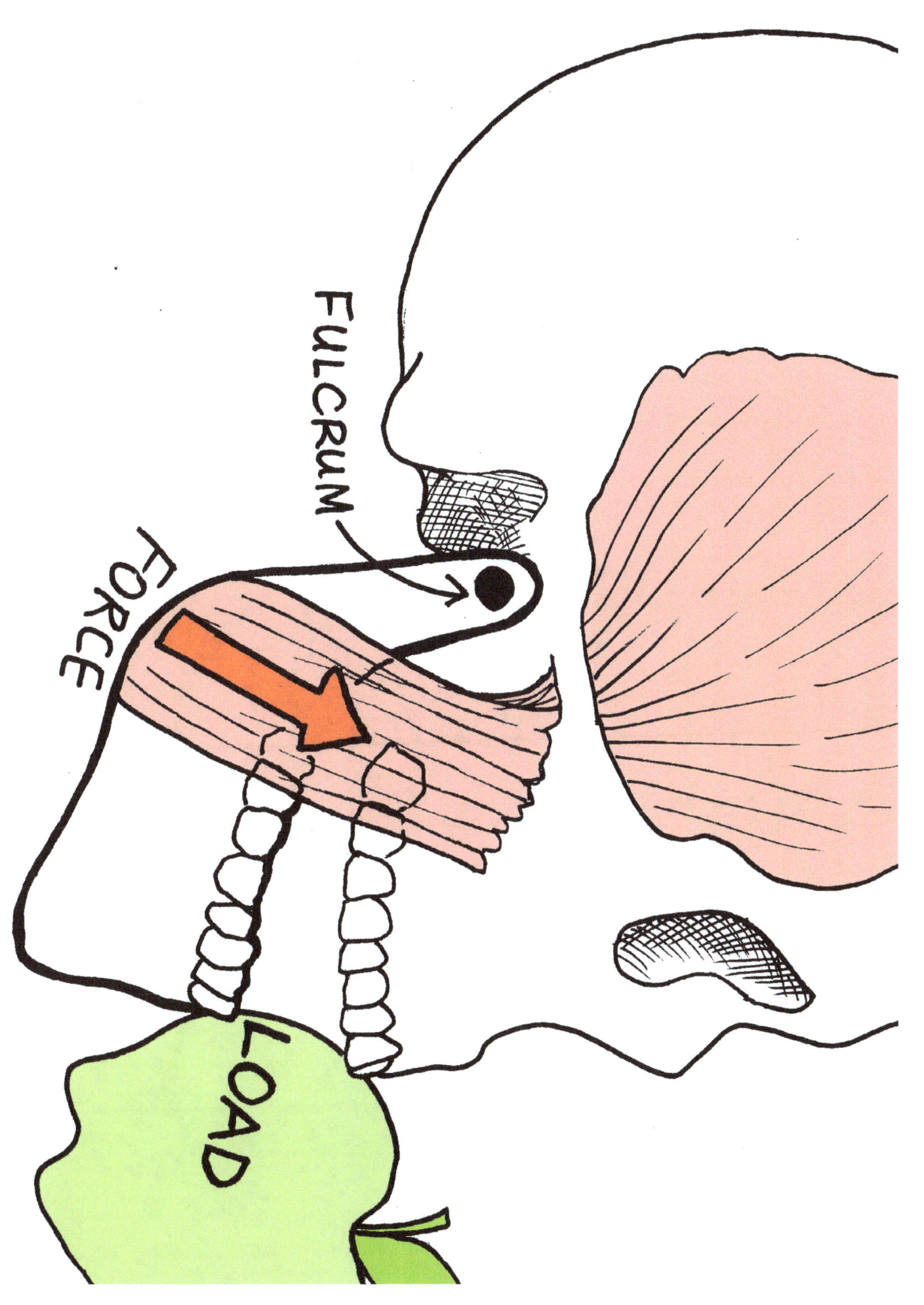

ACTIVITY 2.13: An easy pulley demo without having to buy or make pulleys

You will need: a long piece of rope, two brooms or mops with sturdy handles (or pieces of PVC pipe, or large dowel rods), and three people to do the demonstration
ALTERNATIVE: Try a scaled-down, table-top demo using pencils and string if you don't have much floor space.

1) Tie one end of the rope to one of the handles. Have someone hold this handle tightly, parallel to the floor. This will be person 1 and handle 1.
2) Loop the rope around the second handle once, and have someone hold this handle, also parallel to the floor. This will be person 2 and handle 2.
3) Have these two people stand opposite, facing each other, about 2 meters (6 feet) apart.
4) Have the third person grab the free end of the rope, and (standing very close to person 1) pull on the rope. Person 2 will try to keep the handles from being pulled together. It's like a tug-of-war between person 2 and person 3. Notice how hard or easy it is to keep the handles from being pulled together.
6) Now pass the rope around handle 1 again, then around handle 2 again. Then try the tug-of-war again. Was it easier this time for person 3 to pull the handles together?
7) Make another loop around handle 1 and then handle 2. Try the tug-of-war again. How hard was it to keep the handles apart?
8) Every additional loop you make will increase the mechanical advantage for person 3, because you are making a pulley system.

PENCILS ARE USED IN ORDER TO SHOW THE LOOPING MORE CLEARLY.
Also, the strings are shown very far apart, just for clarity. Your ropes/strings can be much closer together.

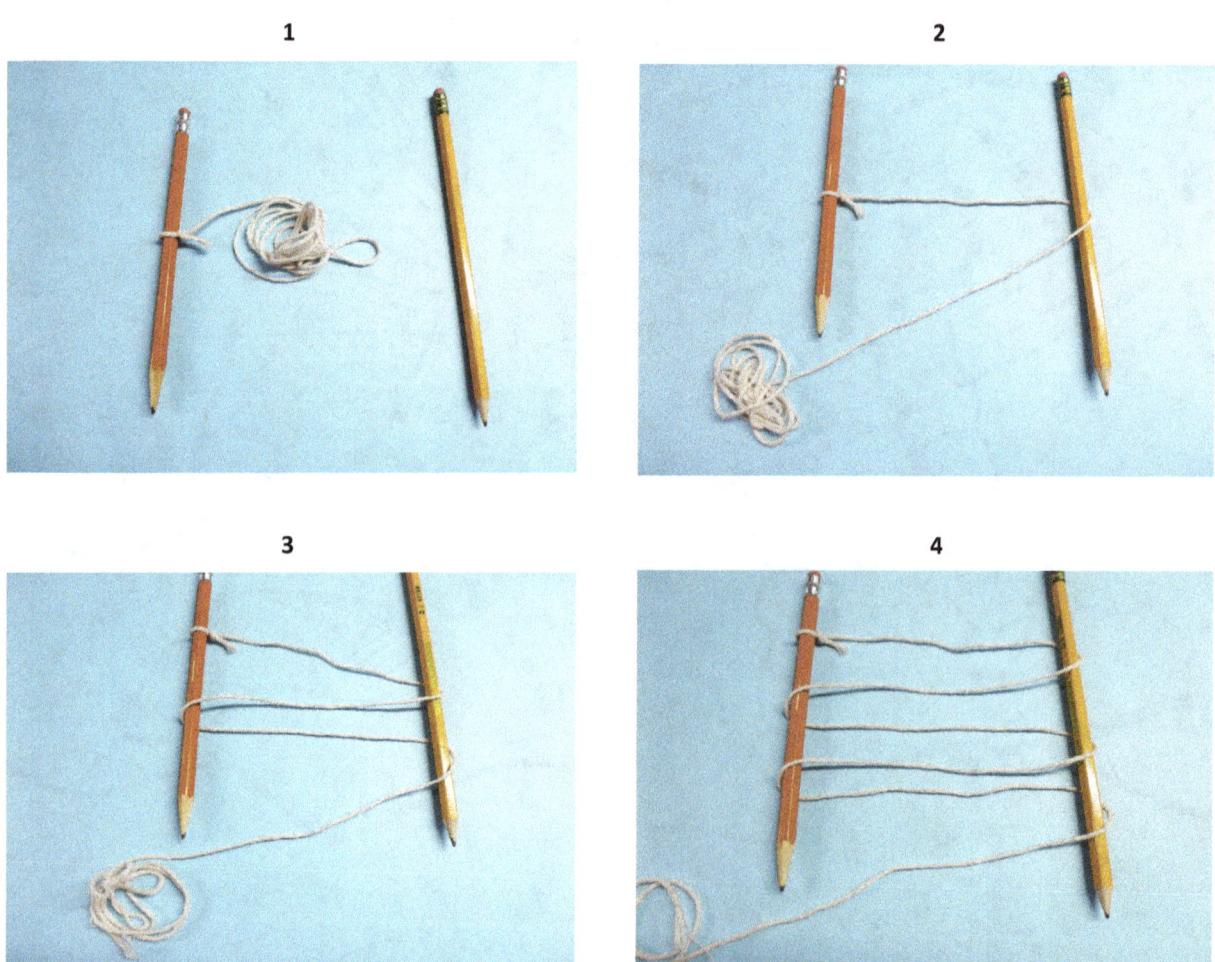

ACTIVITY 2.14   A song about the five simple machines

This song is simpler than the Lever Rap, but if you are working with students who are younger or who have not been introduced to the simple machines previously, you might find this song helpful.
Go to www.ellenjmchenry.com, click on MUSIC, and scroll down until you find it.
You can download a lyrics page as well as listening to the music (both with and without vocals).

The Simple Machines Song

Oh, sing me a song about simple machines and the way, hey, they help up all day,
They let us trade distance for force that we need,
They give us more power and leverage and speed.

The levers and wedges and pulleys and screws, and the inclined planes that we use,
Will lift things and pull things and turn work to play,
And the wheel and the axle will roll them away.

Answers to crossword puzzle, activity 2.10

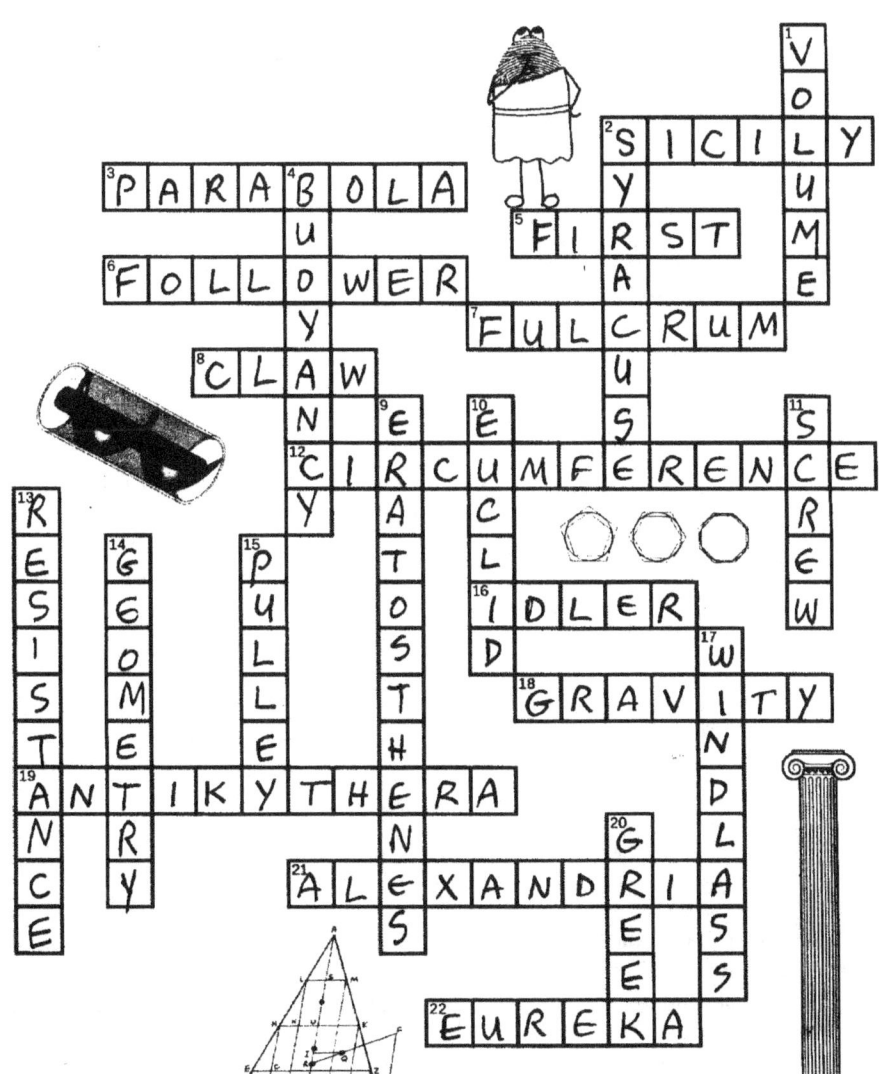

CHAPTER 3:

NOTES FOR ACTIVITY 3.6:

The values my class found: 5 cm 28   20 cm 16   45 cm 10   80 cm 8
Your values should be close to these, but don't have to be exactly the same.
The graph of these values is shown here. Your graph should look similar even if you found slightly different values.

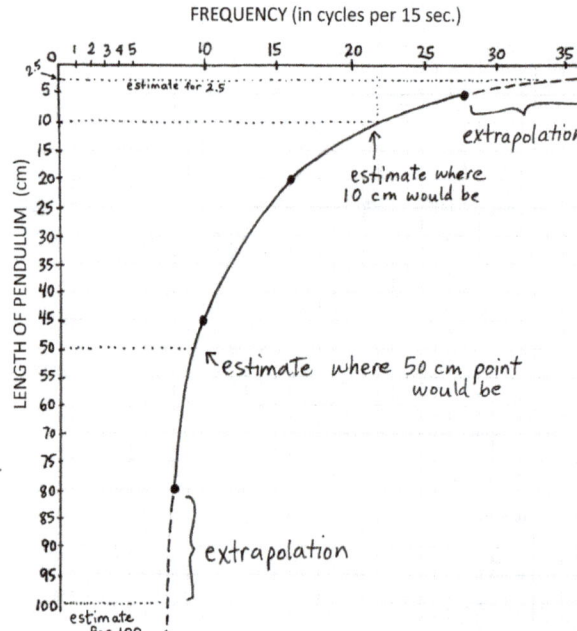

Estimated values for: 125 cm: 5.6   180 cm: 4.6
(These values were calculated using the inverse square law. The students' values might be different, but should be reasonably close to these values.)
If they can't come up with values, just go on to the next section and graph the data points.

These are the frequencies we estimated for:
50 cm: 9   10 cm: 23
Your students might have slightly different answers but they should be close to these values.

Extrapolated values: 100 cm: 7   2.5 cm: 34 or 35

NOTES FOR ACTIVITY 3.10

Here is sample data and a graph by one of my students. Yours won't be identical, but should look similar.

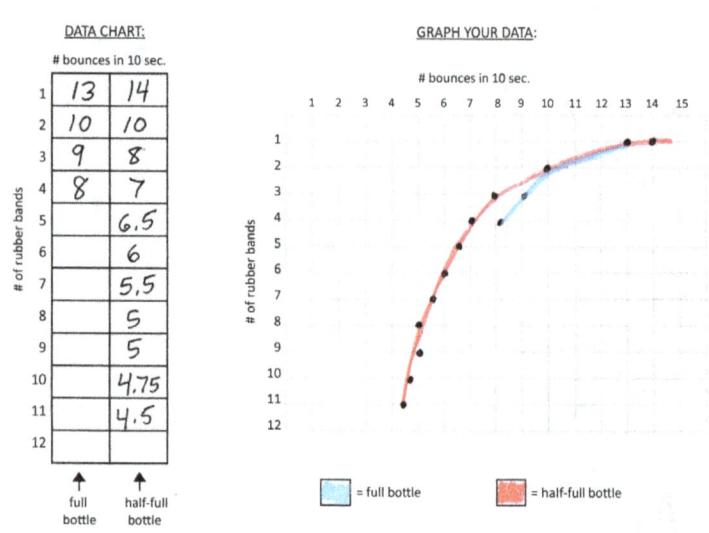

We found it hard to collect data for our large bottle because it was able to stretch 4 rubber bands almost to the floor. If you use smaller rubber bands, you might be able to collect more data.

It is important to use the same size rubber bands for both the large and small bottles. If you use different rubber bands, you can't do a fair comparison.

The students should notice the obvious similarity between this curve and the one they drew in activity 3.6.

Follow-up questions:
1) heavier   2) 10 clips   3) yes   4) Yes. Probably due to the stiffness of the rubber bands. The stiffness is equivalent to friction for a swinging pendulum.   5) fewer   6) yes   7) This is the same curve we saw in activity 3.6.
8) Answers will vary. Main point is to look at graph carefully and use logic to make reasonable estimate.   9) Answers will vary.

The main point of this activity was for the students to understand that all pendulums operate according to the same physical laws, regardless of how they oscillate. Also, all pendulums will produce a graph that has a parabolic curve.

Review 3.11
1) F   2) T   3) F   4) F (Paris)   5) T   6) F   7) F   8) T   9) F   10) T
11-12) tall buildings, bridges, power lines   13) Pisa   14) sine   15) shorten the rod   16) extrapolate
17) resonates   18) oscillate   19) frequency   20) electromagnets
21) c   22) T   23) T   24) d   25) c   26-30) lever, wedge, pulley, screw, inclined plane

ACTIVITY 3.12:   Use a pendulum to calculate the square root of 2

You will need: your pendulum from activity 3.6
NOTE: This activity is challenging and is recommended for advanced students.

1) Tie a knot anywhere in the thread of your pendulum. Hold this knot and time the pendulum. _____ / 15 seconds
2) Fold the thread in half so that the knot touches the middle of the coin. Hold the thread by this
   halfway point and time this half-length pendulum. _____ / 15 seconds
3) Divide the frequency of the full-length pendulum by the frequency of the half-pendulum. ____/____ = _____
4) Use a calculator to find the square root of 2. _____
5) Compare this value for square root of 2 to the value you found in step (4). Are they similar?
6) Try it again using a knot in a different place. Same result?

ACTIVITY 3.13:   Draw some sine waves

You will need: corrugated cardboard, a roll of cash register tape (or make a very long paper strip), a sharp craft knife (such as X-Acto), ruler, glue, pencil, possibly a sharp marker or pen
NOTE: This works best as a 2-person activity. One person holds the cardboard and moves the pencil. The other person pulls the paper tape.

1) Use the patterns on the following page to cut your cardboard pieces. (Printing the patterns on cardstock might be helpful.)

Notice the pieces in the middle (part 3) have the score lines drawn on both sides of the cardboard. You'll score the middle line on the front side, and the two side lines on the back side.

2) Score the lines on these two pieces using the craft knife. Score just the middle line on one side, and just the two side lines on the opposite side. Scoring means you cut through the top layer of the cardboard, so you will get a very clean and crisp fold. Then fold on the score lines so the pieces look like the top one shown here.

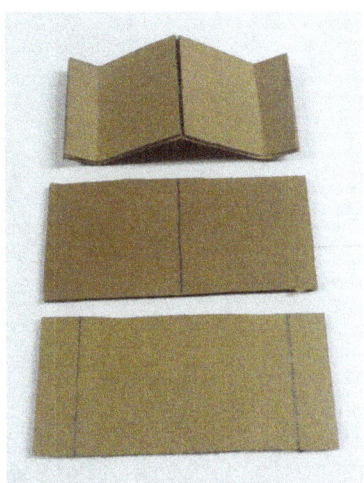

3) Complete the folding process on these parts and glue the middle sections together so they will stand up like this. Then use the craft knife to cut out the slots in the top.

Continued on next page.

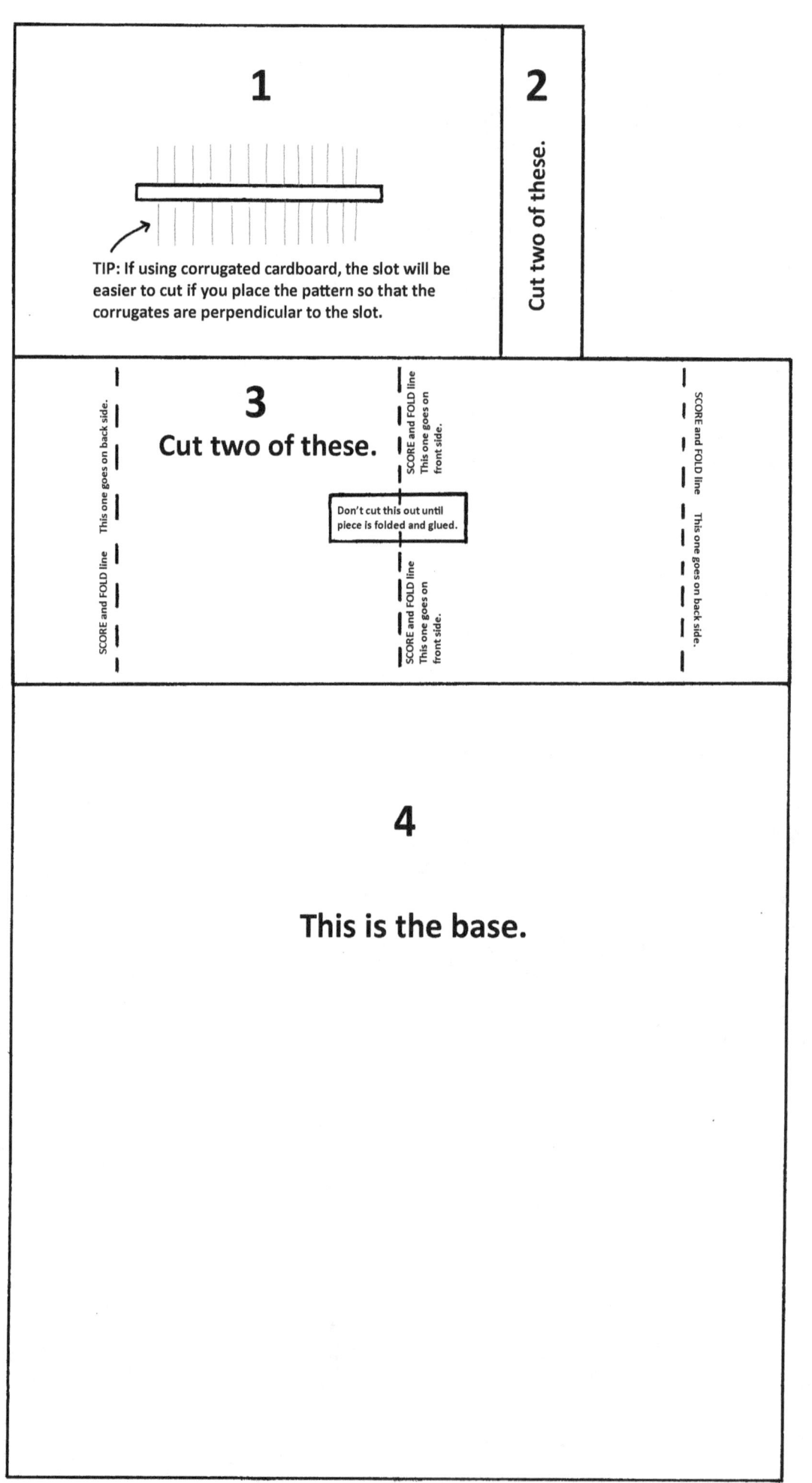

4) You are going to glue these side support in place on the base, but check the spacing first. Place your paper roll between them (as shown) and make sure they are not too close. Once the spacing is right, glue the "feet" of the support in place on the base.

5) Glue the long, narrow spacer pieces (2) on the back of the piece with the long slot (1).

6) Turn the narrow slot piece (1) over and glue it to the base as shown.

7) When dry, use a pencil to put the paper roll in place and feed the end of the paper underneath the narrow slot piece.

8) Now you are ready to draw some sine waves. One person will hold the cardboard device and control the wave pen. Place the pen (or pencil or marker) into the slot. Makes sure it moves up and down easily and marks the paper.
The person who is pulling the paper should now begin pulling the strip out very slowly, making sure to keep it moving at very steady pace. Watch the paper as it comes out.
Do you see a sine wave?

9) Now try some variations.
    FAST PENCIL, SLOW PULLING
    SLOW PENCIL, SLOW PULLING
    FAST PENCIL, FAST PULLING
    SLOW PENCIL, FAST PULLING
    SHORTER PENCIL LINES (don't go to ends of slot)

ACTIVITY 3.14: Make a simple harmonograph

You will need: a wide and flat cardboard box (in the U.S. the Amazon box size 14"x 18"x 3.5" is perfect), another piece of cardboard or foamcore (very flat and wrinkle-free) that is slightly small than the flat top of the box, some lightweight rope, another piece of heavy cardboard (at least 10" by 18"), a clothespin, scissors, craft knife, glue, duct or masking tape, a broom handle, two chairs, some medium-heavy books to put inside the box, a stack of books as a support for your pen holder, clear tape, paper, selection of pens or thin markers

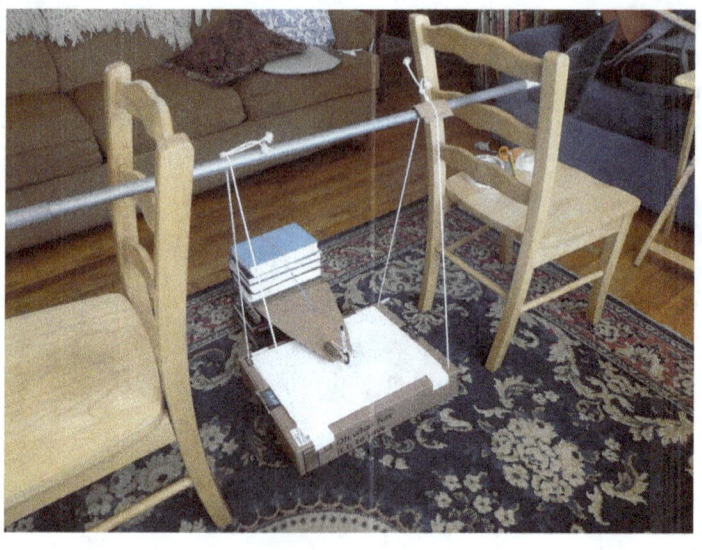

NOTE: There are many ways you can adapt this idea to use supplies that you have around your house. For example, you can use other supports instead of chairs and a broom, your box size doesn't have to be exactly like this one, you can use a heavy board instead of a box, you can change the length or placement of the ropes, you can use you own design for a pen holder, and so on.

Making the platform:
1) Punch holes in the corners of the box so that you can feed pieces of rope through.
2) Cut two pieces of rope that are about 2 meters long.
3) Feed the rope through the holes in the box so that the rope goes across the bottom and out the upper corners, leaving equal lengths coming out the four top corners.
4) While you are still able to open the box a bit, put some books inside the box. This added weight will make the platform swing longer. Remember, the weight of the bob is a deciding factor in how long the pendulum will swing. Heavier bobs make for longer swing times. After the books are inside, tape the box shut.
4) Tie the ends of the rope together.
5) Hang the box from a broom that is stretched between two chairs. (Or use another hanging scheme.)

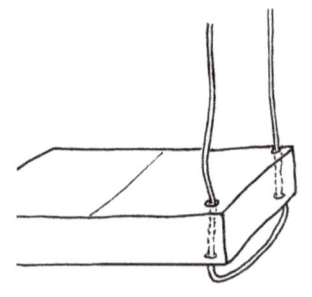

Making the pen holder:
6) Cut a piece of cardboard in the shape shown. The exact shape isn't critical, it just needs to have a wide base and narrower top. The exact size is also not critical. You might need to make yours a little longer if your platform is larger, for example.
7) You'll need to score (dotted line in diagram) near the bottom to create a flap that bends very easily. Make sure the flap is very floppy and not stiff. This flap will be tucked into a stack of books to hold it still.

8) Now you'll need to rig a way to hold the pen. My solution was to use a clothespin glued to a square block made of cardboard. This was then glued to the end of the triangular piece. If you use glue, make sure to let it dry thoroughly before you try to clip a pen into it.

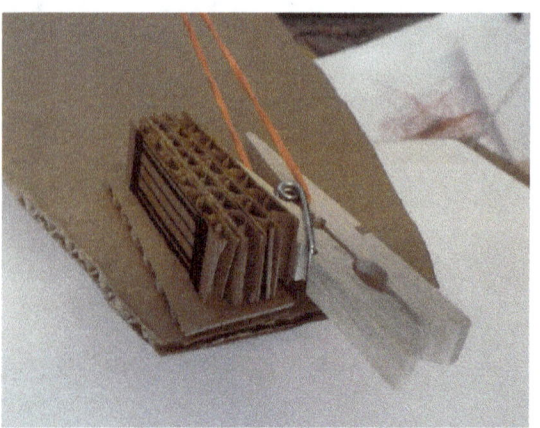

Putting the parts together:
9) Put an extra (very flat) piece of cardboard or foamcore on top of the box and secure with tape.
10) Use the photo as a guide, you'll need to assemble the parts so that the platform has plenty of room to move, the ropes are both the same length, the pen is at just the right height, and the pen is not putting too much pressure on the paper.

NOTE: The pen needs to ride very lightly on the paper. It has to stay in contact with the paper as it moves up and down with the motion of the platform, but if it puts too much pressure on the paper there will be too much friction. Friction from the pen will cause drag on the platform and slow it down. You want the platform to keep moving as long as possible.

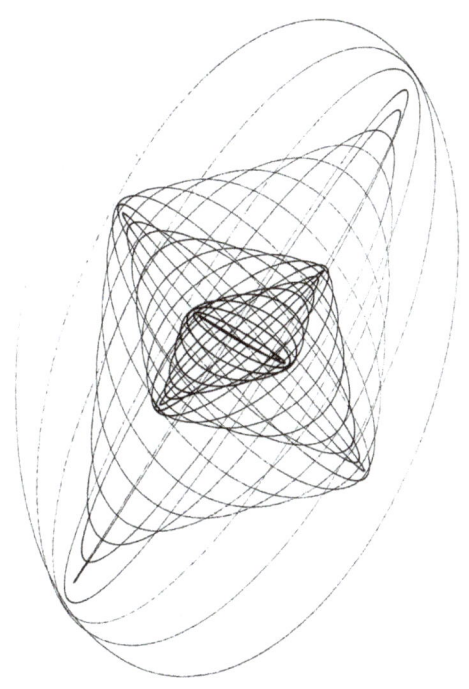

This drawing was made by the set-up in the photo.

You can see my solution to the weight of the pen. I used a chain of rubber bands to pull gently upwards. This isn't the only possible solution. I restricted myself to solutions that did not use anything from my wood shop. I tried to think of something that anyone could rig up. However, you are welcome to try another solution.

11) Place a piece of paper on the platform and secure with a few small pieces of tape. Start the platform moving and then gently lower the pen. Once the pen touches the paper, you can stand back and watch. When the platform runs out of energy and stops, the drawing is done. You can also choose to lift them pen at any time. Also, you might want to switch colors and add another drawing on top of the first one.

With a set-up very similar to mine, you'll get pretty much the same type of shape every time, with small variations. To get much different designs, you'll have to change the way the platform hangs. But this could be a good challenge—to come up with simple ways to alter the motion.

If this project really strikes your interest and you want to make a more permanent one, just do an Internet search for "homemade harmonograph" and you'll find lots of designs.

The cap is on the pen in this photo, but obviously you need to remove pen caps in order to draw.

CHAPTER 4:

ANSWERS to Review word puzzle 4.13:

1) Principia   2) second class   3) Descartes   4) oscillation   5) resonance   6) harmonograph
7) inertia   8) mass   9) force   10) sine   11) frequency   12) Royal Mint
13) x-rays   14) fulcrum   15) bob   16) Pantheon   17) dampen   18) screw
19) advantage   20) parabola

Famous quotes:
:Gravity explains the motions of the planets but it does not explain who set the planets in motion."
"Plato is my friend, Aristotle is my friend, but my greatest friend is Truth."

ACTIVITY 4.16:   Inertia "hat"

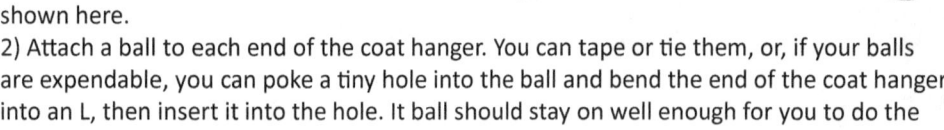

You will need: a coat hanger, two balls of equal size and weight (tennis balls are ideal)

1) Bend the coat hanger into the shape shown here.
2) Attach a ball to each end of the coat hanger. You can tape or tie them, or, if your balls are expendable, you can poke a tiny hole into the ball and bend the end of the coat hanger into an L, then insert it into the hole. It ball should stay on well enough for you to do the activity.
3) Place the middle of the coat hanger on top of your head.
4) Try spinning your head and body quickly to the left or right. Try turning around several times. The inertia hat should stay in place while you turn underneath it. The weight of the balls not only allows to hat to balance (center of mass below the balance point, as we learned in adventure 1), but they also provide inertia.

ACTIVITY 4.17:   Balancing a hammer

You will need: a hammer

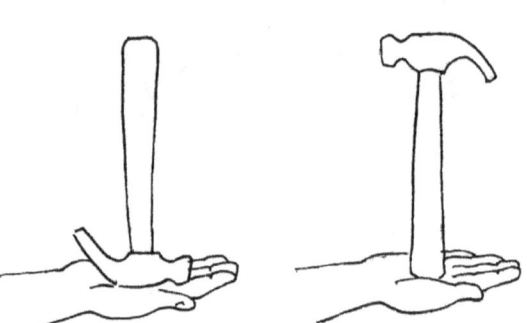

Try balancing the hammer on the palm of your hand, with the head end on your palm and the handle sticking up. Then turn the hammer the other way with the handle on your palm and the heavy head in the air. Which is easier to balance? The heavy metal head has more inertia than the wooden handle. When the top of what you are trying to balance is light and does not have much inertia, it will tip over very easily. However, if the top of what you are trying to balance is heavy, it not only has more mass it also has more inertia. More inertia means it is more resistant to a change in direction, and when you are trying to balance something, a change in direction is exactly what you don't want. So more inertia means it wants to stay in place which is very helpful in the balancing act.

CHAPTER 5:

Activity 5.8: Use a pencil to lift a bottle of rice (using friction)

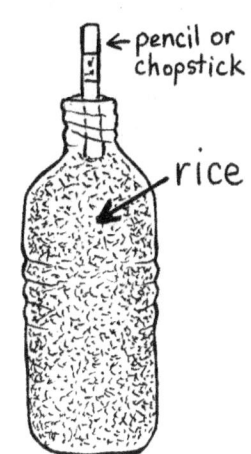

You will need: an empty bottle (clear is best, such as a water bottle), a funnel, rice, a long pencil

1) Use the funnel to fill the bottle with rice. (Make sure the inside of the bottle is dry.)
2) Push the pencil into the bottle.
3) Hold the end of the pencil and gently lift. You should be able to lift the bottle. If not, put the bottle back down and gently tap on the side, allowing the rice grains to settle in. Then try it again.

What is happening here? Each rice grain has a small amount of friction with the pencil, but when you add up all these small frictional forces, they are enough to keep the pencil from sliding out.

CHAPTER 6:

ANSWERS to 6.13 Review questions

1) water clock   2) b   3) Odd Numbers   4) T   5) velocity   6) Jupiter   7) b   8) F
9) F   10) T   11) T   12) d   13) c   14) d   15) F   16) LaPlace
17) Einstein   18) b   19) F   20) F

ANSWERS to 6.15 Math review questions

1) 54/9= 6   2) 8/32= 1/4 or .25   3) 20   4) 28   5) 24/6= 4   6) 28/7= 4, force
7) 30, 20   8) 1.2 kg   9) .6 m   10) 150 g   11) .003 km   12) .4
13) G(44) or 44G   14) 1/9   15) -29.4 m/sec^2 16) 9.8 m, 19.6 m, 29.4 m

BONUS POINT:

NOTES on answers for 6.10
1) We start with M1=5 and M2=5. Then M2 increases to 60. 5x12=60, so M2 increases by a factor of 12.
2) The distance goes from 3m to 9m. The distance increases by a factor of 3 (3d). Therefore the force decreases by a factor of 1 divided by 3^2, which is 1/9.
3) The distance doubled, so d went from 1 to 2 (2d). We have r^2=2^2 which gives us 4 in the denominator, which is 1/4.
4) Here we have 5d, giving us 5^2 in the denominator, which is 1/25.
5) The same because (10)(200)/10^2 = 2000/100= 20/1 = 20, and (5)(4)/1^2 = 20/1= 20.
6) 40x20/10^2 = 8. Thus, 20x10/r^2 must equal 8. 200/r^2=8. 200=8(r^2). 25=r^2. r=5 .

ACTIVITY 6.16: A spaghetti accelerometer

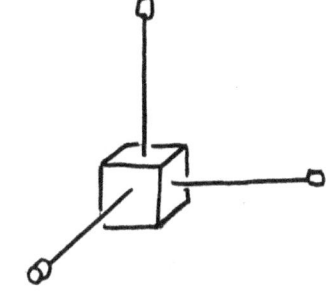

You will need: a wooden block, three pieces of long spaghetti (dry and uncooked), three miniature marshmallows, a drill with a bit that is the diameter of a spaghetti noodle

Directions for this project can be found at:
https://www.exploratorium.edu/snacks/spaghetti-accelerometer

ACTIVITY 6.17: Calculate the value of "g" using a cell phone

You will need: a smartphone, a pencil and paper

There are many YouTube videos about how to do this. They all use basically the same method. Just search YouTube (or another video service) for "measuring g with a smartphone."

ACTIVITY 6.3 : Additional notes, instructions and patterns

If you are not able to construct any kind of ramp, you can do this activity as a video lab and record the data you see on the video. Go to the playlist and look for "Discovering Motion Activity 6.3." You can watch me rolling balls down my 22-foot ramp and use my data.

If you are able to set up a long wooden ramp like I did, go ahead and do so. It can be any length. Just make sure it is very straight (no sagging in the middle). You can use any increments, but it will work out well on the graph if you divide the ramp length into 20-25 equal pieces. Those pieces can be a foot, 20 centimeters, 10 centimeters, 9 inches-- it doesn't matter, as long as the pieces are all the same length.

If you want to make a 2-meter ramp using only paper and your meter sticks, use the following instructions.

You will need: 2 copies of the following pattern page printed onto card stock, your two meter sticks, two short (1 ft/ 30 cm) rulers, roll of masking tape, glue stick, scissors, pencil, stopwatch, marble (Use a steel marble/bearing if you have one.

1) Copy the following pattern page onto cardstock. Make 2 copies.
2) Cut on the darker lines.
3) Fold on the lighter lines so that your strips look like these. —>
4) Overlap the strips by about an inch (2 cm) and glue using a glue stick or white glue. Make sure your overlaps are all going the same way (be mindful of "over" and "under" sides. Avoid tape because this will introduce more additional "speed bumps" over which the marbles will have to roll.

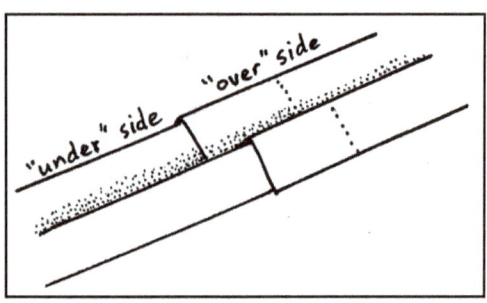

5) Starting with the "under side" on the end (which will be the bottom of your ramp) make marks every 10 centimeters. (You can also do this afterwards, as the last step but if you need your meter sticks to measure, they'll be covered up after step 7.)

6) Put your two meter sticks end to end with a small ruler underneath. Secure with tape. (I use blue masking tape which comes of easily when you go to disassemble.)

7) Put your paper trough on top of your new, very long 2-meter stick. Secure edges with tape.

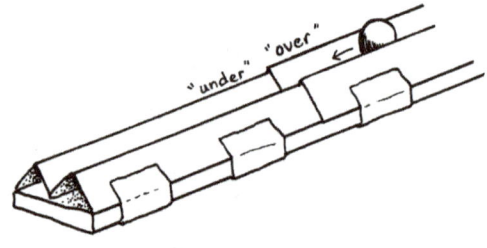
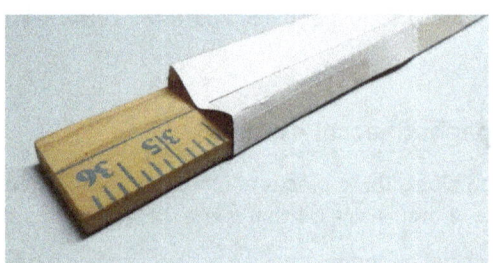

8) You'll want to roll your marbles in the direction shown in the diagram so they go over the seams easily. Your 10-cm marks should be measured up from this end.

HOW TO RUN THE EXPERIMENT:

1) Put the top of your ramp on a book or block so that the start is a few centimeters off the table (or floor). Use your stopwatch to time how long it takes for the marble to go down the ramp. If the time is less than 4 seconds, reduce the height of your ramp. You want the marble to take as long as possible; 5 seconds is even better. However, don't make it so shallow that the marble stops at the joints.

2) You might want to put a metal box, or a hollow box, right at the end of the tracks so that when the marble hits it, it will make a sound. Having a definite sound as the marble hits the end of the ramp might help you get timing stopped at the right point. You might even be able to anticipate when he marble will hit. You want to stop the clock right as the marble hits the end, not a split second afterwards. (Ideally, you'd have the marble hit the stop button on the stopwatch. I tried to rig this up on my ramp, but could not get the ball to hit the button well enough.)

3) Hold the marble at the lowest (10 cm) mark and get your stopwatch ready. Time how long it takes for the marble to go from 10 cm to the end of the ramp. Write this down.

4) Hold the marble at the 20 cm mark and time how long it takes to get to the bottom. Write this down.

5) Continue up the ramp, timing the marble at 10 cm intervals. (If you want to skip an interval here and there, that is okay.) Write down all your results. You will use these results for the graph on page

6) Just as possibly helpful information, here are some numbers I got on my paper ramp. Your numbers won't be exactly the same but they should look somewhat similar.

10 cm= 1.0 sec.	20 cm= 1.3 sec.	30 cm= 1.5 sec.	50 cm= 2.2 sec.	60 cm= 2.4 sec.
70 cm= 2.5 sec.	80 cm= 2.85 sec.	90 cm= 2.95 sec.	100 cm= 3.0 sec	120 cm= 3.5
150 cm= 4.1	180 cm= 4.35	(I also added 5 cm= .75 sec.)		

7) Turn to the graph on page 80. Write "time in seconds" below the bottom of the graph and make 1 second equal to 4 blocks. Write "distance in centimeters" along the "up" axis, and label each block as 10 cm.

8) Plot your data points.

9) Draw a "best fit" curve that shows the general trend of your data. It should be an exponential curve that is very shallow at first, and the gets steeper and steeper as distance increases. (See example below.) Some of your data points might not be on this best fit curve. This often happens with experimental data, due to small errors in equipment, timing, and human error.

10) Consider running the experiment again, but this time with the ramp a little higher or lower. Or maybe you can use a different ball. You can graph these results on the second graph on page 80, or you can put the results on the same graph and just use a different color for the best fit curve line.

11) Finally, watch the video on the playlist labeled "Discovering Motion 6.3 Analysis." The video should also show up if you search for it using those keywords. Even though your numbers aren't the same as the ones on the video, it should still help you to analyze your own data and compare it to what Galileo found.

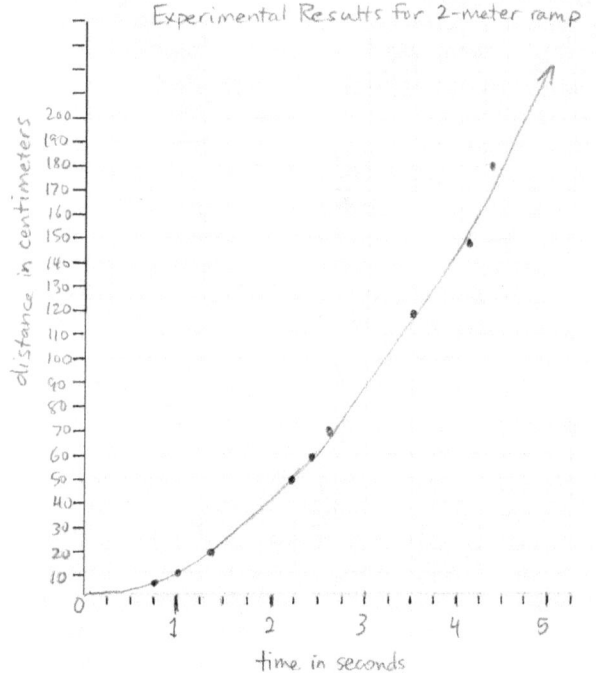

We've seen this curve before, haven't we? It's a parabola. You get a parabola whenever your data goes up exponentially. The equation for the curve would involve an expression like x^2.

CHAPTER 7:

Graph paper for 7.2 is on the next page.

Answers to 7.3:
1) 607.6 N (Multiply 62 kg by 9.8 m/sec².)   2) 600 N (Multiply 20 kg by 30 m/sec².)
3) 50 kg (Divide 490 by 9.8 to get 49.999=50)   4) 9 m/sec² (Divide 490 by 15.)
5) .0784 N (Multiply .008 kg by 9.8 m/sec².)   6) 5,100 N (Multiply 1,700 kg by 3 m/sec².)
7) 588 N (Multiply 60 kg by 9.8 m/sec².)

Answers to 7.4:
Note: If your students don't know how to do unit conversions, this is a good opportunity to introduce the topic. Example shown here on right.

$$\frac{.5 \text{ lb}}{1} \times \frac{.45 \text{ kg}}{1 \text{ lb}} = .225 \text{ kg}$$

1) 2.2 m/sec². (F=1, m= .45 kg. Divide 1 by .45 to get 2.2)
2) 17.7 m./sec² (3 oz.=.25 pound. Multiply .45 kg by .25 to get .1125 kg. So you have 2=(.1125)(a). Divide 2 by .1125.)
3) 5,000,000 N (a=200/10 = 20 m/sec². Then do mass times acceleration: (250,000 kg)(20 m/sec²) = 5 million N.)
4) 12 m/sec² (8 ounces = .5 lb, then multiply .5 times .45 to get .225 kg. Then (2.7)=(.225)(a). Divide 2.7 by .225 to get 12.)
5) 2 kg (Convert .25 km to 250 m. Then find (a) by dividing 250 by 50 seconds. So (a)=5 m/sec². Then 10=5a, so a=2.)
6) 1 N (Multiply .2 kg times 5 m/sec². 10 m in 2 sec. is 5 m in 1 sec.)

Answers to 7.10:
1) 1.5 kg(m)/sec (Multiply .05 kg by 30 m/sec.)
2) 20 m/sec (Convert grams to kg. 145 g = .145 kg. Then (2.9)=(.145)(v). Divide 2.9 by .145 to get 20 m/sec.)
3) 7.5 N (1.5 divided by .2.)   4) .11 seconds (50(Δt)=5.5 Divide both sides by 50. 5.5 divided by 50 is .11)
5) 20 m/sec (3.2=(.16kg)(v) 3.2 divided by .16 is 20.)   6) 20 N (5 divided by .25.)

Answers to 7.13:
1) False, he said things that essentially meant the same thing as this formula, but he did not actually write "F=ma."
2) False, he wrote in Latin   3) newtons, N   4) momentum   5) 1 kg, 9.8 N   6) yes
7) yes   8) 980 N   9) True   10) True   11) barycenter
2) True   13) normal   14) 2 pairs   15) contact, field
16) b   17) large   18) Descartes   19) True   20) no

Answers to 7.14:
1) I   2) J   3) N   4) O   5) A   6) D   7) E   8) C   9) K   10) B   1 1) L   12) F   13) M   14) G   15) H

ACTIVITY 7.15: Ye Old "Balloon Rocket" Activity

This activity is so well-known that I decided not to include it in the text. I figured that probably most of the students using this book have already done this activity. It is a classic activity for elementary grades. It's pretty basic. Blow up balloon, let it go, it travels along the string.

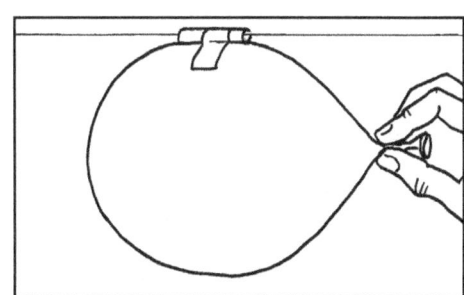

You will need: a balloon, a straw, scissors, tape, fishing line

1) Cut a small piece of straw, about 3 to 4 cm long.
2) Thread the straw onto the fishing line.
3) Stretch the fishing line across a room, and tape the ends securely so that the line stays taught.
4) Cut some pieces of tape and have them ready.
5) Blow up the balloon. Do not tie it shut!
6) Hold the balloon right under the piece of straw and tape the straw to the balloon.
7) Pull the balloon to one side of the string, and then let go.
8) The balloon will "rocket" across the room, along the fishing line.
9) Make sure the students understand that the opposing forces here are the air molecules escaping out the back of the balloon, and the kinetic motion of the balloon traveling along the line.

For 7.2: Make three copies of this pattern page. Trim the ends of two of them so you can make them overlap in a way that gives you a continuous 3-page graph. Use glue stick, not tape, and try to make sure that the overlap goes in the direction that will not create a smooth ride for the cardboard corner that will be gliding over the paper. You don't want to create a "speed bump" that the cardboard corner might get caught on. If you are working from a paperback copy of this book and need a digital pattern to print, you can find it by downloading at www.ellenjmchenry.com/discovering-motion-printables

ACTIVITY 7.16: Experiment with a "Newton's Cradle" toy

You will need: a Newton's cradle, easily purchased from any online science store or from Amazon

Newton's cradle was not invented by Newton himself, only named in his honor because it demonstrates the third law. The metal balls provide very elastic collisions. This toy is a lot of fun, but be aware that you must be very gentle with it. I've found that younger students left unattended will find a way to tangle the strings somehow. The strings are made of fishing line so they are plenty strong for ordinary use, but will break if the toy is misused.

ACTIVITY 7.17: Field trip to elevator to stand on scale and see weight change

You will need: a bathroom scale, an elevator (one that goes fairly high and fairly fast)

1) Enter the elevator, place the scale on the floor and stand on the scale. Notice the weight reading.
2) Press the button to go to the highest floor. (Hopefully, you can get a straight shot up and the elevator won't have to pick up passengers.)
3) Watch the scale very carefully. What happens to the weight reading as you go up?
4) When you get to the top, press the button to go down again.
5) Watch the scale very carefully again. What happens to your weight reading?

On the way up, you will experience additional acceleration (in addition to gravity), so your weight reading should go up.
On the way down, you will experience negative acceleration (which will counter gravity), so your weight reading will go down.

ACTIVITY 7.2: Sample graph

Here is a sample of what one student recorded. Your results won't be the same as these, but this gives you the general idea of what results might look like. If you want to do the experiment a second time, you might use dotted or dashed lines to represent the second ramp height. You can label the lines or add a note to the key.

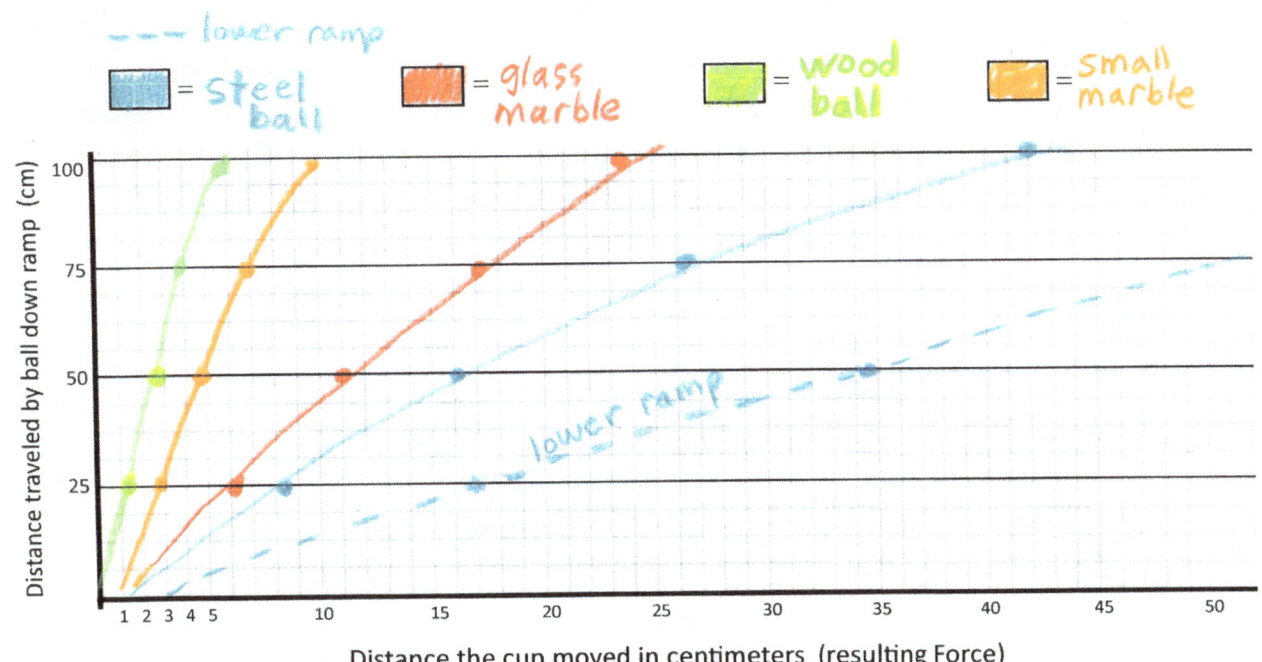

99

CHAPTER 8:

Answers to 8.2:

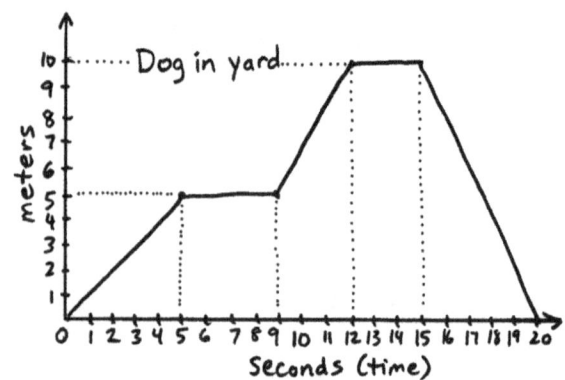

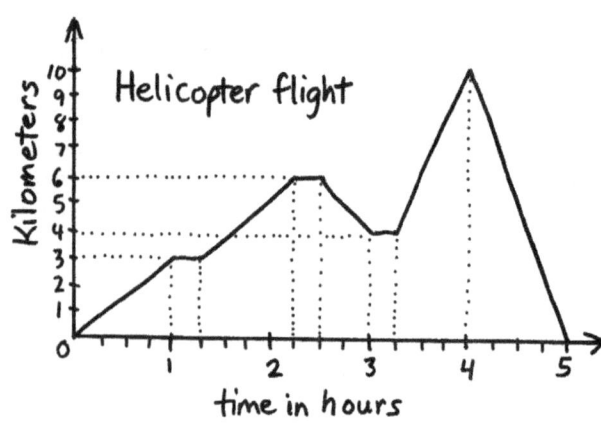

Answer to 8.3:

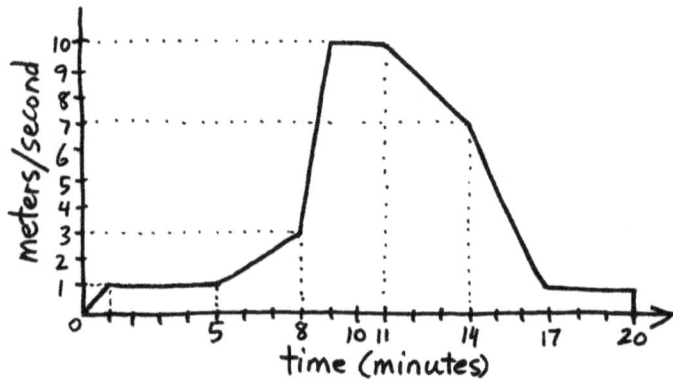

Answers to 8.4:
1) 141.82 meters. (The listed height is 142 meters, so this is very close.)
2) 507.79 meters (The listed height is 508 meters, so our estimate is close.)
3) 11.35 seconds
4) 678.8 meters (The listed height is 679 m. so this is very close.)
5) about 8.8 seconds. (Maybe, if they react quickly and run <u>very</u> fast!.)
6) 323 meters

1) $t = \sqrt{\dfrac{2(142)}{9.8}}$

2) $d = \dfrac{1}{2}(9.8)(10.18)^2$

3) $t = \sqrt{\dfrac{2(632)}{9.8}}$

4) $d = \dfrac{1}{2}(9.8)(11.77)^2$

5) $t = \sqrt{\dfrac{2(380)}{9.8}}$

6) $d = \dfrac{1}{2}(9.8)(8.12)^2$

8.5: Extra photos

End of marble-catching tray

Sample made using tempera paints

Even my high-schoolers enjoyed this activity.

Answers to 8.8:

1) 198.45 m 2) 474.44 m
3) 4 m/sec 4) 10.5 m, so yes he will miss the rocks.

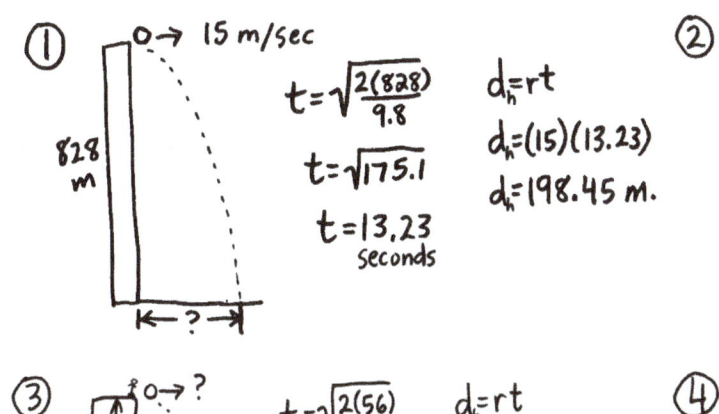

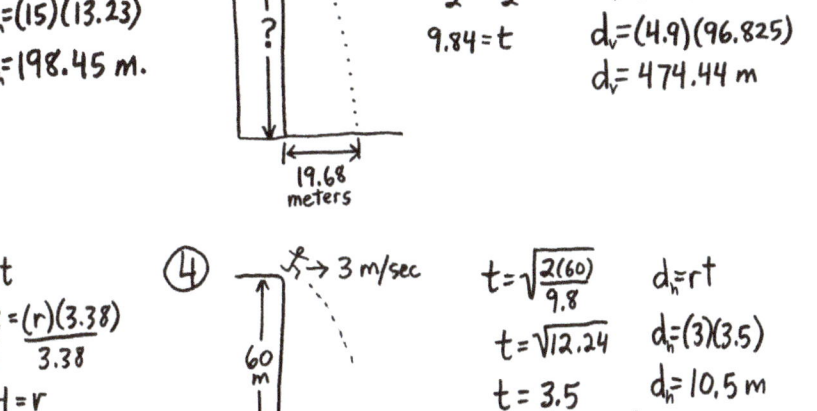

Answers to 8.11:
1) False 2) True 3) Sweden 4) I think, therefore I am. 5) c 6) True 7) magnitude and direciton
8) b 9) True 10) Strasbourg Cathedral 11) False 12) 45 degrees

Answers to 8.12:
1) E 2) G 3) B 4) H 5) C 6) D 7) A 8) F 9) E 10) D 11) G 12)

Sample graph for 8.10 (cotton ball cannon)

This is a sample of student data. Your data won't be exactly the same, but this gives you a general idea of what to expect.

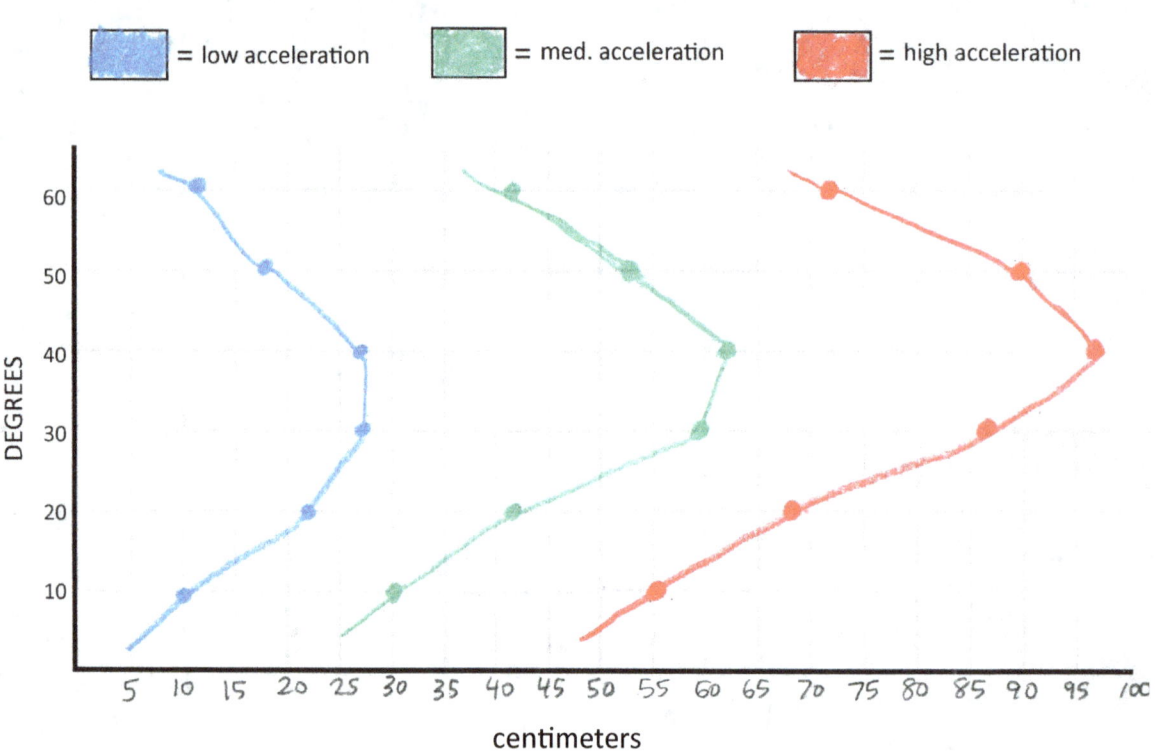

ACTIVITY 8.13: Reaction time experiment (using $d=1/2gt^2$)

You will need: a metric ruler, a stopwatch, a pencil, a calculator, a lab partner

1) The test subject sits in a chair with their arm on a table. Their hand should be off the edge of the table, as shown.
2) The tester holds the ruler exactly at the level of the thumb and fingers.
3) The tester will drop the ruler without any warning. The test subject tries to catch the ruler as quickly as possible using their thumb and first two fingers.
4) Read the number of centimeters the ruler fell before it was caught.
5) You can now use the formula ($d=1/2gt^2$) to figure out the test subject's reaction time. However, it will be easier to use the rearranged form of the formula because we want to find "t." Use the formula in the form shown here on the right, t= square root of 2d/g.
6) Plug in 9.8 for "g," and the numbers of meters the stick fell before it was caught for "d." (Convert centimeters to meters.)
7) Do the calculation and you'll find out how many seconds it took to react.
8) Do this experiment with as many people as you can and compare reaction times. Is there a difference between males and females? Between older and younger? Or is it independent of any obvious factors?

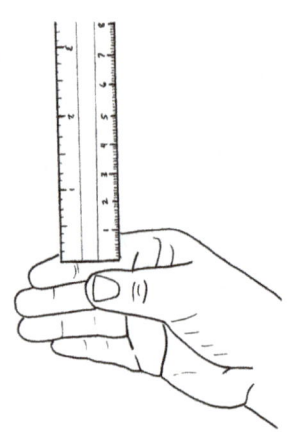

$$t = \sqrt{\frac{2d}{g}}$$

CHAPTER 9:

Answers to 9.21:
1) d   2) b   3) b   4) T   5) F   6) c
7) rabbit or bunny trail   8) magnitude and direction
9) c   10) magnitude or direction   11) T   12) F
13) a   14) c   15) b   16) T   17) d   18) a
19) a   20) T   21) F   22) e   23) right-hand rule
24) F   25) b   26) E   27) D   28) C   29) A
30) B   31) T   32) T   33) cylinder
34) conservation of angular momentum
35) hollow cylinder
36) angular momentum of a point object
37) centripetal force
38) linear momentum
39) moment of inertia
40) angular momentum of an extended object

Answers to 9.22:

ACTIVITY 9.25: Make "moment of inertia" tops

You will need: heavy cardboard, scissors, compass for drawing circles, pennies, poster putty or sticky dots (for adhering pennies to cardboard), stopwatch

<u>How to make the tops:</u>
1) Use the compass to draw four circles of different sizes. (Suggested diameters: 10 cm, 12 cm, 14 cm, 16 cm) The exact sizes are not important, as long as you have at least small, medium and large.
2) Cut out the circles. You'll want to use a good quality large pair of scissors.
3) Use either the tip of the scissors or an sharp craft knife to cut a slit exactly in the center that is as long as the diameter of a penny. Don't widen it very much—you want the penny to fit tightly.
4) Press a penny into the central slot so half of it is showing on each side of the circle. Do this for each circle

<u>What to do with the tops:</u>
1) Using what you know about moment of inertia and the formula for a solid disc ($I = 1/2\, MR^2$), predict which top (small, medium, large) will keep spinning for the longest time and which will stop first. Then use the stopwatch and find out how long it takes for each top to stop spinning. Write down your results. Was your prediction accurate?
4) Now use the poster putty or sticky dots to stick pennies around the center of one top, grouped around the pencil. Give it a spin and time how long until it stops. Compare this to your time without the pennies.
5) Move these pennies to the outside of the disc. Don't add any new pennies, just change their location. Will the top spin for a longer time with this arrangement? Find out.
6) Now add more pennies to the outside. (You might even fill the outer rim.) Will this add to the time of spin? Find out.
7) Try "racing" different combinations. Does the smallest top with pennies on it spin as long as the largest top with no pennies? Make a prediction then see if you are right.

ACTIVITY 9.9 Sample data and graphs

This is just sample data. Your numbers don't have to match these exactly. However, your data will likely be very close to these figures. Your washers might have been a different size than the ones we used, or your string could have been thicker or thinner than ours. Both of these factors would affect your results.

DATA CHART for 1 WASHER

radius	total dist.	# seconds	velocity
48 cm	60 m.	14	4.28
36 cm	45 m.	13	3.46
24 cm	30 m.	11	2.72
12 cm	15 m.	8	1.87

DATA CHART for 2 WASHERS

radius	total dist.	# seconds	velocity
48 cm	60 m.	11	5.45
36 cm	45 m.	10	4.5
24 cm	30 m.	8.5	3.5
12 cm	15 m.	7	2.14

DATA CHART for 3 WASHERS

radius	total dist.	# seconds	velocity
48 cm	60 m.	10	6.0
36 cm	45 m.	8.5	5.3
24 cm	30 m.	7	4.28
12 cm	15 m.	4.5	3.3

DATA CHART for 4 WASHERS

radius	total dist.	# seconds	velocity
48 cm	60 m.	9	6.6
36 cm	45 m.	7.5	6.0
24 cm	30 m.	6	5.0
12 cm	15 m.	4	3.75

10) Add another large washer, making a total of 3 large washers hanging down. Repeat the previous steps, swinging at each radius and counting seconds it takes to do 20 circles. Record your data in the table that says 3 WASHERS.

11) Now add another washer, making a total of 4 hanging down. Repeat time trials and record data on 4 WASHER chart.

12) Now reduce the number of washers to just one. Repeat time trials and record data on the 1 WASHER chart.

13) Now you are ready to graph some data. We will be making two graphs, looking at the data in two ways. The graph on the left asks you to plot number of washers versus the velocity of the small washer. Choose four colors for the sizes of radii.

14) Now for the graph on the right. Choose **one** of your charts above and plot radius in cm versus velocity.

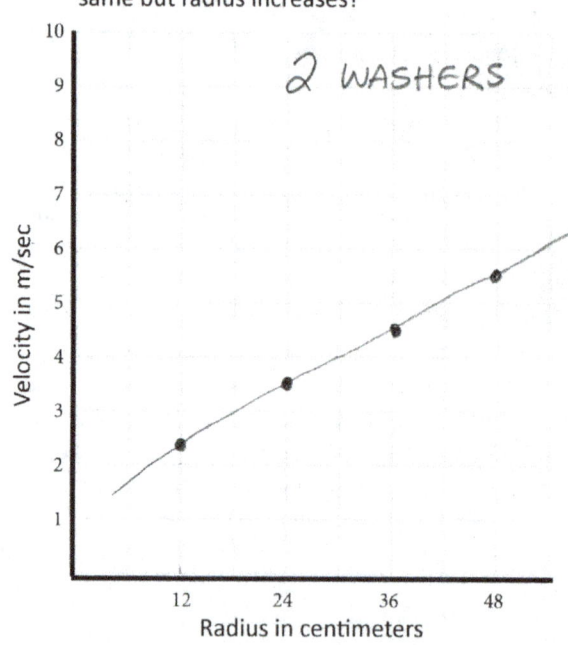

What happens to velocity when F_c increases?

- = 48 cm radius
- = 36 cm radius
- = 24 cm radius
- = 12 cm radius

Velocity of small washer (distance/#seconds) vs. Number of washers (increasing F_c)

What happens to velocity when F_c stays the same but radius increases?

2 WASHERS

Velocity in m/sec vs. Radius in centimeters

ACTIVITY 9.9: Discussion

Questions for graph on left:
1) Student estimates will probably be a little more or less than 7.
2) "When we keep the radius the same, the velocity <u>increases</u> and the centripetal force <u>increases</u>."
3) No, the velocity did not double. The increase was less than double. (In our sample graph, looking at the red line, the increase in velocity between 2 and 4 washers was only approximately 5.5 to 6.5. Double would be 5.5 to 11.)

Questions for graph on right:
4) Speed of the small washer also increases, as radius increases.
5) It probably seemed that the small washer was going faster at smaller radii. However, our data does not support his observation. The small washer was completing more revolutions per second at smaller radii, but the speed at which it was moving through the air was actually decreasing. Our graph shows this very clearly.

Bonus question:
6) When you pull down quickly on the bottom string, you are quickly increasing the centripetal force. Increasing the centripetal force increases the speed of the small washer. This is different from what we were doing in the right hand graph. In that graph we were keeping centripetal force constant.

Extension:
1) Fc= (.001 kg)(5.45 m/sec)(5.45 m/sec)/.48m = .06 N
You might not be working with 5.45 for velocity. Just plug in your velocity number instead of 5.45.

2) Fc= (.001)(3.3 m/sec)(3.3 m/sec)/.12m = .09 N
You might not be working with 3.3 for the velocity. Just plug in your velocity number instead of 3.3.

3) Results will vary depending on what you try.

ACTIVITY 9.14 answers

1) .000125 kg m²/sec   L=(.5m)(.0005kg)(.5 m/sec)
2) 1 meter   .12 kg m²/sec = (r)(.04 kg)(3 m/sec)   .12(r)= .12   r=1
3) 8 m/sec   100 kg m2/sec = (1.25 m)(10 kg)(v)   12.5(v)= 100   100/12.5=8
4) 26 x 10^{17} kg m²/sec   L= (1,000,000 kg)(52,000 m/sec)(50,000,000 m)

ACTIVITY 9.16- 9.18 Discussion

    Size and mass should not make a significant difference in the outcome of the races, though in the case of comparing solid cylinders, we found that size/mass did make a difference. Larger wooden cylinders rolled faster than smaller wooden cylinders. The thickness of a hollow ball will also play a significant role in moment of inertia.
    In general, the distribution of the object's mass is the most important thing. The most important thing to keep in mind during these activities is what we learned in activity 9.15. Keep thinking of tiny juice boxes on sticks inside the cylinders and balls. The pictures of the objects at the **top of page 154** are a great cheat sheet. They show which objects should win races over other objects. The hollow cylinders always lose and the solid balls always win. Pick an object. Anything to the right of it should be faster. Nothing is to the right of the solid ball so it always wins.
    In my class, we found that the mass of the solid balls (glass, wood, metal) did not make much of a difference. Glass and steel marbles of the same size usually tied. Sometimes there would be a split second difference, but if we ran the race several times these small differences did not form a pattern. Each ball had a turn being a split second ahead of the others, so the discrepancies canceled out. We also found that the size of the ball did not make much difference. We raced a small steel marble against a large steel bearing with a diameter 3 times larger than the marble, and they tied every time.
    The reason for larger or heavier solid cylinders being faster than similar smaller cylinders was unclear. We did some Internet research and found conflicting answers, even among physics experts. Our results were very clear: the radius of a wooden cylinder did play a large role in its velocity. However, solid balls of different sizes did not show a large difference. We guessed that it might have something to do with the amount of surface area touching the ramp. In the case of balls, no matter what the radius is, only one small point is touching the ramp. With cylinders, all of the surface area (except the ends) comes into contact with the ramp.

ACTIVITY 9.26: Make your own cycloid ramps

You will need: the pattern on the following page (you can cut it out of the book if you are working from a paperback copy, but you can print it onto card stock if you are working from a digital book), scissors, a sharp craft knife (such as X-Acto), tape, ruler, pencil, white craft glue (PVA glue), corrugated cardboard (or foamcore), several pieces of card stock, some marbles,
Optional: rope about 1.4 inch diameter to use as guide rails on sides of ramps

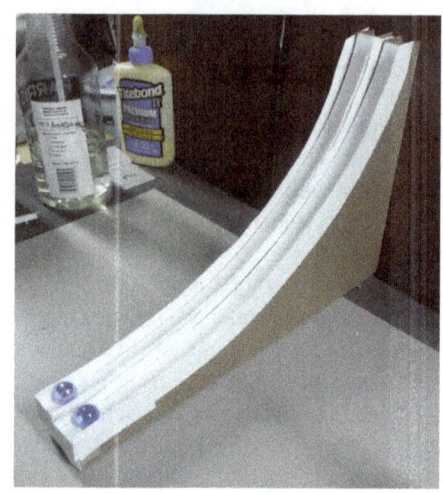

NOTE ABOUT GLUE: I don't recommend using standard Elmer's glue or "school glue," as they are not sticky enough. Elmer's Craft Bond is okay, or you might try Aileen's Craft Glue, or a similar white glue made for crafting. White PVA glue sold outside the U.S. is still pretty good.

How to make the ramps:
1) Print the following pattern page onto card stock If you are working from a paperback copy of this book (not digital) you can download this pattern (along with all the other printables) at **www.ellenjmchenry.com/discovering-motion-printables**
2) Cut out the patterns very carefully. Be patient as you cut and stay right on the lines. You want your curve to be as perfect as possible. Tape the two pattern pieces together along the side with the letters so that the X's, Y's and Z's are next to each other. Also, make sure the entire bottom edge is straight (check it against a flat surface or ruler).
3) Use the pattern to make 4 identical cardboard/foamcore ramp pieces. Cut very carefully! Take your time and be accurate.
4) Now you need to cut some card stock strips. Assuming you'll be rolling marbles down your ramps, the best width is 2 cm. Use your ruler to make half a dozen strips. (Or you can cut these strips with your knife blade instead of scissors.)
5) Cut some small scraps of cardboard/foamcore —pieces that are about an inch by an inch (2 cm x 2 cm) but they don't have to be exactly this size and they don't have to be square. You won't see these when your ramps are finished.
6) Put one ramp piece flat on the table and put some stacks of cardboard pieces in a few places. Glue the scraps together and glue each stack to the ramp piece.

7) Then put the other ramp piece on top and glue in place. While the glue is still soft, Set the ramp up on each edge and make sure the whole thing is straight. Then let it dry. (If it wants to warp, set a medium-weight large book on top while drying.)
8) Repeat these steps for the other set of ramp pieces.
9) When the ramp forms are dry (or dry enough), apply glue along the curved edges and cover with a strip of card stock. If the strip is not long enough, you can cut another short piece for the end. If you want to make your entire ramp have a more finished look, you can put strips along the back and bottom as well.

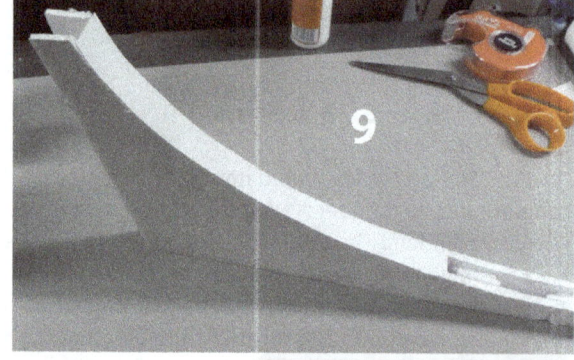

10) Now you'll need to add something along the sides of the ramp to keep the marbles from going off the track. You can cut curved pieces of cardstock to glue along the sides, or you can glue some strips of rope. I used red rope to add a touch of color.
11) Finally, you'll need to put something at the end of the ramp to stop the marbles. You can use a scrap of card stock glued around the end, or a piece of cardboard/foamcore. (See top picture of finished product.)

What to do with the ramps:
1) Hold two marbles at any place along the ramps and let go. They should both hit the end at the same time.
2) You might also want to make a straight ramp that goes from the start to finish points of the curved ramp. Try racing the straight ramp against the cycloid ramp.
3) Also, you could make another curved ramp, but not a cycloid shape. Set this ramp next to a cycloid ramp and race the marbles down both.

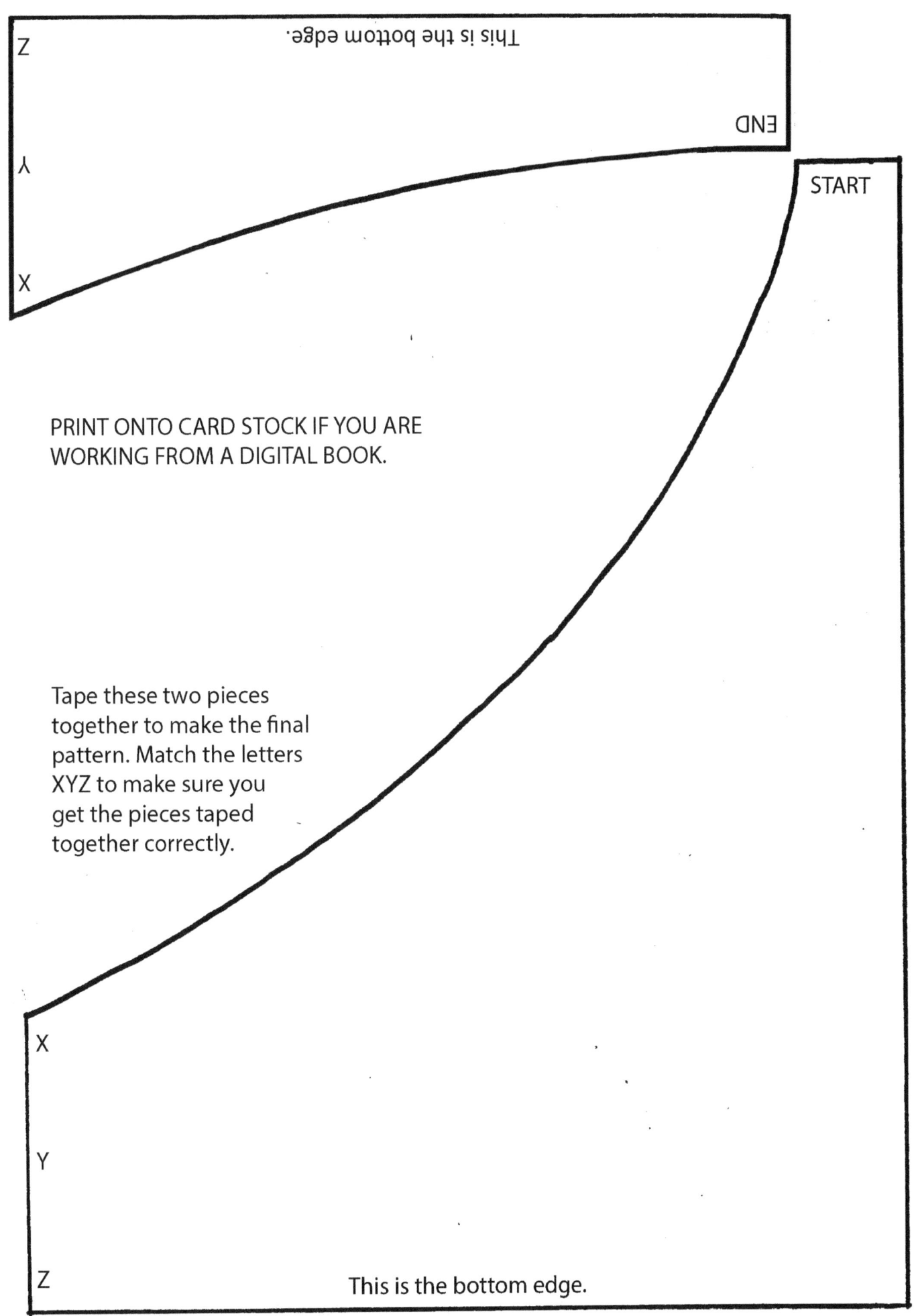

CHAPTER 9: (continued)

ACTIVITY 9.27: A hard-boiled "gyroscope"

You will need: a hard boiled egg (optional: tray of ice)

Spin the hard boiled egg. After a few seconds, it should adjust its position (while still spinning), so that it is up on end. Why does it do this? This action is related to the precession of gyroscopes. The small amount of friction between the egg and the table is the primary cause of this strange rotation of the egg. If you spin an egg on ice, thus reducing friction, it will not stand on end. For a video explanation, see "The Self-Reversing Spin Experiment" by "The Action Lab" channel on YouTube.

ACTIVITY 9.27: Make a homemade "rattleback"

You will need: a plastic spoon, a small amount of clay or "poster tack" dough

NOTE: Plastic rattlebacks can be ordered various places online. They are not expensive.

ALSO NOTE: Watching the suggested video in the previous activity ("The Self-Reversing Spin Experiment" by "The Action Lab") ahead of time will be very helpful.

1) You will need to cut the handle off the spoon and then smooth the end. I did this with my woodworking tools (scroll saw and sander) but if you don't have these tools, you can probably make do with a large pair of scissors and some sandpaper.
2) Place little dots of dough as shown in photo.
3) Spin the rattleback clockwise. If the dough dots are the right size and in the right place, the rattleback will slow down then reverse direction. (The reverse spin will be slow-- don't expect it to spin as fast as it did at first.)
4) If this does not happen, try adjusting the dots. (Watch the suggested video and look how his dots are placed.) Try it again.

CHAPTER 10:

Answers to 10.1:
1) 49 Joules   PE= (5 kg)(9.8 m/sec²)(1 m)= 49 J
2) 2,352,000 J   PE= (80 kg)(9.8 m/sec²)(3,000 m) = 2,352,000 J
3) 10,000 Joules   KE= 1/2(2 kg)(10,000 m/sec)= 10,000 J
4) 51,667.5 J   30 km/hr= 8.3 m/sec   KE= 1/2(1,500)(8.3)(8.3)= 51,667
5) 206,670J   8.3+ 8.3= 16.6 m/sec   KE= 1/2(1,500)(16.6)(16.6)=206,670 J   Answer more than doubles!

Answers to 10.12:
1) potential energy   2) kinetic   3) T   4) 500,000 joules 5) zero   6) 25,000 joules   7) b   8) a   9) b   10) yes
11) d   12) b   13) F   14) force, distance   15) c   16) b   17) T   18) T   19) T   20) a   21) b   22) F
23) sun and planet

Answers to 10.13:
KINETIC
COSINE
POTENTIAL
ENERGY
DISTANCE
WATT
WORK
CONSERVATION

Coaster: Kingda Ka

Draw a free body diagram for each of these situations. (Answers in teacher's section.)

1) A car is sitting in a parking lot and is not moving. What forces are acting on the car?

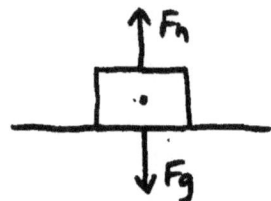

2) A rightward force is applied to a box across a rough surface. The rope is parallel to the ground.

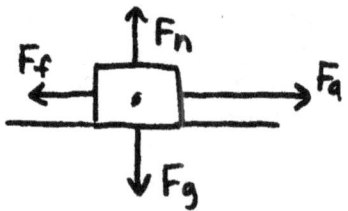

3) A hockey puck is gliding across the ice at a constant velocity. (We'll ignore air resistance and the tiny amount of friction from the ice.)

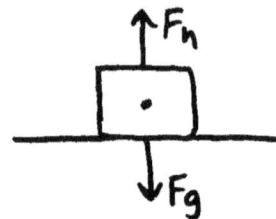

4) An elevator is accelerating upwards.

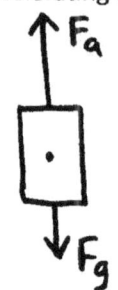

5) A skydiver has opened her parachute and is now descending slowly, moving at a constant velocity. Indicate air resistance with $F_{(air)}$.

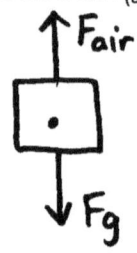

5) A box is sitting on a ramp, but there is enough friction to keep it from sliding.

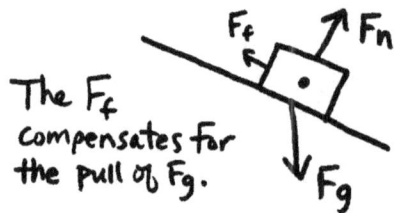

The F_f compensates for the pull of F_g.

7) A ball was thrown into the air. It is slowing down as it approaches the peak of its parabolic flight. (Is its horizontal velocity constant? Review on page 122.)

8) A refrigerator magnet is stuck to the fridge. It is not moving. Show a side view of the magnet and the forces acting on it. Use F_m for the magnetic force. Don't forget the friction between the magnet and the fridge surface!

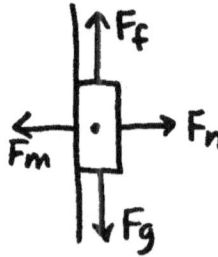

9) A horse is pulling a sled across a rough surface. The angle of the rope is 45 degrees.

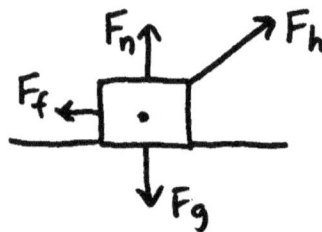

10) A boy is trying to drag his stubborn dog across the kitchen floor, but the dog weighs too much and try as he might, he can't make the dog budge. The leash is parallel to the floor.

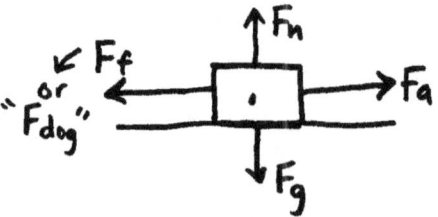

11) Two people are lifting a log, one at each end. They are pulling with equal force. Their two pulling forces must be greater than the force of gravity in order to lift the log. (Remember, the size of the arrows indicates strength of force.)

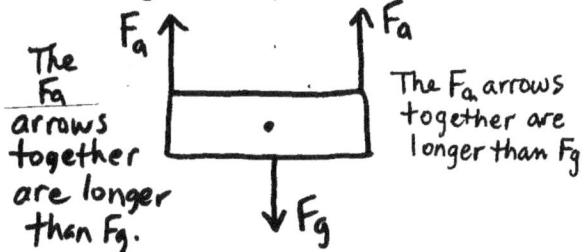

The F_a arrows together are longer than F_g.

The F_a arrows together are longer than F_g.

12) A child is hanging from a horizontal bar in a playground. Show each arm as a force arrow. How long should these arrows be in comparison to the F_g arrow?

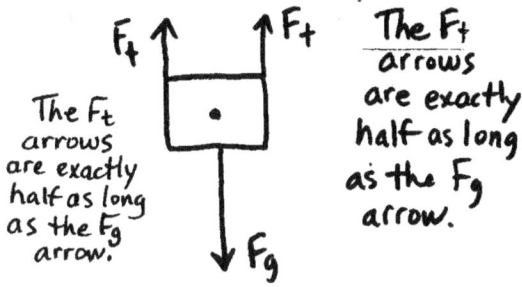

The F_t arrows are exactly half as long as the F_g arrow.

The F_t arrows are exactly half as long as the F_g arrow.

13) A sky diver has not yet opened his parachute. He has not yet reached terminal velocity. Include air resistance.

14) A satellite is traveling through the solar system. It has instruments on board that can keep its velocity at a constant 100 m/sec.

Add some magnitudes (numbers) to the force arrows in these problems:

15) A sailboat is moving to the right with a wind force of 40 N. Water currents are working in the opposite direction with a force of 5 N. (We know what Fg is, and that Fn is equal and opposite so you don't have to label these with numbers.)

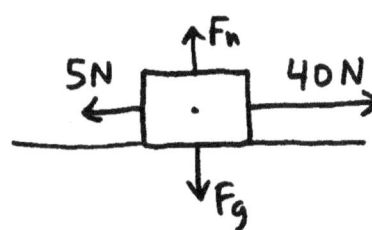

16) A penguin is walking into an Antarctic wind that is blowing with a force of 8 N. He is walking with a force of 9 N. (What will be the total force of the penguin's forward motion?)

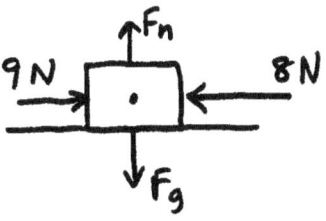

17) A satellite was launched with a force of 2 N. Since it is in outer space, it is not experiencing any air resistance, and therefore continues to accelerate.

18) A crane is lifting a box with a force of 12 N. At the same time, a gust of wind blows the box to the right with a force of 12 N. Use a dotted line to indicate what the combined motion of the crane and the wind will look like.

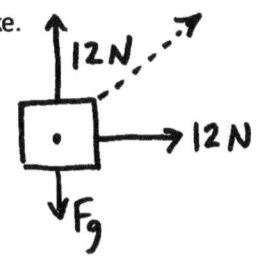

OPTIONAL QUIZZES

(These quizzes are intentionally "light" on math, so they can be used with all students.)

QUIZ for Adventure 1   Name _____

1) TRUE or FALSE? An object's center of mass is always inside the object itself.

2) What do you call the point around which two objects orbit?
   a) center of gravity   b) barycenter   c) epicenter   d) focal point   e) none of these

3) TRUE or FALSE? "Center of mass" and "center of gravity" are the same when we are on earth.

4) Who was the librarian in the Egyptian city of Alexandria?
   a) Archimedes   b) Euclid   c) Eratosthenes   d) Phidias

5) The measurement around the outside of a circle is called its _____.

6) What did Archimedes discover in his bathtub?
   a) buoyancy   b) the Archimedes screw   c) the value of pi   d) the area under a parabola   e) soap

7) TRUE or FALSE? The ancient Greeks did not know that the world is a sphere.

8) Archimedes lived on the island of _____.

9) TRUE or FALSE? If the center of mass of an object is not over its base, it will fall over.

10) Which one of these authors did Archimedes NOT read?
    a) Aristotle   b) Plato   c) Socrates   d) Copernicus   e) Euclid

11) Which of these letters is the only one that you could balance on the head of a pin?
    (We are assuming that the letter is lying flat, parallel to the floor.)
    a) L   b) O   c) C   d) D   e) Y   f) U   g) V

12) A balanced object becomes (more/less) stable if its center of mass is lowered. (circle one)

Name three practical applications of center of mass:

13) _____

14) _____

15) _____

QUIZ for Adventure 2   Name _____

Calculate the mechanical advantage of these levers:

1) _____ ◯_____△_____ 2) _____ ◯_____△_____
          8 cm   40 cm   33 ft.   11 ft.

3) "In a first class lever, the fulcrum's _____"

4) Where is the load in a second class lever? _____

5) "If you want some speed to swat or bat, the _____ class lever is where it's at."

6) Which type of lever is a pair of scissors?   a) first   b) second   c) third

7) Which type of lever is a wheelbarrow?   a) first   b) second   c) third

8) The principle of the lever is that you can trade _____ for a gain in force.

9) In a second class lever, if you want to make your work easier, where should you position the load?
   a) closer to the fulcrum   b) farther away from the fulcrum

10) Something that is made of more than one simple machine is a _____.

11) Which one of these is NOT a way to calculate mechanical advantage?
    a) radius of wheel/radius of axle   b) radius of follower/radius of driver
    c) length of lever/length of fulcrum   d) length of force arm/lenght of resistance arm

12) TRUE or FALSE?   An idler gear has no effect on mechanical advantage.

13) If a driver gear has 12 teeth and its follower has 24 teeth, this will result in multiplication of: _____

14) The wheel of a windlass is 60 centimeters. The shaft with the rope wound around it has a diameter of 15 cm. Will the mechanical advantage of the windlass be enough for you to have 5 times your ordinary lifting power? _____

15) What did Archimedes use pulleys for?
    a) lifting buckets of water   b) lifting ships   c) lifting logs   d) lifting houses

16) Archimedes lived in the city of _____.

17) TRUE or FALSE?   To calculate the mechanical advantage of gears you can use either the number of teeth or the radius of the gears-- it doesn't matter which.

18) TRUE or FALSE?   A mechanical advantage of less than 1 means that your work is much easier.

19) Which pulley provides no mechanical advantage?
    a) fixed pulley   b) movable pulley

20) Which part of the rope is not helping to bear the weight of the iron block?

a) A
b) B
c) C
d) D

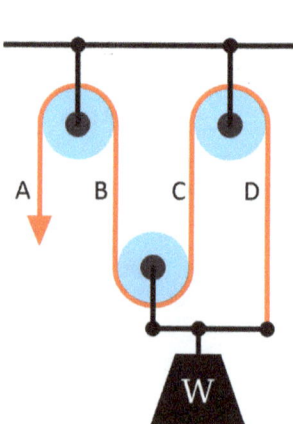

QUIZ for Adventure 3   Name _____

1) What did Galileo's father want him to become?
   a) doctor   b) lawyer   c) musician   d) artist   e) scientist   f) priest

2) What determines how fast a pendulum swings?
   a) the weight of the bob   b) the initial force   c) the length of the rod   d) the size of the swing

3) TRUE or FALSE? Pendulum swings always have exactly the same period no matter how high or low the bob goes.

4) The size of a pendulum's swing is called its:
   a) frequency   b) period   c) amplitude   d) cycle

5) What happens when you make the bob heavier?
   a) The pendulum swings for a longer period of time before it comes to rest.
   b) The frequency of the pendulum increases.
   c) The frequency of the pendulum decreases.
   d) The pendulum comes to rest more quickly.

6) What type of clock did Galileo use to time his pendulums? _____

7) What do we call the "law" of pendulums that Galileo discovered? _____

8) This law (in previous question) says that if we make a pendulum 4 times as long, its frequency will decrease by:
   a) 1/8   b) 1/4   c) 1/2   d) 1

9) Why did Foucault hang a pendulum from the ceiling of the Paris Pantheon?
_____

10) When something swings at its natural frequency, we call this: _____

Name two places you might find tuned mass dampers:

11) _____   12) _____

13) When something goes in and out, or back and forth, or up and down, in a steady rhythm, we call this:
   a) resonance   b) oscillation   c) pendulum motion   d) extrapolation   e) variation

14) When you extend the lines on a graph, going beyond your collected data, this is called: _____

15) When you graph the up and down motions of a pendulum, you can produce a curve called:
   a) cycloid   b) parabola   c) oscillation   d) sine wave

QUIZ for Adventure 4   Name _____

1) The Latin word "inert" means: a) not moving   b) incapable   c) stiff   d) lazy

2) What scientist/philosopher was considered the ultimate authority until Galileo's time?
   a) Archimedes   b) Aristotle   c) Euclid   d) Plato

3) TRUE or FALSE? Isaac Newton was born into a fairly wealthy family.

4) Newton developed a lifelong interest in chemistry while he was attending a boarding school (as a young teen). This might have been because he lived with a man who was _____.
   a) an apothecary   b) a doctor   c) a fireworks manufacturer   d) the inventor of the Periodic Table

5) Who said "I think therefore I am"? _____

6) Which one of these did Newton not study?   a) light   b) motion   c) gravity   d) astronomy   e) electricity

7) What was the name of Newton's famous book? _____

8) TRUE or FALSE? Newton broke tradition by writing in English.

9-11) Fill in these blanks.

A body at rest will _____; a body in motion will _____

unless _____ acts upon it.

12) What is a measure of inertia?   a) weight   b) mass   c) resistance   d) force

13) Which one changes if you travel from the earth to the moon?   a) mass   b) weight

14) TRUE or FALSE? A spring scale can measure in both Newtons and grams.

15) About how much does a pineapple weigh?   a) 1 gram   b) 10 grams   c) 100 grams   d) 1 kg

16) TRUE or FALSE? There isn't any way to measure mass without a scale that relies on gravity.

17-18) Mass is the measure of an object's resistance to a _____ in _____.

19) Where did Newton work during the last two decades of his life? _____

20) Why was Newton not as brilliant in his later years?
   a) He suffered from severe fatigue due to old age.
   b) He had suffered from heavy metal poisoning.
   c) He suffered from a lifetime of poor nutrition.
   d) He contracted small pox, which left him weak.

QUIZ for Adventure 5   Name _____

1) TRUE or FALSE? Friction always opposes motion.

2) What is the force you feel as you drag a heavy box across the floor?   a) static friction   b) kinetic friction

3) Which one of these will register the highest force on a spring scale?
   a) a box dragged with its largest, flattest side on the table
   b) the same box dragged with its short and narrow top side on the table
   c) the same box dragged with its long and narrow side on the table
   d) they will all register the same force

4) Something that has a rough surface has a very (high/low) coefficient of friction. (circle one)

5) Which one of these probably has the lowest coefficient of friction?
   a) glass   b) sandpaper   c) wood   d) metal   e) rubber   f) concrete

6) When is static friction at its maximum?
   a) when the box is not moving
   b) when you are pushing the box across the floor
   c) when you are pushing the box but it has not moved yet

7) TRUE or FALSE? At the microscopic level, friction involves some electrical attraction between atoms.

8) The strength of a force is called its: a) force   b) mass   c) magnitude   d) mechanical advantage

9) Kinetic means _____.

10) What does "normal" mean in physics? _____

11) What is the normal force?
    a) The force of gravity pulling you towards the ground
    b) The force of the ground pushing up on your feet
    c) The force of your weight
    d) The force of your feed pushing on the ground

Match the descriptions with the free body diagrams.

12) _____ The box is going up.
13) _____ The box is moving on a frictionless surface.
14) _____ Two people are pushing the box with equal strength in opposite directions.
15) _____ Someone is pushing on the box but it is not moving.

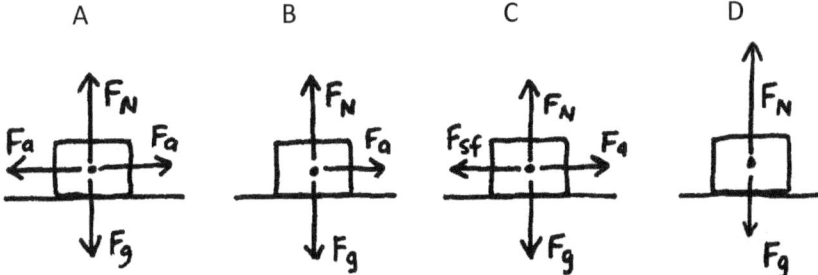

16) Technically, motion is the result of:
    a) kinetic forces   b) unbalanced forces   c) direction and magnitude   d) equilibrium

QUIZ for Adventure 6   Name _____

1) TRUE or FALSE?   Falling is just a very fast version of rolling downhill.

2) What kind if timer did Galileo use for his ramp experiment? _____

3) TRUE or FALSE? Galileo found that heavier balls produced different mathematical patterns than lighter balls.

4) Acceleration measures:
   a) the velocity of an object
   b) the rate at which velocity is increasing
   c) the rate at which velocity is changing
   d) how fast a car is traveling at a certain point

5-6) Vectors have both _____ and _____.

7) Where would gravity be stronger: mountains or valleys? _____

8) What is the acceleration due to gravity on earth? _____

9) TRUE or FALSE? When you are traveling at a very fast speed, an accelerometer will be swinging backwards.

10) TRUE or FALSE? A skydiver's terminal velocity would be the same on the moon as it is on earth.

11) TRUE or FALSE? The shape of an object has no effect on terminal velocity.

12) Which number is larger?   a) little g   b) big G

13) Who actually invented the device that (indirectly) measured the value of big G?
    a) Cavendish   b) Michell   c) Newton   d) Galileo

14) Who was the first person to propose that gravity is a giant "field" that is warped by giant objects?
    a) Newton   b) Einstein   c) LaPlace   d) Michell   e) Cavendish

15) Which one of these is the correct formula for the Universal Law of Gravitation?

   a) $F_g = \dfrac{G(m_1)(m_2)}{r^2}$ b) $F_g = \dfrac{g(m_1)(m_2)}{r^2}$ c) $F_g = \dfrac{G(m_1 - m_2)}{r^2}$ d) $F_g = \dfrac{G(m_1)(m_2)}{r}$

16) TRUE or FALSE? You and the earth both have gravity and are pulling on each other.

17) TRUE or FALSE? If an astronaut was in a rocket that was accelerating at 9.8 m/sec², it would feel to him the same as normal gravity.

18) TRUE or FALSE? Einstein proposed that time isn't affected by gravity.

19) TRUE or FALSE? Einstein said that gravity isn't a force.

20) Name a branch of science that doesn't line up with relativity: _____

QUIZ for Adventure 7   Name _____

1) TRUE or FALSE? Because of earth's greater mass, it pulls harder on the moon than the moon pulls on the earth.

2) Does F=ma work in zero gravity? _____

3) Newtons are the units used to measure _____.

4) How much does a 20 kg child weigh in newtons? _____

5) How much force is needed to accelerate a 1,500 kg truck at 2 meters per second? _____

6) TRUE or FALSE? The earth has more inertia than the moon.

7) The center of gravity around which two objects orbit is called the _____.

8) A nerf gun shoots a dart with a force of 1 newton. If the dart weighs 50 grams, how much acceleration will the dart experience? _____

9) To prevent something from breaking when it falls, do you want Δt to be large or small? _____

10) When the mass of an object doubles and acceleration stays the same, what happens to the resulting force?
   a) It stays the same.   b) It doubles.   c) It triples.   d) It increases by the square of 2, which is 4.

11) TRUE or FALSE? Conservation of momentum occurs in all collisions.

12) TRUE or FALSE? According to Newton's Third Law, the reaction occurs first, then the reaction occurs.

Definitions:

13) The measure of how fast something is going in a particular direction: _____

14) The measure of how fast something is moving: _____

15) The measure of how resistant something is to a change in velocity: _____

16) The measure of how fast velocity is increasing: _____

17) Mass multiplied by velocity gives you: _____

18) Force times the time interval in which the force is applied: _____

19) The measure of how large or small something is: _____

20) Measuring mass while in a particular gravitational field: _____

WORD BANK FOR DEFINITIONS: (There are 2 extra words that you won't use.)

acceleration, conservation, impulse, force, mass, magnitude, momentum, speed, velocity, weight

QUIZ for Adventure 8   Name _____

1) TRUE or FALSE? A ball dropped straight down from a tower will hit the ground sooner than a ball thrown out horizontally from the tower.

2) What is "the amount something moves in a certain direction"?
   a) distance   b) impulse   c) velocity   d) displacement

3) TRUE or FALSE? If you multiply both sides of an equation by the same thing, both sides will remain equal.

4) What angle will give a cannon the longest range?
   a) 15 degrees   b) 25 degrees   c) 45 degrees   d) 75 degrees   e) 90 degrees

5-6) Velocity and acceleration are vectors. What two things does a vector have?

   _____ and _____

7) TRUE or FALSE? Any launched object (baseball, cannonball, etc.) will <u>always</u> follow a parabolic path (even if it is only half of a parabola).

8) TRUE or FALSE? When you analyze the parabolic motion of a ball, you can separate the motion in the x direction (horizontal) from the motion in the y direction (vertical).

9) TRUE or FALSE? It is impossible to graph a circle on a Cartesian grid.

10) If a ball is dropped from a tower and takes 5 seconds to hit the ground, how tall is the tower? _____
    (Use 10 for the acceleration due to gravity, instead of 9.8)

The displacement-time graph on the right shows the path of a turtle in a zoo. The time starts a sunrise and ends 15 hrs. later at sunset. The turtle's starting point is his little house.

11) Letters "a" and "c" both show a 3-hour time periods. During which period did the turtle travel faster? _____

12) How much time did the turtle spend sitting still? ____ hours

13) Did the turtle return to its house? _____

14) Does this graph show acceleration? _____

Match the problem with the formula you need to solve it:

15) ___ A 10-kg ball is rolling along at 2 m/sec. What is its momentum?

16) ___ A beetle is crawling along at the rate of 2 cm per second. How many cm can it crawl in one minute?

17) ___ Find the strength of the gravity between the sun and Jupiter.

18) ___ If a falling apple hits the ground in 1.2 seconds, how high is the branch it fell from?

19) ___ A 50-kg satellite is launched with a force of 1,000 newtons. What will be its acceleration?

20) ___ If a rock is dropped off a 50-meter cliff, how long will it take for it to hit the ground?

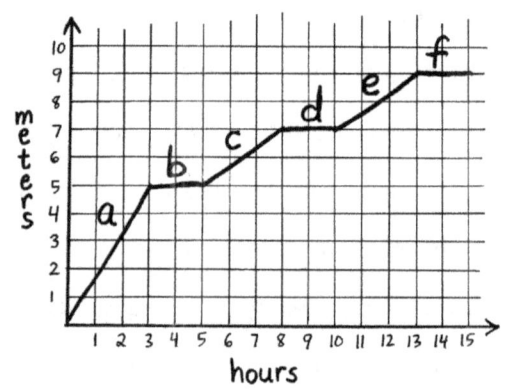

Ⓐ $F=ma$ Ⓑ $d=rt$ Ⓒ $p=mv$

Ⓓ $d=\frac{1}{2}gt^2$ Ⓔ $t=\sqrt{\frac{2d}{g}}$ Ⓕ $F_g=\frac{G(m_1)(m_2)}{r^2}$

244

QUIZ for Adventure 9   Name _____

1) TRUE or FALSE? An orbiting moon is constantly changing direction.

2) A change in acceleration can be due to a change in velocity or a change in _____

3) TRUE or FALSE? It is not possible for a human to experience constant acceleration.

4) Which one of these is NOT a vector quantity?
    a) acceleration   b) velocity   c) speed   d) angular momentum

5) Which one of these is NOT something that can provide centripetal force?
    a) gravity   b) friction   c) string   d) velocity

6) What is the maximum number of G forces that you'll experience on a roller coaster?
    a) 3   b) 5   c) 8   d) 10

7) What do you call a line that touches a circle at just one point? _____ line

8) What would happen to the moon if gravity suddenly disappeared?
    a) nothing   b) It would hit the earth.
    c) It would keep going in its orbit.   d) It would fly off along a straight-line path.

9) What rule do you use to find the direction of angular momentum? _____

10) The direction of angular momentum is always _____ to the direction of spin.
    a) parallel   b) perpendicular   c) opposite   d) unrelated

11) When you use $F_c=mv^2/r$ to find centripetal force, your answer will be in _____. (what unit?)

12) A hammer thrower is whirling his hammer in a circle. If the rope is 2 meters long, the velocity of the weight is 5 meters per second, and the mass of the weight is 2 kg.
    What is angular momentum? _____ (Remember to include the unit of measurement.)

13) TRUE or FALSE? The same formula can be used to calculate angular momentum for any object, regardless of whether it is a disc, a sphere, or a ball on a string.

14) Why did Foucault invent the gyroscope?
    a) To prove that spinning objects have angular momentum.   b) To see moment of inertia of the earth.
    c) To see the turning of the earth.   d) To make better navigational instruments for boats and planes.

15) TRUE or FALSE? If the earth shrinks, its rotational speed will slow down.

16) If an ice skater pulls in her arms, she will go: a) faster   b) slower   c) the same

17) A solid cylinder and a hoop had a race to the bottom of a ramp. Who won? _____

18) What letter is used to represent moment of inertia?   a) L   b) ω c) I d) θ

19) TRUE or FALSE? Increasing the length of a wrench will increase its torque.

20) The rotational equivalent of "force" is: a) moment of inertia   b) torque   c) angular momentum

QUIZ for Adventure 10   Name _____

1) Which one of these is NOT a factor in determining gravitational potential energy, PE?
   a) mass   b) height   c) velocity   d) acceleration due to gravity (g)

2) Where does a roller coaster car have its maximum potential energy?
   a) at the top of the first hill   b) at the bottom of the first hill   c) at the top of the second hill

3) TRUE or FALSE? Some of the coaster car's potential energy will be lost due to friction.

4) The equation for calculating potential energy is PE = _____.

5) What energy unit is used for all forms of energy? _____

6) Force is measured with this unit: _____.

7) What is the formula for kinetic energy, KE?
   a) mv^2/r b) mgh c) $1/2gt^2$ d) rmv e) MR^2 f) $1/2(mv^2)$

8) TRUE or FALSE? Energy cannot be created or destroyed.

9) For an ideal loop in a physics lab, the ball has to be _____ times higher than the radius of the loop.
   a) 2.0   b) 2.5   c) 2.7   d) 2.9

10) What Greek letter is used to represent angles? _____

11) What units are used to measure work? _____

12) How much work is done when a force of 15 newtons is applied through 2 meters? _____

13) Who invented the steam engine (originally called the atmospheric engine)?
    a) Newcomen   b) Watt   c) Joule   d) Newton

14) Who coined the word "horsepower"?   a) Newcomen   b) Watt   c) Joule   d) Newton

15-16) How do you find the cosine of an angle?

   You divide the _____ by the _____.

17) What unit is used to measure power? _____

18) Which one of these did Watt NOT contribute to make an improved steam engine?
    a) a centrifugal governor
    b) a chamber for turning water into steam
    c) planet and sun gears
    d) flywheel
    e) separate chamber for cooling steam
    f) automatic valves

19) What was the first use for steam engines? Pumping water out of _____.

20) Energy is "the ability to _____."

Answers to quizzes

Adventure 1
1) F   2) b   3) T   4) c   5) circumference   6) a   7) F   8) Sicily
9) T   10) d   11) e   12) more   13-15) race cars, athletics, astronomy, other answers possible

Adventure 2
1) 5   2) 1/3   3) in the middle   4) in the middle   5) third class   6) a
7) b   8) distance   9) a   10) complex or compound machine   11) c   12) T
13) force   14) no, only 4 times   15) b   16) Syracuse   17) T   18) F   19) b   20) a

Adventure 3
1) a   2) c   3) F   4) c   5) a   6) water clock   7) inverse square law   8) c
9) To show the rotation of the earth   10) resonance   11-12) bridges, high wires, skyscrapers
13) b   14) extrapolation   15) d

Adventure 4
1) d   2) b   3) F   4) a   5) Descartes   6) e   7) Principia   8) F
9) remain at rest   10) remain in motion   11) an external force   12) b   13) b
14) T   15) d   16) F   17) change   18) velocity   19) Royal Mint   20) b

Adventure 5
1) T   2) b   3) d   4) high   5) a   6) c   7) T   8) c   9) moving   10) perpendicular
11) b   12) D   13) B   14) A   15) C   16) b

Adventure 6
1) T   2) water clock   3) F   4) c   5) speed (or magnitude)   6) direction   7) valleys   8) -9.8 m/sec^2
9) F (assuming your fast speed is constant)   10) F (the moon has no atmosphere)   11) F   12) a   13) b   14) c
15) a   16) T   17) T   18) F   19) T   20) quantum mechanics or SED physics

Adventure 7
1) F   2) yes   3) force   4) 196 N   5) 3,000 N   6) T   7) barycenter   8) 20 m/sec^2
9) large   10) b   11) T   12) F   13) velocity   14) speed   15) mass   16) acceleration
17) momentum   18) impulse   19) magnitude   20) weight

Adventure 8
1) F   2) d   3) T   4) c   5-6) magnitude and direction   7) T   8) T   9) F   10) 125 meters
11) a   12) 6   13) no   14) no   15) C   16) B   17) F   18) D   19) A   20) E

Adventure 9
1) T   2) direction   3) F   4) c   5) d   6) b   7) tangent   8) d   9) righthand rule   10) b
11) newtons   12) 20 N   13) F   14) c   15) F   16) a   17) cylinder   18) c   19) T   20) b

Adventure 10
1) c   2) a   3) T   4) mgh   5) joules   6) newtons   7) f   8) T   9) c   10) theta θ
11) joules   12) 30 J   13) a   14) b   15-16) adjacent side, hypotenuse   17) watts   18) b
19) mines   20) do work

BIBLIOGRAPHY

BOOKS and PRINTED ARTICLES:

On the Life of Galileo Galilei; Viviani's historical account and other early biographers. Edited and translated by Stefano Gattei. Princeton University Press, 2019. ISBN 978-0-691-17489-1
I used the sections written by Vincenzo Viviani and Noccolo Gherardini.

Exploring Creation with Physics by Dr. Jay Wile. Published by Apologia Educationals Ministries, Inc. www.apologia.com ISBN 978-1-932012-42-2

"Hunting the White Elephant; When and how did Galileo discover the law of fall?" by Jurgen Rnn, Peter Damerow, Simone Rigeer and Michele Camerota. Published by the Max Planck Institute for the History of Science, 1998. ISSN 0948-9444.

WEBSITES and ONLINE BOOKS:

How does a lever work?
https://owlcation.com/stem/Simple-Machines-How-Does-a-Lever-Work

Is a screwdriver a lever?
https://web.physics.ucsb.edu/~lecturedemonstrations/Composer/Pages/28.12.html

How a wheel is a series of infinite levers
https://owlcation.com/stem/How-Do-Wheels-Work-The-Mechanics-of-Axles-and-Wheels
https://en.wikipedia.org/wiki/Wheel_and_axle

How the jaw is a lever
http://ffden-2.phys.uaf.edu/104_2012_web_projects/kjersten_williams/Page_3_Physics_of_Chewing.html

How do gears work?
https://en.wikipedia.org/wiki/Gear

Archimedes
http://archimedespalimpsest.org/about/history/archimedes.php
https://www.hellenicaworld.com/Greece/Technology/en/ArchimedesGears.html
https://www.youtube.com/watch?v=D75q3vsbID8&t=284s
https://en.wikipedia.org/wiki/Archimedes
Simon Whistler: https://www.youtube.com/watch?v=55_QQRDXlW0

How Foucault started his pendulums
https://www.youtube.com/watch?v=sWDi-Xk3rgw&t=11s (Sixity Symbols, Univ. of Nottingham)

The French Pantheon
https://en.wikipedia.org/wiki/Panth%C3%A9on

Newton
https://www.youtube.com/watch?v=OK1bCqkn6Vk ("Biography")
https://en.wikipedia.org/wiki/Isaac_Newton
https://en.wikipedia.org/wiki/Philosophi%C3%A6_Naturalis_Principia_Mathematica
https://www.youtube.com/watch?v=perbFN0ztl4 (how to pronouce Principia)
https://en.wikipedia.org/wiki/Great_Recoinage_of_1696

Inertia definition
https://en.wikipedia.org/wiki/Inertia

Falling objects calculation
https://sciencing.com/calculate-force-impact-7617983.html
https://www.airitas.com/the-physics-behind-the-real-danger-of-falling-tools/
https://www.calculatorsoup.com/calculators/physics/gravitational-potential.php

Friction
http://scienceline.ucsb.edu/getkey.php?key=167
https://www.youtube.com/watch?v=fo_pmp5rtzo (Crash Course Physics)
https://www.tribology-abc.com/abc/history.htm
https://phys.org/news/2016-05-leonardo-da-vincithe-systematic-friction.html
https://www.sciencedirect.com/science/article/abs/pii/S0043164816300588
https://www.youtube.com/watch?v=RIBeeW1DSZg (Organic Chemistry Tutor channel)
https://simple.wikipedia.org/wiki/Coefficient_of_friction
https://www.youtube.com/watch?v=QwuldyEP9Jk (explaining coefficients of friction-- Flipping Physics)

Gravity
https://www.youtube.com/watch?v=7gf6YpdvtE0 (Crash Course Physics)
https://www.youtube.com/watch?v=nR9nE1TalZc (edwardCurrent)
https://en.wikipedia.org/wiki/Schiehallion_experiment (gravity of nearby mountain detected)
https://www.youtube.com/watch?v=PY36XOafp58 (Calculating 9.8 from universal law of gravitation. Prof. Matt Anderson)
https://www.youtube.com/watch?v=7ymoYwItyMw (Pattern youtube channel- big G)
https://en.wikipedia.org/wiki/Newton%27s_law_of_universal_gravitation
https://www.youtube.com/watch?v=Ki5g0CpRj1I (UNSW Physics)
https://www.physicsclassroom.com/class/circles/Lesson-3/Newton-s-Law-of-Universal-Gravitation
https://www.sciencemusings.com/albert-einsteins-most-happy-thought/

Normal force
https://www.youtube.com/watch?v=fRQq4_ry9-Q

Calculating G and Cavendish experiment
https://www.youtube.com/watch?v=xoam-au5A1U (Doc Schuster)
https://www.youtube.com/watch?v=2PdiUoKa9Nw (Sixty Symbols)
https://www.youtube.com/watch?v=MbucRPiL92Q (Rich Lund)
https://www.youtube.com/watch?v=SjCEhdvnj6A (Jim Al-Kahlili)
https://en.wikipedia.org/wiki/Gravitational_constant

Is Gravity a real force?
https://www.youtube.com/watch?v=EmrZ3ZaXmS4&t=14s (Sabine Hossenfelder)
https://www.youtube.com/watch?v=NblR01hHK6U&t=87s (PBS Spacetime)
https://www.youtube.com/watch?v=XRr1kaXKBsU (Veratsium)
https://www.youtube.com/watch?v=OTMELHUAzSM (MinutePhysics)

Quantum mechanics versus Einstein's explantion of gravity
https://www.youtube.com/watch?v=Ov98y_DCvRY (Sabine Hossenfelder)

Acceleration due to gravity
https://www.youtube.com/watch?v=GDUdUumkv0o (Flipping Physics)

Relativity
https://wpsu.pbslearningmedia.org/resource/phy03.sci.phys.energy.sprelativity/einsteins-special-theory-of-relativity/
https://www.youtube.com/watch?v=tzQC3uYL67U (Arvin Ash)

Descartes and Cartesian graph
https://wild.maths.org/ren%C3%A9-descartes-and-fly-ceiling
https://www.britannica.com/biography/Rene-Descartes

Projectile motion
http://galileo.rice.edu/lib/student_work/experiment95/paraintr.html
https://www.hindawi.com/journals/ijmms/2020/9695053/ (Harriot the English Galileo)

Conservation of momentum
https://en.wikipedia.org/wiki/Momentum
https://www.youtube.com/watch?v=uQ1UMHXyFrM (Bozeman Physics Essentials 49)
https://www.youtube.com/watch?v=w2zQJ8JMlBA (GPB Education)
https://www.youtube.com/watch?v=iDUaPp6bELw (PhysicsHigh)
https://www.physicsclassroom.com/class/momentum/Lesson-2/Momentum-Conservation-Principle

Impulse Force
https://www.youtube.com/watch?v=ph48Xwj_eS8 (Bozeman Physics Essentials 50)
https://www.youtube.com/watch?v=gv8nvpqo5aU (Bozeman Physics Essentials 62)

Definition of force
https://www.youtube.com/watch?v=fiT2R88Zt58 (Flipping Physics)

Misconception about Third Law applying to graivty and normal force
https://www.youtube.com/watch?v=wmmjfbl7zG4

Collisions
https://www.youtube.com/watch?v=8ko3qy9vgLQ (Bozeman)

Cycloids
https://www.irishtimes.com/news/science/the-curved-history-of-cycloids-from-galileo-to-cycle-gears-1.2347025
https://en.wikipedia.org/wiki/Cycloid
https://nnu.whdl.org/sites/default/files/resource/academic/Chavez%252C%2520Natalee%2520HP%2520Paper.pdf

Moment of Inertia
https://www.youtube.com/watch?v=v5vicPCvHsU (Flipping Physics)
https://www.youtube.com/watch?v=lNx0yPdl960 (Flipping Physics)
https://www.youtube.com/watch?v=uyU25DdONjo
https://probingphysics.com/understanding-moment-of-inertia-with-real-life-examples/
https://physics.stackexchange.com/questions/440946/cylinder-vs-cylinder-of-double-the-radius-roll-down-an-incline-plane-which-one

Centripal force
https://en.wikipedia.org/wiki/History_of_centrifugal_and_centripetal_forces
https://en.wikipedia.org/wiki/Coriolis_force
https://en.wikipedia.org/wiki/Centripetal_force
https://www.youtube.com/watch?v=SQX22VVmRPs (Organic Chemistry Tutor)
https://www.physicsclassroom.com/mmedia/circmot/ucm.cfm
https://geteducationbee.com/centripetal-acceleration-formula/

G suit and G force training
https://en.wikipedia.org/wiki/G-suit
https://en.wikipedia.org/wiki/High-g_training

Angular momentum
https://www.youtube.com/watch?v=4GgJ6EgAmFg (Organic Chemistry Tutor)
https://www.youtube.com/watch?v=MULe4xv3lVk (Bozeman)
https://arxiv.org/pdf/1511.07748 (A Historical Discussion of Angular Momentum and its Euler Equation)

Kinetic and Potential Energy
https://www.youtube.com/watch?v=DyaVgHGssos (Organic Chemistry Tutor)
https://www.youtube.com/watch?v=zDcf7eEaP0M (Bozeman)
https://www.youtube.com/watch?v=zt0ov2C5svw (Ben Ryder)

Marey
https://en.wikipedia.org/wiki/%C3%89tienne-Jules_Marey
https://www.youtube.com/watch?v=Xs87FijgVaA (Vox)
https://www.youtube.com/watch?v=Q02SAf_eUmU (The Olympic Museum)
https://www.youtube.com/watch?v=BIKwns3y2_w (complete videos by Marey)
https://skullsinthestars.com/2013/07/26/cat-turning-the-19th-century-scientific-cat-dropping-craze/

Gyroscopes
https://en.wikipedia.org/wiki/Precession
https://en.wikipedia.org/wiki/Gyroscope
https://en.wikipedia.org/wiki/Foucault%27s_gyroscope

Torsional pendulums
https://en.wikipedia.org/wiki/Balance_spring
https://www.physics.louisville.edu/cldavis/phys298/notes/torpend.html
https://farside.ph.utexas.edu/teaching/301/lectures/node139.html
https://www.youtube.com/watch?v=6iucHadtnX4
https://www.youtube.com/watch?v=GGO2I_sN8hI
https://www.youtube.com/watch?v=T_GtrM-5AJo

Work and energy
https://en.wikipedia.org/wiki/Work_(thermodynamics)
https://www.youtube.com/watch?v=w4QFJb9a8vo&t=3s (Crash Course Physics)
https://www.mathsisfun.com/physics/energy-potential-kinetic.html

Joule
https://en.wikipedia.org/wiki/James_Prescott_Joule
https://www.youtube.com/watch?v=fKoOD29qISs (Kathy Loves Physics & History channel)
https://www.britannica.com/biography/James-Prescott-Joule

Roller coasters and Loop-the-Loops
https://www.heartofconeyisland.com/sea-lion-park-coney-island.html
https://www.youtube.com/watch?v=3Kzl2suBE2w (Vox video on looping coaster history)
https://mathsciencehistory.com/2020/01/19/physics-of-the-flip-flap-rollercoaster/
https://www.youtube.com/watch?v=SZ5QIesKy48   (Math and Physics Tutor channel)

Work and power
https://www.youtube.com/watch?v=_MR1Dp8-F8w (Organic Chemistry Tutor channel)
https://www.youtube.com/watch?v=w4QFJb9a8vo (PBS Crash Course)
https://www.youtube.com/watch?v=pmOXi-My6ZI&t=5s (Bozeman)
https://www.youtube.com/watch?v=qcb9GTdNmIs (Matt Anderson)
https://www.youtube.com/watch?v=KCHEBwNLlQ4 (Step by Step Science channel)

Watt and steam engine
https://www.youtube.com/watch?v=fsXpaPSVasQ (Real Engineering channel)
https://www.youtube.com/watch?v=Caqf4hQBYBI   (BBC Earth Science)
https://www.youtube.com/watch?v=UVBq27luj8A (Engineering Education)
https://www.youtube.com/watch?v=3apcSyFUeOs (The History Guy channel)
https://www.youtube.com/watch?v=GMgP-4O99qU (Branch Education)
https://www.youtube.com/watch?v=ecFTXRDm_vA (IET)
https://www.youtube.com/watch?v=QltRwiu4U2Q ("My bit of history" channel)
https://en.wikipedia.org/wiki/Newcomen_atmospheric_engine
https://en.wikipedia.org/wiki/Horsepower

www.ingramcontent.com/pod-product-compliance
Lightning Source LLC
Chambersburg PA
CBHW082044250426
43661CB00080B/2735